The Middle East

D0226415

The Middle East has been the focus of international attention for a number of reasons. Historically it was a central meeting place of the three Old World continents. More recently, it has become the world's premier geopolitical flashpoint for two major reasons: (i) the establishment of the state of Israel centrally in the Arab core of a region which is overwhelmingly Islamic, and (ii) the realization by a few Middle Eastern states that, between them, they possess the world's major reserves of petroleum.

Although the focus is on geopolitics rather than regional geography, the aim of the book is to provide a reorientation, while maintaining an essential continuity with W. B. Fisher's book of the same name, the seventh edition of which was published in 1978. A comprehensive account is provided of the physical and human geography of the region and, in this, certain sections from Fisher's book have been retained. Based on the geography, a geopolitical assessment is then made of the states of the Middle East and of the major issues, such as water and conflict, which illustrate the interplay of geography and politics.

With comprehensive illustrations, this is a lively and much-needed book, based upon a well-respected work, providing an excellent synopsis of the complexities that make the Middle East such an intriguing and important global region.

The Middle East

Geography and geopolitics

Ewan W. Anderson

London and New York

DS
44.96
A 53
2000

LIBRARY
BUNKER HILL COMMUNITY COLLEGE
CHARLESTOWN, MASS. 02129

First published 2000
by Routledge
11 New Fetter Lane, London EC4P 4EE

Simultaneously published in the USA and Canada
by Routledge
29 West 35th Street, New York, NY 10001

Routledge is an imprint of the Taylor & Francis Group

© 2000 Ewan W. Anderson

The right of Ewan W. Anderson to be identified as the Author of
this Work has been asserted by him in accordance with the
Copyright, Designs and Patents Act 1988

Typeset in Galliard by Florence Production Ltd, Stoodleigh, Devon
Printed and bound in Great Britain by MPG Books Ltd, Bodmin

All rights reserved. No part of this book may be reprinted or
reproduced or utilized in any form or by any electronic,
mechanical, or other means, now known or hereafter invented,
including photocopying and recording, or in any information
storage or retrieval system, without permission in writing from
the publishers.

British Library Cataloguing in Publication Data
A catalogue record for this book is available from the British Library

Library of Congress Cataloging in Publication Data
Anderson, Ewan W.
  The Middle East : geography and geopolitics / Ewan W. Anderson.
    p. cm.
  Revision of: The Middle East / W.B. Fisher. 1978.
  Includes bibliographical references and index.
  1. Middle East – Geography. 2. Geopolitics – Middle East.
  I. Fisher, W. B. (William Bayne). Middle East. II. Title.

DS44.96.A55 2000+
956–dc21   99–049755

ISBN 0–415–07667–6 (hbk)
ISBN 0–415–07668–4 (pbk)

To the immortal memory and,
as he would have considered particularly appropriate,
specifically to MTF

Contents

List of figures

Preface

It was a particular honour and pleasure for me to be invited by the publisher
to reorientate and revise what has been for long a standard book on the
region, *The Middle East* (1978) by W. B. Fisher. It was an honour because
Professor Fisher was instrumental in bringing me to Durham to head the
Geography Department in the College of the Venerable Bede and later to
become a member of the University Geography Department. It was a pleasure because, once a member of Professor Fisher's Department, my career
was rapidly set on Middle Eastern affairs. Initially, my research focused upon
hydrology and geomorphology within the region and later my major concern
switched to geopolitics. As a result, I have been fortunate to have worked
in most of the countries in the region on subjects which were predominantly
geographical but ranged from cartography to child care.

Professor Bill Fisher belonged to the last generation of complete scholars
of the subject, equally at home in physical and human geography. Fisher
himself had an abiding interest in meteorology and climate which paralleled
in depth his fundamental understanding of so many aspects of Middle
Eastern culture. I felt it important to retain in this book his writing in these
key areas.

As the term Middle East is at least as much geopolitical as geographical,
it seems wholly appropriate to recast the book in a geopolitical rather than
its original regional format. Although, as ably demonstrated by Claval (1998),
regional geography is not dead, the Middle East tends to be regarded particularly in geopolitical terms. So many defining events in international relations
have occurred within the region that it is reasonable to allude to the states
as actors. As a result, Chapters 10 and 11, concerned respectively with the
states of the region and the key aspects of Middle Eastern geopolitics, have
replaced the regional section in *The Middle East* (Fisher 1978). Of the other
chapters, the Prologue, 1, 5, 9, 12 and Epilogue are either new or completely
rewritten. Chapters 2 and 4 retain very little of the original volume while
Chapters 3, 6, 7 and 8 have been updated with certain additions but are
substantially based upon the original. Using a page count, something under
30 per cent of the book is recognizable as the work of Fisher and it is hoped
that this is sufficient to provide the required continuity.

In writing this volume, I would like to express gratitude to many current and former students of the University of Durham, most notably: Dr Edward Twiddy, James Carnegie, Jenny Reeves, Gareth Stansfield, Dr 'Rajan' Puvanarajan and Sahab Farwaneh. I have been privileged to work with these colleagues. In addition, it is a particular pleasure to thank two of my sons; Greg Anderson for the Prologue which so effectively sets the book in its new context and Liam Anderson for his help and advice throughout the preparation of the volume. Greg illustrates the point very clearly that although the term may not have been used, the geopolitical approach can be traced back to very early times. For her thought and meticulous effort in editing the volume, I am most grateful to Rosemary Baillon and for typing and encouragement I am as ever indebted to my sister, Bid Austin. The other feature of the book is the maps, many based upon those used in *The Middle East* (Fisher 1978) but all redrawn. For his patience, attention to detail and for all the interest he has shown, I should like to thank Ian Cool. Finally, I must attribute the inspiration for this volume to Bill Fisher and the errors to me alone.

Abbreviations and acronyms

AD	Anno Domini
AE	actual rates of evapotranspiration
AIOC	Anglo-Iranian Oil Company
AOC	Arabian Oil Company
ARAMCO	Arabian–America Oil Company
BC	Before Christ
bcm	billion cubic metres
BP	before the present era
BP	British Petroleum
CENTO	Central Treaty Organization
CFP	Compagnie Française des Pêtroles
CIA	Central Intelligence Agency
DFLP	Democratic Front for the Liberation of Palestine
ED	electro-dialysis
EEZ	Exclusive Economic Zone
ENI	Ente Nazionale Idrocarbui
ENOSIS	the political union of Cyprus and Greece, as an ideal or proposal
EOKA	National Organization for Cypriot Struggle (*Ethniki Organosis Kipriakou Agonos*)
ESCWA	Economic and Social Commission for Western Asia
EU	European Union
FAO	Food and Agriculture Organization (United Nations)
FMA	Foreign Military Assistance
FMF	Foreign Military Financing
FSU	Former Soviet Union
GAP	South-East Anatolian Project
GCC	Gulf Co-operation Council
GDP	gross domestic product
GIS	geographical information systems
GNP	gross national product
IAEA	International Atomic Energy Agency
ICBM	intercontinental ballistic missile
ICJ	International Court of Justice

ILA	International Law Association
IMF	International Monetary Fund
IPC	Iraq Petroleum Company
IR	international relations
ITC	Intertropical Convergence Zone
KDP	Kurdish Democratic Party
KDPI	Kurdish Democratic Party Iran
MINURSO	UN Peacekeeping forces in Western Sahara
MSF	Multiple Stage Flash
NATO	North Atlantic Treaty Organization
NPT	Nuclear Non-Proliferation Treaty
OAPEC	Organization of Arab Petroleum Exporting Countries
OAS	Organization of American States
OAU	Organization of African Unity
OPEC	Organization of Petroleum Exporting Countries
PCP	Palestine Communist Party
PE	potential rates of evapotranspiration
PFJ	Polar Front Jetstream
PFLP	Popular Front for the Liberation of Palestine
PFT	Polar Front Theory
pH	acidity/alkalinity scale
PKK	Kurdish Workers Party
PLO	Palestine Liberation Organization
PNA	Palestinian National Authority
PUK	Patriotic Union of Kurdistan
RO	reverse osmosis
R/P	reserves to production
SABIC	Saudi Basic Industries Corporation
SPLA	Sudan People's Liberation Army
SPLM	Sudan People's Liberation Movement
STJ	Subtropical Jetstream
SUMED	Suez–Mediterranean oil pipeline
TAP Line	Middle East–Mediterranean oil pipeline
TJS	Tropical Jetstream
UAE	United Arab Emirates
UK	United Kingdom
UN	United Nations
UNCLOS	UN Conference on the Law of the Sea
UNDOF	UN Peacekeeping Forces in Syria/Israel
UNEFI	UN Peacekeeping Forces in Egypt/Israel 1956–67
UNEFII	UN Peacekeeping Forces in Egypt/Israel 1973–79
UNESCO	UN Educational, Scientific, and Cultural Organization
UNFICYP	UN Peacekeeping Forces in Cyprus
UNIFIL	UN Peacekeeping Forces in Lebanon
UNIKOM	UN Observer Mission in Iraq/Kuwait

UNIMOG	UN Observer Mission in Iran/Iraq
UNOGIL	UN Observer Mission in Lebanon
UNRWA	UN Refugee and Works Agency
UNSCOM	UN Special Commission to Iraq
UNTSO	UN Observer Mission in Palestine
UNYOM	UN Observer Mission in Yemen
US	United States
USA	United States of America
USSR	Union of Soviet Socialist Republics
V/A	volume to area ratio
VCD	vapour compression desalination

Prologue

Geopolitics: a prehistory

The term 'geopolitics' is commonly used, by specialists and non-specialists alike, as a kind of short-hand for 'politics on the global level'. It is a contention of this book that the popular usage of 'geopolitics' is unnecessarily narrow, since it fails to take full advantage of the rich analytical possibilities presented by the word's two components. In the present work the term will be used to refer to the contribution of geography, here considered in all of its physical, historical, economic, strategic and social dimensions, to political decision-making on all levels, from the local to the supranational.

'Politics' derives from the Greek adjective *politikos*, one of a range of words which cluster around the noun *polis*. Originally *polis* meant 'citadel', the area of raised ground around which Greek towns and cities were generally situated. By the archaic period (*c.* 700–480 BC), this meaning was extended to include the entire area covered by an urban settlement and, subsequently, that portion of territory beyond the walls which the city held under its political sway. Now meaning something closer to our idea of 'city-state', the word *polis* underwent a further broadening of its semantic range by the beginning of the classical age (*c.* 480–320 BC) so that it could also be used to refer to the community which made up its citizen body.

Thus, suggesting at once a physical, a political and a social entity, the term generated a variety of related words, ranging from *politeia* ('political regime', 'polity') and *politeuma* ('institutions of the state'), to *polites* ('citizen'), the ultimate root of our own word 'polite'. A further derivation was the adjective *politikos*, which, by virtue of its primary meaning, 'relating to the *polis*', could be used with considerable semantic latitude to connote anything from 'civil', 'public' or 'constitutional', to 'urbane', 'secular' or 'communal'. It is in fact this latter sense which underlies Aristotle's (*Politics* I.1253a) famous observation that 'man is by nature a *politikon zoon*'. Here, he means less that humans are inherently 'political animals' than that they are creatures specially designed by nature for the communal life of the *polis*. In any case, the Greek use of the neuter plural form of this adjective as an abstract noun, namely *ta politika* (literally 'matters relating to the *polis*'), gives us our own word 'politics', though clearly much of the broader resonance of the original Greek word has been lost in the process.

In early Greek cosmogony, the creation of life on earth was seen as the product of sexual congress between the god Ouranos and the goddess Ge (or Gaia), personifications respectively of the sky and the earth. And throughout the course of Greek antiquity, the word *ge* would always refer in some sense to 'the earth's surface', whether suggesting '(dry) land' in contrast to sea, an individual's '(native) land', or the 'soil' tilled by farmers. Like *polis*, the word also provided productive roots (*ge-* and *geo-*) for use in compounds, such as *georgos* ('farmer', literally 'worker of the soil'), *geometria* ('land measurement', 'geometry'), and, of course, *geographia*.

Meaning essentially 'description of the earth's surface', *geographia* was first established as a field of enquiry in the sixth century BC by the Ionian Greeks who inhabited the coastal cities of Asia Minor. While the range of topics covered by this field varied somewhat over time, its boundaries broadly correspond to those of the modern field of geography, embracing study of the full range of human and physical phenomena which are found on, or have some kind of impact on, the earth's surface. As the field became increasingly well established during the archaic and classical periods, the word *geographia* also came to be used to refer to the two concrete products of geographical enquiry, namely the 'map' and the 'geographical treatise'.

Though derived from Greek roots, the term 'geopolitics' is of course a modern coinage, in current use for barely a hundred years. However, as used in this book, it suggests a mode of enquiry which is arguably of much older pedigree. As I will try to show in this Prologue, the importance of the interplay between geography and politics at various levels was long recognized and studied by the Greeks themselves, and state political and economic interests in fact provided the momentum for many of the more important advances made in geographical research during classical antiquity. I begin with a brief overview of the history of Greek geography.

Intimations of an emerging geographical consciousness are present in even the very earliest Greek literature, as seen, for example, in the fairly elaborate frames of topographical reference employed in the works of Homer (*c.* 750–700 BC) and Hesiod (*c.* 700 BC). Formal speculation about the layout of land and sea on the earth's surface, and about the place of the *oikoumene* ('inhabited world') within this scheme, began during the sixth century among the Greeks of Ionia, a narrow strip of land along the Aegean coast of Asia Minor. The first map was credited to Anaximander of Miletus, who seems to have conceived of the earth as a kind of drum. Around the turn of the fifth century, a fellow-Milesian, Hecataeus, refined Anaximander's claims in his *Description of the Earth* (*Periegesis* or *Periodos Ges*), one of the earliest known treatises on a geographical subject, though only fragments of it have survived. Hecataeus' findings were the product of observations made during travels around the Mediterranean and Black Sea. But it was his older contemporary Scylax of Caryanda who appears to have begun the long Greek tradition of this style of enquiry with his account of a more extensive voyage, one which supposedly took him as far afield as the Indus valley and the isthmus of Suez. What survives of the text assigned to him by the ancients dates however only to the fourth century BC.

Recognition of a spherical earth is first found among the fragments of the philosopher Parmenides of Elea (born *c.* 510 BC), though it may have been postulated somewhat earlier by Pythagoras of Samos (born *c.* 550 BC). After further research in the fourth century by scholars such as Eudoxus of Cnidus (*c.* 390–340 BC), Aristotle, and his student Dicaearchus, who worked largely from astronomical observation, it was generally understood that the world known to the Greeks occupied only a portion of the earth's northern hemisphere. However, during the aftermath of Alexander's conquest of the Persian Empire and the dawning of the Hellenistic age, a more scientific, mathematically grounded approach to geographical enquiry was beginning to emerge, associated particularly with the new intellectual centre of Alexandria. Without doubt the most significant figure in the development of this new approach was the polymath Eratosthenes of Cyrene (*c.* 285–194 BC), a chief librarian at Alexandria, who much improved on previous attempts to systematize the fixing of co-ordinates of longitude and latitude and was able to produce a remarkably accurate calculation of the earth's circumference.

Investigation along these lines would be pursued at Alexandria and elsewhere well into the Roman era, though it never entirely displaced the earlier tradition of coastline observation and geographical description. Indeed, the most enduring product of this latter tradition would not be written until the time of the Augustan principate (31 BC to AD 14). The monumental *Geographia* of Strabo towers over all previous works in this genre with its exhaustive description of the human and physical world both within and beyond the boundaries of the Roman Empire. At the same time, the work also offers its Greek readers a highly ambitious synthesis of history and geography in its attempt to identify the geographical imperatives which encouraged the creation of the Augustan world order.

Meanwhile, the younger tradition of a mathematically grounded geography would reach its culmination in the *Geography* of Claudius Ptolemy (*c.* AD 130–180), which aims to provide sufficient data for the construction of a map of the world stretching from Ireland to the Malay peninsula. Inaccuracies inevitably abound in the work, especially in those sections which concern the world beyond the confines of the Roman Empire; a southern land mass is said to link Africa with China, while Scotland is placed at a right angle to the east coast of England, and Sri Lanka is imagined to be some fourteen times its actual size. Yet, with its remarkable detail and mathematical precision, the *Geography* made advances in the science of cartographic projection which would not be superseded until the sixteenth century.

For the purposes of this Prologue, the most striking feature of Greek geography is that its history appears to be informed throughout by a recognition of geography's political implications. This can be seen first of all in the way that new departures in geographical research were often encouraged, directly or indirectly, by the economic and political interests of states. At the same time, this recognition is also reflected in the work of historians like Thucydides and Polybius and geographers like Strabo, who frequently draw on distinctively 'geopolitical' forms of analysis to trace patterns of cause and effect in political and military history.

It is surely no coincidence that the earliest maps and geographical writings sprang from the city-states of Ionia on the coast of Asia Minor. In fact, it has long been recognized that these developments were conceived expressly to facilitate the extensive commercial and colonial ventures pursued by the Ionian Greeks, initiatives which would have brought them into repeated contact with regions outside the Greek *oikoumene*, such as North Africa, Babylonia and the lands surrounding the Black Sea, as well as with non-Greeks in the neighbouring states of Lydia and Caria and, later, with the Persians.

Significantly, the geographer Hecataeus was among those consulted by the Milesian Aristagoras in *c*. 500 BC when the latter was planning an attempt to overthrow the Persian hegemony which had held sway over Ionia since the fall of the Lydian kingdom in *c*. 546 BC. Knowing all too well the vast extent of the power and resources available to the Persian king Darius, Hecataeus, according to Herodotus (5.36), wisely advised Aristagoras to abandon the whole idea. Unfortunately, this advice was ignored. Aristagoras meets an ignominious end (5.126) and the revolt is crushed soon thereafter, with disastrous consequences for the Ionians (6.31–32); their cities are looted and burned, their temples are razed to the ground, the most beautiful of their girls are taken away to serve as royal concubines, and the most handsome of their boys are castrated.

Meanwhile, on an earlier occasion, it appears that Darius himself had actually commissioned the famous voyage of Scylax down the Indus to the Indian Ocean. The king's motives for despatching this mission were not, however, purely scientific, as Herodotus goes on to relate (4.44). Once armed with the knowledge gained from Scylax's travels, Darius promptly 'subdued the Indians and made use of this sea'.

These few examples suggest that an awareness of geography's political ramifications informed the work of even the earliest Greek geographers. Unfortunately, all too little survives of their texts, and we have to look elsewhere to find systematic attempts to apply this awareness to the study of human affairs past and present in the form of a recognizably 'geopolitical' mode of analysis. For the earliest extended efforts in this direction we should turn to the work of Herodotus and Thucydides, the two great historians of the fifth century.

The primary task which Herodotus set himself in his sprawling and often highly entertaining *Histories* was to explain how the Greeks came to fight and defeat the Persians in what became known as the Persian Wars (480–479 BC). Significantly, having dismissed prevailing popular accounts, which sought to trace the mutual hostility back to events in the legendary past, Herodotus (1.6) begins his own explanation by focusing not on a particular event, but on a specific geographical area, the narrow strip of land along the Aegean coast of Asia Minor occupied by the Ionian city-states which marked the interface between the Greek world and the non-Greek kingdoms to the east. As the evolving narrative makes clear, it was a series of major political upheavals affecting this region – its subjection first to the Lydian and then to the Persian Empire, and its subsequent revolt against the latter under Aristagoras – which, in the eyes of Herodotus, set in motion the chain of events leading to the Persian Wars. What we have here, in other words, is an ancient anticipation of modern 'flashpoint' analysis.

We saw a little earlier how, when planning his revolt against Persia, Aristagoras of Miletus had sought out the counsel of a geographer, his fellow-Milesian Hecataeus, only to ignore his advice with fateful consequences for the Ionian Greeks and for himself. This consultation is in fact representative of a larger pattern or theme in the *Histories*, one which recurs at key points in the Herodotus' account of the sequence of events which led to the wars with Persia. In each case, an influential actor plans a major military venture, encouraged typically by his own greed, emotions and ambition or by the misinterpretation of dreams and oracles. He then consults with a knowledgeable advisor about the wisdom of the plan. After evaluating the situation, the latter states his objections and recommends abandoning the venture. The advice is invariably ignored and disaster predictably ensues.

In terms of Herodotus' larger narrative purposes, the most significant instances of this theme concern the decision of the Lydian king Croesus to attack the Persian army (*c.* 546 BC), that of Aristagoras to revolt against Persian hegemony (*c.* 500 BC) and, most consequential of all, that of the Persian king Xerxes to invade Greece in 480 BC. For our purposes, the most noteworthy feature here is the fact that the advisor's objections in all three instances are stated from a distinctively 'geopolitical' perspective. We have already looked at the advice given by Hecataeus to Aristagoras, and one further example will suffice to illustrate the point.

According to Herodotus, Xerxes' decision to invade Greece was motivated primarily by a desire for revenge against the Athenians for their successful resistance to a Persian attack ten years earlier at Marathon, an attack which was itself conceived in part as a punitive measure against the Athenians for their support of the Ionian revolt in *c.* 500 BC. During his long and suitably dramatic account of the process by which Xerxes arrived at his momentous decision, Herodotus counterposes the sage objections of his uncle Artabanus. The intervention fails, and Xerxes goes on to mobilize an army and fleet of vast proportions for the expedition. But when the Persian hordes are just about to cross from Asia to Europe by means of a pontoon bridge which has been specially constructed over the Hellespont, Artabanus makes a last-minute attempt to pull his nephew back from the brink. He opens his appeal (7.47–49) by claiming that 'the two mightiest powers in the world' are against them, namely 'the sea and the land'. He then proceeds to elaborate on this rather cryptic observation, beginning with an explanation of the sea's hostility:

> So far as I know there is not a harbor anywhere (in Greece) big enough to receive this fleet of ours and give it protection in the event of storms: and indeed there would have to be not merely one such harbor, but many – all along the coast by which you will sail. But there is not a single one; so I would have you realize, my lord, that men are at the mercy of circumstance, and not their master.

> Now let me tell you of your other great enemy, the land. If you meet with no opposition, the land itself will become more and more hostile to you the further you advance, drawn on and on; for men are never satisfied by success. What I mean is this – if nobody stops your advance, the land itself – the mere distance growing greater and greater as the days go by – will ultimately starve you.
>
> (Trans. Aubrey de Selincourt, Herodotus (1996)
> *The Histories*, Harmondsworth: Penguin Books)

Both of these observations turn out to be prophetic, though Xerxes again chooses to ignore his uncle's advice and proceeds headlong with an expedition that will end in humiliating failure.

What, if any, are the implications of the recurrence of this theme at key points in narrative? It is hard to avoid the conclusion that, for Herodotus, the long sequence of events which led ultimately to the Persian Wars was driven, at least in part, by the failure of a series of powerful political figures to take geographical factors into account before they embarked on major, and ultimately disastrous, initiatives. That Herodotus himself had a profound interest in geography and the work of pioneers like Hecataeus is clear enough from the numerous anecdotes and digressions relating to the subject which punctuate and, at times, threaten to overwhelm his narrative. That he also had some appreciation of geography's political implications is revealed by the kind of arguments he chooses to put in the mouths of 'informed' individuals like Artabanus, arguments which give him some claim to be considered perhaps the first serious student of 'geopolitics'.

When we then turn to Thucydides and his *History*, an account of the exhausting war between the Athenian Empire and the Peloponnesian League which lasted from 431 to 404 BC, we find a considerably more disciplined and sequential narrative of events, along with an altogether more sophisticated application of geopolitical analysis. Thucydides' identification early in his text of Corcyra (modern Corfu) and Potidaea (on the northern Aegean coast) as key 'flashpoints', where the conflict of Athenian and Corinthian interests helped to precipitate the outbreak of the Peloponnesian War in 431 BC, still provides the basis for modern interpretations of the war's immediate origins.

In the opening section of the *History*, Thucydides attempts to reconstruct the history of Greece from prehistoric times down to the mid-fifth century with a view to demonstrating that no earlier conflict could have been as significant as the Peloponnesian War. Working from paltry fragments of mostly unreliable information, Thucydides nevertheless manages to produce a remarkably lucid and coherent picture of Greek socio-political evolution, one grounded in insightful analysis of the historical interplay between politics, economics and geography. The following passage, where he attempts to explain why Attica had historically enjoyed greater political stability than other areas of Greece, is particularly characteristic (1.2):

> Where the soil was most fertile, there were the most frequent changes of population, as in what is now called Thessaly, in Boeotia, in most of the Peloponnese (except Arcadia), and in others of the richest parts of Hellas. For in these fertile districts it was easier for individuals to secure greater powers than their neighbours: this led to disunity, which often caused the collapse of these states, which in any case were more likely than others to attract the attention of foreign invaders. (Meanwhile) Attica . . ., because of the poverty of her soil, was remarkably free from political disunity, (and) has always been inhabited by the same race of people.
>
> (Trans. Rex Warner, Thucydides (1972) *History of the Peloponnesian War*, Harmondsworth: Penguin Books)

The text of the *History* appears to be unfinished, and, as it stands, reaches its climax in the disastrous Athenian expedition to Sicily of 415–413 BC. For Thucydides there was an almost tragic inevitability to this event, being in his view the ultimate product of years of mismanagement and arrogant miscalculation by the Athenians since the death of Pericles in the early stages of the war. In this particular instance, miscalculations based on a lack of geographical knowledge were to play a major role in the Athenians' eventual undoing, as the historian makes clear at the very beginning of his account of the expedition (6.1): 'They were for the most part ignorant of the size of the island and of the numbers of its inhabitants, both Hellenic and native, and they did not realise that they were taking on a war of almost the same magnitude as their war against the Peloponnesians.' This then is further developed in Thucydides' subsequent account of the debate in the Athenian assembly which led to the fateful decision, the underlying scheme in many ways recalling the pattern we observed earlier in Herodotus. Here, the decision-making process is similarly characterized as a triumph of greed, ambition and emotion over a more pragmatic caution based on sound assessment of the geopolitical situation, the former case being represented by the charismatic young politician Alcibiades, the latter by Nicias, the 'elder statesman' of Athenian domestic and foreign policy. Among the more effective arguments offered by Nicias are the following (6.10–11):

What I say is this: in going to Sicily you are leaving many enemies behind
you, and you apparently want to make new ones over there and have
them also on your hands. . . . [I]t will certainly not stop our enemies
from attacking us immediately, if in any part of the world any considerable forces of our own should suffer defeat. . . . [T]his is no time for
running risks or for grasping at a new empire before we have secured
the one we have already. . . . [E]ven if we did conquer the Sicilians,
there are so many of them and they live so far off that it would be
very difficult to govern them. It is senseless to go against a people who,
even if conquered, could not be controlled, while failure would leave us
much worse off than we were before we made the attempt.

(Trans. Rex Warner, Thucydides (1972) *History of the
Peloponnesian War*, Harmondsworth: Penguin Books)

The fact that the narratives of both Herodotus and Thucydides represent the
failure of major political initiatives as being the result, at least in part, of
inadequate geographical knowledge suggests that a keen interest in geography's political implications had already emerged among informed observers
in Greece by the end of the fifth century, an interest which can be traced
back ultimately to the geographical enquiries pioneered by the sixth-century
Ionians. The next great phase of geographical enquiry came, unsurprisingly,
in the era following Alexander's conquest of the Persian Empire, which
dramatically expanded the horizons of the Greek world. As in the case of
sixth-century Ionia, state economic and political interests provided much
of the impetus for these new directions in research.

For the Greek and Macedonian rulers who divided up the conquered territories, the provision of precise geographical information was obviously of
enormous value in their efforts to control and exploit their kingdoms. Here,
a precedent had been set by Alexander himself, who employed teams of
surveyors, known as 'bematists', to measure out distances between key points
in his new empire and submit published reports on the forms of human, animal
and plant life they encountered along the way. In a similar spirit, Seleucus I
Nicator, who inherited the eastern portions of the empire after serving as a
general under Alexander, later sent one Megasthenes to visit the Mauryan
emperor Chandragupta at Pataliputra on the Ganges, the result being the *Indika*,
the single most influential work on India produced in classical antiquity.

But the information yielded by these and other state-driven research projects
was not just of interest to the rulers of the new kingdoms. It also opened
up new avenues of enquiry for academic figures like Eratosthenes, himself
an employee of the Ptolemaic court in his capacity as chief librarian at
Alexandria. In particular, it helped to make possible the development of
the more scientifically grounded approach to geography that was one of the
hallmark intellectual achievements of the Hellenistic age.

Following the entry of Rome into the affairs of Greece in the later third
century, the dominant geopolitical perspective is further transformed from

the regional to the 'global'. As far as we can tell, the first individual to apply
this new perspective in a systematic study of recent political developments
was Polybius of Megalopolis, the last of the major Greek historians. Written
in the mid-second century BC, the overall purpose of his *History* is to explain
(largely to his fellow Greeks) how in less than 53 years (220–167 BC) the
Romans had 'succeeded in bringing under their rule almost the whole of
the inhabited world, an achievement which is without parallel in human
history' (1.1). As in the texts of the two earlier historians, geopolitical analysis
plays a central role in Polybius' *History*, though the scale of the geopolitical
environment discussed would have been unimaginable to Thucydides, let
alone to Hecataeus. After all, as Polybius observes (1.3), following Rome's
defeat of Carthage at Zama in 202 BC, the known world and its history
become, as it were, 'an organic whole', since 'the affairs of Italy and of Africa
are connected with those of Asia and Greece, and all events bear a relationship and contribute to a single end'.

But the most impressive sustained work of geopolitical analysis produced
in antiquity is unquestionably the *Geographia* of Strabo, a colossal endeavour
spread over seventeen books. Writing during the reign of Augustus, Strabo,
much like Polybius, sought to promote mutual understanding between Greeks
and Romans, the result being in many ways an eloquent apology for the
new Augustan order, which, in the author's view, had brought civilization
to the known world. As it stands, the work is also the most comprehensive
ancient statement we possess of the importance of geography in the shaping
of political outcomes. Whether describing how the river system of France
had encouraged contacts between the region's native peoples, while also
making it vulnerable to the influence of outsiders (4.1.4), or how the location and character of the Italian peninsula had equipped it especially well as
a base for world domination (6.4.1), Strabo repeatedly displays a profound
understanding of the historical interplay between geography and politics. It
is no surprise, therefore, when, near the beginning of his work (1.1.16), he
explicitly recommends geographical study to all those generals and statesmen
who aim to influence world events, especially those who would 'bring cities
and peoples into a single empire'.

As we noted earlier, it would be left to Claudius Ptolemy to bring the
tradition of Greek geography to its intellectual culmination, the eight books
of his *Geography* representing the ancient world's most detailed and enduring
attempt to project the human and physical features of the known world onto
the surface of the globe. Though this work is less explicitly 'geopolitical'
than that of Strabo, marking as it does a return to the kind of mathematically driven geography pioneered in Hellenistic Alexandria, there was by now
little need to justify such an endeavour, since the practical value of geographical research, especially to those concerned with political dominion, had
already long since been firmly established.

From this brief survey of ancient Greek interest in the interplay of politics and geography, I would offer two firm conclusions. First of all, it seems

safe to say that geopolitics is hardly a new science. From the sixth century BC down to the time of the Roman Empire, the political implications of geography came to be deeply appreciated by decision-makers and systematically explored by intellectuals of various kinds. Second, it is striking how Greek geopolitical analysis was able to adjust its perspective over time to a political environment that was in a state of constant change, whether confronting an energetic if precarious world of small independent city-states, or, centuries later, the relative certainties, however unpleasant, of a world unified and governed by Rome. The post-war twentieth-century world, with its multifarious and continually shifting patterns of political interests, is of course infinitely more complex than anything imaginable in antiquity. But so long as meaningful political decisions are being made at the local, national, international and supranational levels, it is surely incumbent upon the modern science of geopolitics to engage with this complexity and elucidate the geographical contribution to political outcomes at all levels, in other words, to show the kind of adaptability to changing political realities that was the hallmark of its ancient predecessor.

Greg Anderson
Departments of History and Classics
University of Illinois, Chicago

1 Introduction

Since the Second World War, the Middle East more than any other global region has been the focus of international attention. This high profile has been evident for a number of reasons. The Middle East has a central location in the World Island at the meeting place of the three Old World continents. Situated at the major crossroads of global cultures, it is hardly surprising that the Middle East has been the scene of conflict throughout time.

During the modern period, two key factors have become superimposed upon this long-established pattern to ensure that the Middle East remains the world's premier geopolitical flashpoint. The first factor was the establishment of the state of Israel centrally in the Arab core of a region which is overwhelmingly Islamic. The second factor was the realization that between them certain Middle Eastern states possessed the world's major reserves of petroleum. Since petroleum is generally considered to be the most strategic commodity and the one upon which the only truly global industry has developed, the unique geopolitical significance of the Middle East is clear. Over the medium term at least, there is likely to be an increasing dependence for oil upon a region which is seen as among the most volatile. However, suffice it to say that there are probably as many misperceptions as there are accurate perceptions about the Middle East.

Misperceptions arise because of the innate mystique of a region apparently dominated by two things foreign to the Eurocentric viewpoint: an extreme climate characterized by aridity and Islam. Furthermore, until recently, for a variety of reasons, very few from the Western world had travelled at all extensively throughout the Middle East. Even today, only Turkey, Egypt, Israel and to a far lesser extent Jordan among the countries of the region have developed tourism on any scale. Even in those four states, most visits are limited to a relatively small number of globally acclaimed sites. Thus, very few people in the West are familiar with the Middle East and able to make rational judgements from experience. To the many, the region remains a problem, possibly the problem, in international relations.

Given the reality of the diversity within the Middle East, a more fundamental issue is the fact that, even among experts, the exact definition of the region remains unclear. From the Great Age of Discovery in the fifteenth century, it had increasingly become customary to distinguish between the Near East and the Far East. The Near East comprised essentially the eastern Mediterranean and adjacent lands while the Far East was everything east of India. The Indian subcontinent was in terms of trade and military strategic thought the centre of the British Empire and was, other than its most northerly and westerly approaches, accepted as a separate entity located in the East but not the Near East or Far East. Thus, the area in the middle, the Middle East, was almost by default that lying between the Indian subcontinent and the Near East or Levant. Geographically, therefore, there is some logic in the designation of what is now considered the Mashreq as the Middle East. However, this area accounts for only part of the modern region, the basis for which was geopolitical rather than geographical.

Although the term Middle East probably had currency in the British India Office during the 1850s (Beaumont *et al.* 1988), for its introduction into accepted terminology credit is normally given to Alfred Mahan, the American geopolitical historian. The focus of the Middle East as he envisaged it was the Persian–Arabian Gulf. It has been variously interpreted that Mahan saw this area as 'middle' in an east–west sense but also possibly in a north–south sense, essentially between the Russians to the north and the British to the south.

It was, however, with the dawn of the age of global warfare that the Middle East was distinguished as crucial in grand strategy. In the First World War, the operational area of the Mesopotamia Expeditionary Force was considered the Middle East while that of the Egyptian Expeditionary Force was characterized as the Near East. Until the end of the First World War therefore, geographical and geopolitical definitions tended to coincide.

Between the wars, the Middle Eastern Command and the Near Eastern Command of the Royal Air Force were amalgamated under the title of the former. This precedent was followed by the Army which, at the beginning of the Second World War in defence of the Suez Canal in particular, established its GHQ Middle East in Cairo. Effectively, a military province stretching from Iran to Tripolotania was created and named Middle East. The importance of the region, and in particular its political and economic life resulted in the appointment of a Minister of State and the development of an economic organization, entitled the Middle East Supply Centre. The Centre was originally British but later Anglo-American and thereby the term Middle East became generally recognized not only in the Eurocentred world but also in the home of its originator.

After the Second World War, the territorial designation adopted by the military authorities continued and Middle East became the standard term of reference, exclusively used in numerous government publications summarizing political events, territorial surveys and schemes of economic development. An additional possible explanation for its incorporation into official terminology has been advanced in *The Middle East* (1978) by Fisher who suggests that

France had strong military claims in an official Near East theatre of war, but fewer in a Middle East, which was therefore much employed as a term and extended as a geographical concept when the situation of France *vis-à-vis* Britain became equivocal in 1940–41. This idea may or may not be valid but it is clear that over the post-war years, the term Near East, so redolent of Balkan Europe in the nineteenth century, has faded from common usage. Its connotation in geographical terms was always vague if rather more exact than that of Middle East.

Throughout the vicissitudes of the Cold War, the Middle East remained central in the concerns of East and West. Both the USA and the Soviet Union developed close relationships within the region while some states, such as Egypt, changed sides. As a result, the term Middle East has become sanctioned by use world-wide, including within the region itself. Nonetheless, while the geopolitical concept of the Middle East has become relatively clear-cut, the geographical boundaries do not enjoy universal recognition. In the USA, the State Department has compromised by using the term Near and Middle East. In the US military, the region is divided between three Commands, the European, the Middle Eastern and the African. However, it is reasonable to conclude that the major problem of delimitation concerns the states of North Africa. As a result, in many volumes the term 'Middle East and North Africa' is used as a unit.

As with regional geography in general, there is little difficulty in defining the core, it is the boundaries which cause problems. The Mashreq is tightly integrated into the core but the Maghreb stretches westwards as far as Mauritania. Neither geographically nor geopolitically could Mauritania be considered Middle Eastern. Therefore, an appropriate division would seem to be between the states predominantly of the desert and those of the Atlas Mountains and beyond. Libya is accepted as a state of the Maghreb but it would appear to have more in common with Egypt including shared aquifers. Other issues of definition concern Sudan, the southern part of which is clearly more closely associated with Central Africa than the Middle East. Nonetheless, Sudan is a Nile Valley state and there is historical and geographical integration with Egypt. A small but highly significant area of Turkey is technically within Europe but few would doubt that the culture of the state places it firmly within the Middle Eastern setting. The other contentious issue concerns Cyprus because of its dominant Greek connection. However, the island is located geographically within the Middle East and, since partition, its links with Turkey have been strengthened.

Therefore, in this volume, following the pattern established in *The Middle East* (Fisher 1978), the Middle East is taken to include all the generally accepted states together with Libya, Sudan and Cyprus (Figure 1.1). However, it should be noted that in discussing physical landscapes and in particular historical developments, there are several useful terms which predate the modern mosaic of states and, where appropriate, these have been retained. They include Levant, Mesopotamia and Asia Minor.

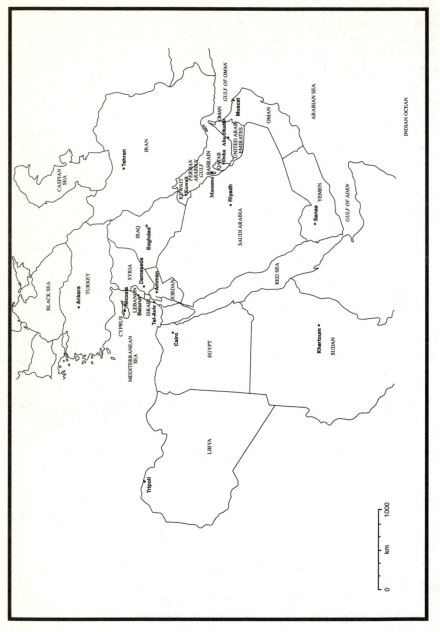

Figure 1.1 Middle East states and capitals.

These terms, current in earlier periods, provide a reminder of the early importance of the Middle East in which the Egyptian, Sumerian, Babylonian and Assyrian civilizations flourished. Within the region also arose three great monotheistic religions: Judaism, Christianity and Islam. A fourth even older, Zoroastianism, arose nearby and spread into the region. The fusion of cultures and civilizations within the Middle East results from the fact that the region was successively part of the Persian, Greek, Roman, Arab, Mongol, Tartar and Turkish Empires (Peretz 1978). As might be expected from its location, the Middle East has been central in human affairs.

By delimiting the region as extending from Libya to Iran and Turkey to Sudan it is possible to postulate the existence of a natural region designated the Middle East on geographical grounds. The region has an unusual characteristic climatic regime which has induced highly distinctive human responses and activities. According to Fisher (1978) in *The Middle East*, the common elements of natural environment and social organization are sufficiently recognizable and strong to justify treatment of the Middle East as one single unit. However, despite the overall unity of climate and culture, the Middle East, when examined in detail on a smaller scale, reveals enormous diversity. There is frequently a juxtaposition of harsh desert environment and intensively cultivated rich agricultural land or extreme modernity and ancient tradition or extraordinary wealth and grinding poverty.

The modern Middle East is central between Europe, Asia and Africa but it can now be seen as intermediate between the developing and the developed world. The Middle East is on the North–South boundary. Through Islam with its 1.2 billion adherents, it has links to most countries in the world while geographically it abuts on to more different cultures than any other. As a result of oil, it is closely aligned with the three great economic regions: the USA, the European Union (EU) and Asia Pacific. Whether viewed geographically or geopolitically, the Middle East is central and exercises a global influence.

2 Land structure and form

The landscape comprises an underlying structure and a surface form. In the Middle East, large areas of which are totally unencumbered by human artifacts or even vegetation cover, the relationship between structure and form is more apparent than in probably any other major region of the world. The principal fault patterns can be clearly seen, the geological characteristics are relatively obvious and the effects of the major geomorphological processes are clear-cut. However, underlying this clarity is complexity. The foundations have been laid by the tectonic history, the main events of which are on a macro-scale. Given the vast extent of the timescale involved and the global nature of the movements, the finer detail of this must remain to an extent conjectural. On the mosaic of land masses resulting from tectonic activity, the geological sequence has been laid. Through emergence and submergence, drier periods and wetter periods, the stratigraphy varies in time and location throughout the Middle East. In the context of tectonics, these would be considered meso-scale differences. The emergent land mass has then been moulded by geomorphological processes to produce the present landscape. Many of these processes operate on what is, comparatively, a micro-scale. Thus there are broad differences in scale between the events which have occurred in tectonics, geology and geomorphology.

There are also differences in the current state of knowledge about these subjects. Interest in the economic aspects of geology has tended to relegate the more theoretical questions of origins to a secondary role. This situation, however, has improved in recent years particularly following the elaboration of the theory of tectonics, which has special applicability to the Middle East. An improved appreciation of the locations of major faults, earthquakes and volcanic activity has allowed the edges of the plates to be more accurately identified. With regard to the geology, the depth of knowledge across the region varies considerably. In areas with a potential for hydrocarbons, drilling has been relatively intense and there is as a result abundant information about the underlying stratigraphy. This has been supplemented by the growing and pressing requirement for geological data concerned with aquifers. In areas where these interests do not obtain, knowledge is more sparse. However, the situation is being rectified as a result of enhanced survey procedures using, for example, satellite imagery. Furthermore, fundamental research has focused on certain key areas in the Middle East such as the Red Sea Basin and its extension northwards through the Jordan Valley.

Geomorphological research presents rather different problems. To understand the operation of processes in the development of landforms, it is normally necessary to establish monitoring programmes. Much of the Middle East, particularly the hyper-arid areas, is inhospitable for sustained research. Again, improved approaches to surveying have enhanced the situation but many key geomorphological events occur on a small scale and there is no real substitute for fieldwork. For example, while broad patterns of erosional change can be established from aerial photographs or satellite images, the measurements of rates of erosion and the relationship between these and the processes involved can only be established by meticulous field procedures. It remains true that there are less textbooks on arid zone geomorphology than there are journals on glaciology published annually. Essentially, tectonics, geology and geomorphology are intimately related. For instance, tectonic movements can, through emergence and submergence, control the geological succession and through lateral movements induce folding. The thrusting up of fold mountains and the development of faults affects the shape of the landscape and geomorphological processes. Geology, as a result of the permeability, structures and resistance to erosion of rocks, influences the development of landforms.

Tectonics

The basis for the fundamental movements which result in major global features, particularly the distribution of land and sea, is plate tectonics. The surface of the Earth comprises a series of plates, some very large and some relatively small. The movement of these plates results in events at their edges which produce change sufficiently discernible for the broad location of those edges to be identified. The movements take place in the crustal layer and are generated by convection currents produced as a result of differential heating within the mantle.

There are three categories of plate edge effects. Plates may be forced apart by upwelling from the mantle and this movement is termed 'spreading'. Alternatively, plates may move together, the result being that one plate is compressed against another. One plate is forced below the other and its lower advancing edge is then consumed in what is termed a 'subduction zone'. Third, two plates may move laterally with respect to each other producing transform faults. Clear-cut examples occur of all three in the Middle East. The central trough of the Red Sea is a zone of spreading while to the north-east, at the other end of the same plate, the Taurus and Zagros mountains indicate a subduction zone. The Jordan Valley represents a transform fault. While there is some variation in interpretation, the conventional large-scale model suggests that for much of its evolution Arabia was part of the African continental plate, the regions of what are now the Taurus and Zagros mountains representing the more active leading edges of that plate (Figure 2.1). The African and Arabian plate drifted northwards to collide with the Eurasian plate, crushing as it did so the sediments laid between the two plates in the Tethys Sea, a geosyncline formed during Triassic–Jurassic times. The Tethys Sea was a vast geosyncline or downfold in which great thicknesses of sedimentary deposits had been formed as a result of erosion of the land plates to the north and south. As these sediments accumulated, they trapped the organic matter which eventually gave rise to the occurrence of hydrocarbons. At the time of the collision, the northern part of Iran lay at the southern edge of the Eurasion plate and the Zagros belt represents the main area originally occupied by the Tethys Sea. Following this at the same time, structural stresses occurred in southern Arabia which resulted in faulting along the median line of the Red Sea and the Gulf of Aden. This tension was related to spreading and resulted in rifting and the eventual formation of the Red Sea and the Gulf of Aden.

The chronology of these events has been described by Shannon and Naylor (1989). During the Precambrian, Arabia was part of the Gondwanaland super-continent. From that time until the Late Palaeozoic, there were three orogenic events corresponding to plate collisions together with a number of epeirogenic events. The Palaeozoic closed with the main phase of the Hercynian orogeny. During the Mesozoic the Tethys Sea developed and the Mesozoic ended with the onset of the Alpine–Himalayan orogeny. Synchronous arching of

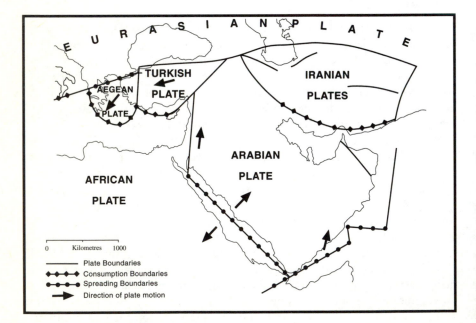

Figure 2.1 Middle East: tectonic plates.

western and southern Arabia resulted in faulting, rifting and collapse which occurred during the Tertiary. By the beginning of the Tertiary the Tethys Sea had been reduced to a residual trough line. The Arabian plate is held to be active today, evidenced by further elevation of the Zagros, continued widening of the Red Sea trough and activity along the Jordan transform fault.

Within this broad pattern of events, other smaller scale activity occurred. By the late Mesozoic, one or more Iranian plates, of smaller size, had separated. Around these plates the Tethys sediments were later compressed to give the present-day local basins or micro-plates of Iran and Asia Minor with their surrounding fold garlands. The effects of the Red Sea and Gulf of Aden spreading was to detach the Arabian plate from the African plate. This not only resulted in the continuation of movements present today but affected the plates composing Asia Minor and the Aegean Sea. A Turkish plate comprises most of central Asia Minor and an Aegean plate is located immediately to the west. As a result of the extreme pressures resulting from the continued northerly motion of the Arabian plate, these two small plates are being thrust westward with transform faulting along their northern edges and consumption of their southern edges against the African plate.

Due to this tectonic evolution, it is possible to distinguish broad regional differences within the Middle East. In *The Middle East* (1978), Fisher identified two major zones:

1   central and southern parts, which in the main are continental plate areas, interrupted only by the fault structures resulting from spreading or transform faulting; and
2   the zone where moving plates impinge strongly, due to the northward movement of the Arabian and African plates.

Shannon and Naylor (1989) recognize three broad geotectonic areas. To the south and west is the Arabian Shield comprising Pre-Cambrian massives and separated from the Nubian Shield of Africa by the faulting of the Red Sea depression. To the north and east of the Arabian Shield is an area termed the Stable Shelf. Like the Arabian Shield, this area remained basically stable historically, dipping gently towards the basin centre in the north-east. This area can be subdivided into an interior 'homocline' which immediately borders the Stable Shelf and to the north-east again, an 'interior platform'.

Geology

The structural evolution of the Middle East with plate movements, faulting, volcanicity, emergence and submergence together with drier and wetter climates has produced a complex geology (Figure 2.2). There are marked variations in the sequence according to location and therefore a general picture can only be obtained through simplification and generalization. The

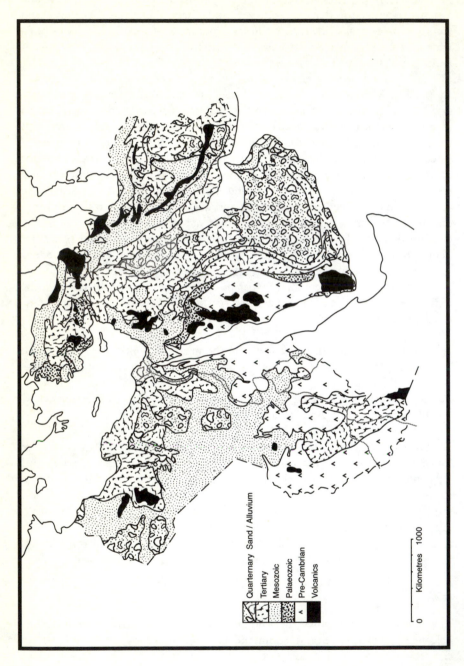

Quaternary Sand / Alluvium
Tertiary
Mesozoic
Palaeozoic
Pre-Cambrian
Volcanics

0   Kilometres   1000

Figure 2.2 Middle East: generalized geology.

oldest rocks are Pre-Cambrian which form a crystalline basement outcropping in North Africa and various parts of the Arabian Peninsula. In general terms, the overlying rocks are sedimentary and become younger from south to north. Intermingled with the sediments are igneous intrusions and volcanic features including large-scale larva flows. Good examples of the latter occur in eastern Jordan and the Jebel Druze in Syria. The complexities of the facies can be judged from the generalized relationships illustrated by Shannon and Naylor (1989) which provide a section from south-west to north-east extending from the Carboniferous to the Quaternary. Figure 2.3 provides a simplified version. It is apparent that limestones dominate the column. Emergence and submergence are illustrated by the alternating successions of sandstones and conglomerates followed by shales and then limestones. Located within this pattern but at no time extending across the entire region are deposits of gypsum and anhydrite. Similarly concentrated in basins but occurring principally during the latter Jurassic and Mid- to Late-Cretaceous are the important bituminous shales.

The variations according to location can be illustrated by profiles taken at different stages across the section. At the south-western end, much of the sequence is missing but that which occurs is almost equally divided between limestones and sandstones, conglomerates, siltstones, shales and marls. A profile through the centre of the section shows a preponderance of shales and marls over limestones but also includes the important successions

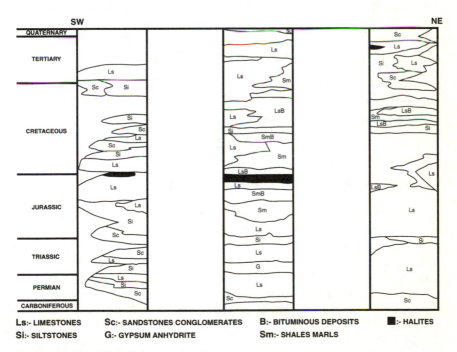

Ls:- LIMESTONES   **Sc:-** SANDSTONES CONGLOMERATES   **B:-** BITUMINOUS DEPOSITS   ■:- HALITES
Si:- SILTSTONES   **G:-** GYPSUM ANHYDRITE   **Sm:-** SHALES MARLS

Figure 2.3 Simplified facies relationships.

of halytes, gypsum and anhydrite and bituminous shales. A profile of the north-eastern end of the section is dominated by limestone until the Mid- to Late-Cretaceous. From then until the Quaternary much of the succession is missing and that which occurs is predominantly composed of siltstones, sandstones and conglomerates.

The key elements of the geology are well summarized by Burdon (1982) into five categories. The Precambrian basement comprises crystalline rocks which outcrop notably in the Tibesti and Ahaggar Ranges in the Sahara, the mountains of the Eastern Desert of Egypt, western Saudi Arabia, Yemen and, to a lesser extent, in Sinai and around Aqaba. Epicontinental sediments laid upon the basement comprise, in the Palaeozoic, predominantly sandstones of various kinds. In the later Palaeozoic and throughout the Mesozoic, these become increasingly intercalated with and eventually dominated by limestones and other marine sediments. In the region of the former Tethys Sea, deposits of terrestrial origin, eroded from the littoral continents, are interdigitated with sediments of direct marine origin. As the Arabian plate approached Eurasia, the Tethys Sea became narrower and shallower and Tertiary evaporite deposits become of importance. The plate movements disturbed the vast thicknesses of sediments laid in the Tethys Sea, the disruption caused varying from the gentle folds of the Jebel Akhdar in Libya to the intense folding and thrusting apparent in the Zagros Ranges and the mountains of northern Oman. During the same period, the disruptions resulted in the Tertiary volcanic outpourings of basalt which are associated with the lines of rifting. They occurred not only in Syria and Jordan but also in the Jebel Marra of Sudan, and the Jebel Soda and Jebel Haroudj of Libya. The fifth element identified by Burdon is the Quaternary cover. This comprises the sand, predominantly sand seas and gravel planes. Of more limited extent and more specific in location are lacustrine deposits, inland and coastal *sebkha* and the outwash gravel terraces which fringe the mountains. Figure 2.4 provides a summary of the main structural elements.

Geomorphology

Landforms are the product of processes operating for varying lengths of time upon the structure of the landscape. With so many possible variables it is hardly surprising that Thornbury (1954) lists among the cardinal principles of geomorphology that landscapes tend towards complexity rather than simplicity. At first sight, given the abundance of flat or nearly flat surfaces in the Middle East, the region might seem to provide an exception to this general rule. However, not only does the basic structure present wide variations and complications but there are major controversies surrounding the efficacy of the geomorphological processes involved. The area is very large and detailed research is limited.

A key point is the extent to which current landforms reflect present geomorphological processes rather than those which obtained during past climates.

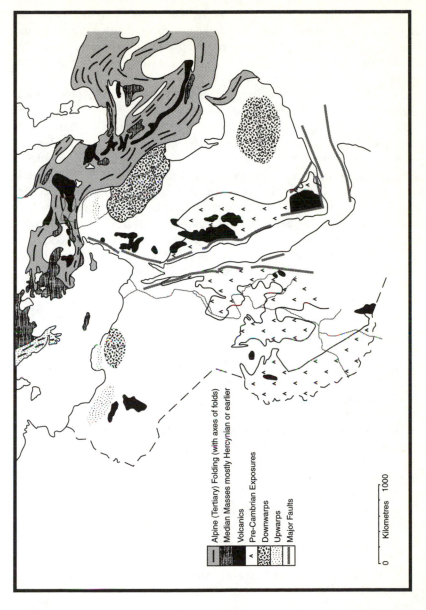

Alpine (Tertiary) Folding (with axes of folds)
Median Masses mostly Hercynian or earlier
Volcanics
Pre-Cambrian Exposures
Downwarps
Upwarps
Major Faults

0   Kilometres   1000

Figure 2.4 Middle East: structural elements.

At the present time, the most obvious processes are aeolian but questions must be raised about their effectiveness in producing major erosional landforms other than through deflation. The effects of run-off generated by the characteristically short duration intense storms can be far more devastating, particularly in a landscape largely unprotected by vegetation cover. However, for large parts of the Middle East, the vast plains covered in lag gravels would seem to be relict landforms, virtually unaffected by present processes. This discussion is developed further with an analysis of the major landforms, particularly the constructional features, by Cooke *et al.* (1993).

Controversy also surrounds dating, although the situation is being alleviated by modern techniques. Since there are outcrops in the Middle East ranging from the Pre-Cambrian to the Quaternary, there are almost any number of time sequences in which landforms might have developed. Over such lengths of geological time, no variation in structure or process can be considered insignificant. However, recent research suggests that landform development is episodic, involving climatic changes in time-scales of anything from decades to centuries (Lancaster 1992). Evidence indicates that the morphology of modern dune systems may be less than 2,000 years old. Attempts continue to relate eustatic and tectonic changes to such features as erosion surfaces.

A full coverage of Middle Eastern geomorphology would include consideration of a number of processes which, while they produce elements of the landscape, are not in any sense typical. For example, the assemblage of features produced by marine processes could occur in a wide range of other environments. For different reasons, despite the presence of what is reputed to be the second largest cave in the world in northern Oman, Karst development is omitted. Assuming at some stage the provision of sufficient water, Karstic landforms depend only upon the presence of appropriately structured limestone. In the Middle East, the key determinant of landforms is present-day aridity or hyperaridity combined with the effects of previous pluvial periods. The preferred terminology would be arid zone geomorphology in that it occurred in what is now the Arid Zone. The distinctive assemblages of landforms are those comprising:

1   slopes, varying in angle from mountain to plain, with their covering and mass movement features;
2   wadi networks in mountains, foothills and plains; and
3   dunefields or sand seas with their variety of aerodynamically shaped constructional landforms (Figure 2.5).

Owing to the general lack of moisture, subaerial weathering rates tend to be very slow. Therefore, apart from the steepest, slopes are generally covered in deposits ranging in size from boulders to fine gravel. Finer material is constantly removed by deflation but extremely fine 'desert dust' tends to infiltrate the surface. The general covering of detritus protects the surfaces from both weathering and erosion.

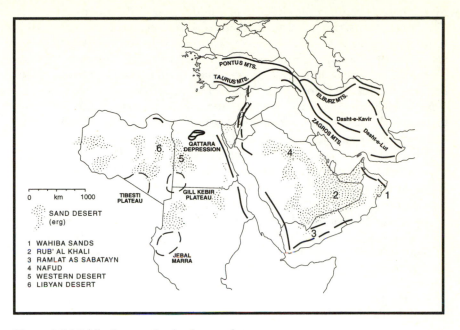

Figure 2.5 Middle East: major landscape elements.

Mass movement processes, triggered predominantly by the occasional heavy rainstorm, range from rock creep and debris avalanche to landslides. At the micro-level, the diurnal rhythm of heating and cooling together with, in favoured micro-habitats, wetting and drying, produce various forms of creep. In contrast, surface run-off in such an unprotected environment can be highly effective in both moving surficial material and eroding the surface itself.

Across the Middle East, fluvial features range from perennial rivers in the north in Turkey to the almost permanently dry wadis of much of the Arabian Peninsula to the south. In many of their features, both extremes are similar but there are substantial differences. These are indicated by a consideration of the *wied*, characteristic of Malta, which is located between the two extremities and appears as a hybrid (Anderson 1997). The form of the *wied*, its long and cross profiles, together with the visual impact of its vegetation cover accords with that of a river valley. The process, resulting from the extreme irregularity of flow and a hydrograph which mirrors that of the wadi, shows an affinity with the fluvial features in the arid south. In a study of four major rivers in western Turkey, Eisma (1978) identified four features common to the lower valleys:

1   hillside alluvial fans;
2   a flat alluvial valley floor some 5–10 m above the present river bed;
3   near the mouth, a broad floodplain; and
4   along the coast, lagoons, beach ridges and dunes.

All of these features are in evidence in Middle East wadi systems.

One question which arises is whether, as with rivers, wadi features can be related to a general sequence of river deposition and erosion. As yet, there is insufficient research to produce a chronology which will allow the approximate dating of wadi terraces and fans which would establish a relationship between the wadi systems of the Middle East and facilitate correlation with climatic and eustatic change. For the eastern Mediterranean, the following sequence of deposition and erosion put forward by Vita-Finzi (1969, 1973) allows comparison:

1   deposition of older valley deposits [Fill I] 20,000–10,000 BP;
2   erosion and down-cutting with the deposition of deltas 10,000–2,000 BP;
3   deposition of younger valley deposits [Fill II] 2,000–300 BP; and
4   erosion and increased delta-building 300–0 BP.

This chronology would seem to indicate increased erosion is related to drier conditions. The basis for the argument is that, with increasing aridity, rainfall tends to occur in powerful short bursts which, in the absence of anything like a complete vegetation cover, is highly erosive. In wetter periods, rainfall totals may be higher but downpours tend to be less intense and the land surface is protected by vegetation. Evidence from the USA indicates strongly that climatic changes play a major role in controlling patterns of erosion and aggradation in arroyos or wadis (Lancaster 1992).

Field research is limited but a detailed description of the current understanding of aeolian processes and landforms is provided by Pye and Tsoar (1990). Largely as a result of computer simulations, problems of fieldwork have been modified and theory has caught up with wind tunnel and experimental techniques (Anderson *et al.* 1991). Sand accumulations occur in many parts of the Middle East ranging from the linear dune fields of the northern Negev researched by Tsoar (1983) to the vast sand seas of Libya and western Egypt. In the Arabian Peninsula, the Rub' al Khali, with an area of some 560,000 km^2 and a total absence of surface water is predominantly composed of sand. The dunes, an assemblage of barchans, seifs, star and compound dunes in places reach a height of 300 m. In the Wahiba sand sea of Oman, an area of approximately 20,000 km^2 the dune types can be broadly separated regionally. In the north are the high, north–south aligned lineage dunes, towards the south is an area of transverse and crescentic dunes, and round the edges are the peripheral dunes often varying in character between the two (Jones *et al.* 1988). Difference in age are apparent in that the high linear dunes are themselves virtually stable but are capped by currently active barchans. According to Warren (1988), in the Rub' al Khali, Nafud and Wahiba sand seas aridity and dune formation occurred between 20,000 and 9,000 BP. As with wadis, there is increasing evidence that modern sand seas have accumulated episodically with climatic and eustatic changes controlling patterns of sedimentation (Lancaster 1992).

Consideration of sand seas introduces the significant issue of man as an agent of erosion. Attempts to limit the effects of many morphological agents such as marine waves and running water have been extremely costly or abortive but in the case of moving sand there have been important successes. The control of blown sand is fully considered by Watson (1990) who uses the Jafurah sand sea of eastern Saudi Arabia as a case study. In that area it is estimated that the annual drift is some 30 m^3 per metre width. The four main approaches considered are:

1   the promotion of deposition away from the problem area by ditches, barriers and fences or vegetation belts;
2   enhancement of sand transportation within the problem area by aerodynamic streaming, surface treatment or panelling (e.g. over a road surface);
3   the reduction of sand supply by surface treatment, fences or vegetation; and
4   the deflection of moving sand by fences, barriers or tree belts.

Variations in the geomorphology of the Middle East can be illustrated with reference to Egypt (Said 1990). The Western Desert is characterized by plateaux and escarpments with large-scale depressions and a vast sand sea dominated by linear dunes. The Eastern Desert is a rugged mountainous region minutely dissected by wadis. The east is essentially a rock desert, the west a gravel and sand desert.

The Red Sea system

Using the Red Sea and its associated features, it is possible to indicate how tectonics, geology and geomorphology are inter-related. According to Crossley *et al.* (1992), in the Late Oligocene the onset of rifting occurred and the Gulf of Aden and southern Red Sea were inundated by marine waters. At this stage, the northern part of the Red Sea was probably a largely continental rift. By the Early Miocene, continued rifting had established marine conditions throughout the system. However, episodic isolation of the Red Sea system leading to evaporite deposition in some basins began in Mid-Miocene and large deposits of salt had been accumulated by the end of the Miocene. In the Pliocene, maritime conditions were re-established and carbonate build-ups occurred in shallow water areas. The escarpments bordering the Red Sea which first developed during the onset of rifting were uplifted from the Miocene to the Quaternary, the evidence for this being drawn from clastic sediment. Conversely, there was subsidence of basin floors which appears to have been particularly rapid during the period of salt deposition. These movements resulted in the eruption of basalts from the axial trough along the median line of the Red Sea and volcanism elsewhere, particularly in the Afar triangle.

Geodetic surveys reveal magnetic anomalies along the axial trough and these together with the fact that it is filled with basalt, provide firm evidence that its genesis resulted from sea floor spreading. The magnetic lineation seems to indicate that the spreading occurred some 5–6 million years ago at the southern end and appears to be advancing from south to north (Said 1990). It would be reasonable to conclude from this that there was a pre-rifting stage during the Oligocene and that the main trough formed during the early to Mid-Miocene.

A different approach has been adopted by Crossley *et al.* (1992) who, from the synthesis of a range of data sources, produced a series of palinspastic reconstructions. The research drew on studies made on the amount and timing of movement on the fault together with work on the orientation of movements. The resultant pre-rift and drift configuration predicted by the reconstructions was found to be largely in agreement with the geological data presented (Beydoun 1970, 1982). Furthermore, the implications of the sequence of rift and drift for periods of accelerated and decelerated compression along the northern margin of the Arabian plate is in broad agreement with the structural data. In particular, Hempton (1987) postulated that the Late-Miocene termination of extension across the Red Sea and Gulf of Aden was due to the collision of the Arabian plate with the Eurasian plate to the north.

Thus, in several ways the geology helps confirm the tectonic history. The relationship between stratigraphy and rifting is clearly set out in Coleman (1993) in which the sedimentary history is divided into five sections:

1   basement sediments;
2   pre-rift sediments;
3   syn-rift sediments;
4   evaporite sediments; and
5   post-evaporite sediments.

There have been no systematic geomorphological accounts of the Red Sea Basin but certain elements of the geomorphology can be cited to illustrate the relationship with geology and structural history. For example, the Red Sea shelves show features which indicate emergence in the north and submergence combined with rapid deposition in the south. South of latitude 21°N the shelf merges imperceptibly with the coastal plain which broadens to a width of more than 50 km and in certain areas is covered by recent lava flows. North of 21°N the shelf becomes narrower and is interrupted by sharp topographic breaks, while the coastal plains themselves are narrower and display raised terraces representing older shorelines (Brown 1970).

However, more compelling evidence comes from the examination of the Red Sea islands of which there are three distinct types (Coleman 1993). At the southern end of the Red Sea is a median line series of volcanic islands: Jebal Tair, the Zubayr group, the Zukur group and the Hanish islands. These consist mostly of pyroclastic material and Jebal Tair is considered to be still active. Further surveys have revealed that there are many more small eruptive structures in the axial zone and further spreading could therefore lead to more volcanic features.

To the north of these islands and situated on the shelves are, on both sides of the Red Sea, a series of low-lying islands. These comprise marine sediments, evaporites and coral reef limestone, in no case, other than through faulting, exceeding 20 m above sea level. Both the Dahlak islands on the western side and the Farasan archipelago on the eastern display scallop-shaped coastlines providing evidence of the former location of salt domes which have since been dissolved by the seawater. Thus there is a major difference between the volcanic islands in the axial trough and the carbonate-based islands on the shelves.

A third category is represented by Zabargad Island. This consists of both mantle and lower-crust metamorphic rocks and rises some 235 m above sea level. Geologically it is composed of peridotites and amphibolites and is covered by tertiary marine sediments. It is suggested, from the geology, that this island represents a 'core complex' exposed during extentional faulting attendant upon the opening of the Red Sea.

3   Climate

The countries of the Middle East between them share a number of different climatic regimes. As in geomorphology in which there is no typical assemblage of landforms, so in climate there is not one characteristic data set for pressure, temperature, precipitation and related variables. Furthermore, not only are there climatic variations from area to area within the region but it is clear that there have been major changes through time. Indeed, climatic change has become a focus of research over recent decades although the emphasis has been on Europe and North America.

With its location at the crossroads of the world, most aspects of the geography and the geopolitics of the Middle East are subject to a wide range of influences. The climate of the region is no different, being transitional between the dry climates and the warm, temperate, rainy climates as defined by Köppen (1931). In the classification of Köppen, still regarded as the standard in the field, five major climatic regimes are distinguished, corresponding broadly to vegetation types. These are identified by the letters A, B, C, D and E, each of which is subdivided on the basis of temperature and rainfall.

The coastal fringe of eastern Libya and parts of Egypt, a small area of Yemen, the Levant coastlands, eastern and north-eastern Iran together with Turkey comprise the area classified within the warm, temperate, rainy climatic regime. The remainder of the Middle East, considerably the larger area, falls within the Köppen category of dry climates. Within the warm, temperate, rainy climates, designated C, three climatic types can be distinguished:

Cw: dry winter climate: the wettest month of summer has at least ten times as much rain as the driest month of winter;
Cs: dry summer climate (Mediterranean): the wettest month of winter has at least three times as much rain as the driest month of summer, which itself has less than 30 mm of rainfall;
Cf: climate with no dry season: even rainfall throughout the year.

To this may be added a further lower case letter to indicate a finer subdivision:

a: hot summer: mean temperature of the hottest month above 22°C and more than four months over 10°C;

b: warm summer: mean temperature of hottest month below 22°C but more than four months over 10°C.

Csa, the classic Mediterranean climate, is found along the Mediterranean coastal areas of the Middle East including Cyprus and the Turkish Black Sea coastal region. A more extreme subdivision, Csb is predominant throughout western and north-western Iran. Cfb is limited to the upland interior of Turkey, the Anatolian Plateau.

The dry climates are distinguished as:

BS: steppe climate;
BW: desert climate.

The B climatic regime is recognized by a combination of low rainfall and a wide range of temperature. The most effective division between BS and BW has been the source of considerable discussion and is examined later in this chapter. The exact designation is again modified by subsidiary lower case letters:

h: hot, dry climate: mean annual temperature above 18°C;
k: cool, dry climate: mean annual temperature below 18°C.

To the south of latitude 30°N, the Middle East is predominantly classified as BWh. The BSh climatic subdivision is found in northern Libya and north-western Egypt, in belts inland of the mountain chain on either side of the Red Sea, in an area immediately inland from the mountains of Oman and the United Arab Emirates (UAE), throughout much of Iraq and, other than a coastal and a central belt, throughout most of eastern Iran. The BSk subdivision is found only in the north-east of Iran.

Despite this variety, as with geomorphology, it is possible to identify distinctive aspects of climate. In the Middle East, the distinctive and predominant climatic types are steppe and desert. The perception of the region is of aridity and in fact parts of the Middle East are considered the most arid world-wide.

In terms more amenable to scientific enquiry, the desert (BW) and steppe (BS) climatic types of Köppen can be designated, respectively, arid and semi-arid. Many approaches have been made to distinguishing between the two (Agnew and Anderson 1992), the most significant being those based upon climatic indices such as that of Köppen (1931) and those employing the water balance method developed by Thornthwaite (1948, 1954). Additionally, using the formula for evaporation produced by Penman (1948), a map of the arid lands was interpreted for UNESCO (1977). The formulae for each of these three methods are as follows:

Köppen (1931)

Arid boundary: $P/T < 1$
Semi-arid boundary: $1 < P/T < 2$
where P = mean annual precipitation (cm); T = mean annual temperature
(°C) (after Yair and Berkowicz 1989).

Thornthwaite (1948, 1954)

Arid boundary: I_m = < -66.7
Semi-arid boundary: $-33.3 > I_m > -66.7$
where I_m = $100[(P/P_e) - 1]$; P = mean annual precipitation (mm);
P_e = mean annual potential evapotranspiration (mm) (after Mather 1974).

UNESCO (1977)

Arid and hyper-arid boundary: P/ET_p = less than 0.20
Semi-arid boundary: P/ET_p = 0.50–0.20
where P = mean annual precipitation (mm); ET_p = mean annual potential
evapotranspiration (after UNESCO 1977).

While aridity or even hyper-aridity are characteristic, a simplified and generalized climatic pattern can be identified for the region as a whole. There are
four key determinates of climate (Trewartha 1954):

1 latitude;
2 the distribution of land and sea;
3 specific meteorological systems, such as depressions or upper air controls;
   and
4 altitude.

In general, temperatures and aridity increase from north to south. The pattern
is of course complicated by altitude in that much of the land to the north
is mountainous and lower temperatures and higher precipitation would be
expected. Certainly low winter temperatures and snowfall occur in both
Turkey and Iran but never in Saudi Arabia. This pattern is modified by the
maritime influence of the Mediterranean and to a much lesser extent by
the Red Sea and the Persian–Arabian Gulf. Near the coast, extremes of
temperature are modified while there is commonly an increase in humidity.
Conversely, the effect of continentality, with enhanced temperature ranges,
can be seen in the Fezzan of Libya, upper Egypt and central areas of the
Arabian Peninsula.

The effect of meteorological systems can be seen in the control of climate
exercised by the interplay between lower air pressure systems and upper air
jetstream movements. Winter in the region is characterized by the passage

of depressions from the Mediterranean and unsettled weather is particularly prevalent when the Polar Front Jetstream (PFJ) exhibits a pattern of major oscillations facilitating the latitudinal transfer of cold air. The more southerly positions of the jetstream can result in heavy rainfall. In the spring, the Subtropical Jetstream (STJ) begins to move northwards from the tropic and precipitation is reduced. Thus, by May the Mediterranean depressions are replaced by ridges of high pressure and the Mediterranean, particularly at its eastern end, is associated with a subtropical upper tropospheric high pressure. At the same time, the Indian monsoonal system begins to move northwards and the resultant complex set of jetstreams extends over southern Arabia. The effect of this is that vorticity is induced, drawing in moist air from the Indian Ocean that gives the summer rainfall in Ethiopia, the southern Sudan and the mountains of Yemen.

The fourth control is altitude and the effects of this are most obvious in the mountains of Lebanon and Yemen. In both cases areas of relatively high rainfall occur in otherwise arid conditions. In the case of Yemen, there is a semi-permanent mist belt.

The more detailed description of the climatic elements which follows is, in its essentials, retained from *The Middle East* (Fisher 1978). Fisher was himself a trained meteorologist and took particular pleasure in climatic description. Therefore it is felt that these next sections in particular help retain the flavour of his original volume. The fact remains that the Middle East is an area still deficient in climatic data and research. For example, in the whole of the Badia of eastern Jordan there is only one meteorological station. However, increasing interest in water resources and rangeland development have led to a number of research programmes and the use of portable automatic weather stations has, in recent times, supplemented data.

Pressure

Summer

Intense heating of the southern part of Asia gives rise to the well-known monsoonal low pressure area of the Indian subcontinent. This low pressure zone, shallow in that it does not extend into the upper atmosphere, is however a permanent feature of the months June to September, and has a major centre over north-west India and west Pakistan, prolonged as far as south-east Iran and the Gulf of Oman. The strongly developed wind system to which it gives rise affects not only southern Asia but much of the Middle East, since the low pressure zone is prolonged over the Persian–Arabian Gulf towards Iraq and Syria, with the development of a minor but extremely persistent low pressure centre over the island of Cyprus. This minor low lasts again throughout the summer months and markedly affects pressure and wind distribution in the Middle East. The other major factor is the seasonal extension of the Azores high which intensifies and pushes north-eastwards across north-west Africa as far as Libya and Egypt (Figure 3.1). The Cyprus low develops primarily as the result of differential heating of sea and land with a marked development of local convergent uplift. For a small island, Cyprus shows remarkably high summer temperatures, probably enhanced by its basin-like topography.

The main result of these pressure conditions is to draw air from the north and east southwards over much of the Middle East. A further contributory element would seem to be the presence of the STJ over Asia Minor and the Caspian. Lateral and vertical eddies are generated with an anticyclonic circulation that both reinforces the northerly surface and wind pattern, and also contributes to atmospheric stability, producing marked inversion. To the north of the STJ, cyclonically circulating eddies tend to develop and this can be held to explain the tendency for summer rain on the eastern Black Sea coasts.

The well-developed northerlies, termed Etesian Winds or *Meltemi*, develop consistently over the Aegean, and as far south as the coasts of Libya and Egypt, where they are strongly reinforced diurnally by the sea-breeze effect. Further east, however, over Israel, Lebanon and Syria, the prevailing winds are strongly south-western, influenced by the Cyprus low pressure and again reinforced by a sea-breeze effect which reaches as far inland as Damascus, Amman, and even Palmyra. This air, though northerly, is continental and from the interior of Eurasia rather than from Europe, and so it is dry, being warmed at low levels as it moves south, and at highest levels by the inversion

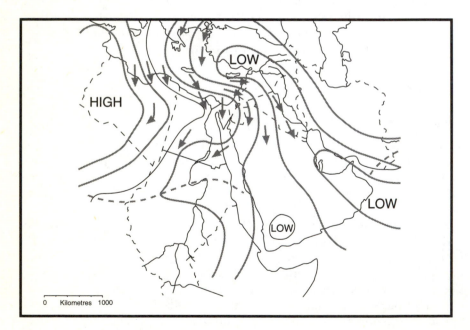

Figure 3.1 Pressure conditions in summer.

previously noted. Thus, there is entire absence of rainfall, except for two special instances: the north-east Black Sea coastlands and the coastlands of southern Asia Minor between Antalya and Antakya where slight surface convergence towards the Cyprus low, a sea track and the presence of high mountains fronting the sea would appear to produce a very slight summer rainfall confined to the coast.

In the south of the Middle East, conditions are more complex. There is the upper jet system, the Tropical Jetstream (TJS), that develops in June to August and blows from the east strongly and at heights of 10–15 km. Below this at low levels, there is the transgression of southern Hemisphere Trade Winds across the Equator into the northern Hemisphere where they are deflected to become south-westerlies. This is in response to the northern migration of the Hyetal Equator and, correspondingly, the STJ that in winter lies over northern Egypt. Consequently, in summer there are convergent winds at low levels: dry northerlies and north-easterlies, and humid south-westerlies off the Indian Ocean, which between them produce what is termed the Intertropical Convergence Zone (ITC). Above this is the seasonal high-level easterly, with considerable disturbance and eddying that is held to produce in a more developed form immediately further east, the monsoonal 'cell' of the northern Indian Ocean. Where topography produces mechanical uplift, as in the plateaux of Ethiopia, the Yemen and south-western Saudi Arabia, considerable summer rainfall occurs from the damp southerly maritime stream over-running the drier northerly currents. This may be held to explain why there is very little or no summer rainfall at sea level along the Red Sea and Gulf of Aden coasts, but considerable rainfall, as much as 1,000 mm, on the high mountains inland. However, in providing such an explanation, it is as well to emphasize the fact that the Middle East remains as yet one of the least well-provided regions for upper air reports.

The composite pressure systems, as outlined, tend to be permanent features from June until September. Relatively minor variations of surface pressure occur from time to time, with slight deepening or falling of the systems. However, in general, quasi-stability of the pressure situation remains so that regularity in surface wind patterns and in weather conditions is the chief characteristic.

Winter

In the lower atmosphere, high pressure covers the interior of Asia, and extensions of this may reach as far as Iran. In Asia Minor, because of its elevation and consequent low temperatures, a second, much smaller and rather more intermittent anticyclone may form. However, unlike the larger Siberian 'high' it can disappear from time to time.

Over the entire Middle East, the westerly STJ becomes very well established. This is at high velocity, with its axis over the southern Mediterranean coast, which is especially favourable for the formation of depressions in the lower atmosphere. Winter is therefore characterized by a succession of disturbed cyclonic conditions, broken from time to time by a temporary build-up of high pressures over Asia Minor and the mainland of south-west Asia. At the same time, the easterly TJS, characteristic of July in the extreme south, is no longer present, hence the convergent air flow is much less marked and there is very little rainfall in the southern Sudan and southern Arabia.

Well-developed depressions may pass from the Atlantic, via north-west Europe, Spain or north-west Africa, into the Western Mediterranean basin. Rejuvenated by contact with the sea and maintained or even steered by the upper jet streams, the depressions continue by way of the Mediterranean as far as Armenia, Iran and the Persian–Arabian Gulf to Pakistan. A few move southwards along the Red Sea to bring slight winter rainfall to the coastlands.

Coincident with this cyclonic activity and on average distinctly more frequently, new low-pressure systems develop and greatly intensify within the Mediterranean basin. It would appear that the impact of jet streams on major mountain masses is especially favourable for the development of 'lee depressions': vortices or eddies in the lee of major hill massifs such as the Atlas or northern Apennines. Also, cyclogenesis can take place in the Gulf of Sirte and over Cyprus which has important effects on the weather of Libya, Egypt and the Levant.

This means that the Mediterranean in winter has its own pronounced system of weather. Sometimes fully formed depressions may continue on a track from Europe or the Atlantic, but often a change or new development takes place.

Furthermore, frontal systems associated with depressions are not necessarily the same in the Mediterranean as in north-west Europe. Warm sectors in particular, which, in higher latitudes, are associated with extensive layer cloud, high humidities and much precipitation, may well in the Mediterranean be almost cloudless and quite dry.

A further point of difference arises in the relative shallowness and small extent of Mediterranean depressions. Whereas Atlantic depressions may cover half the entire ocean, with a minimum pressure approaching 960 mbar, disturbances in the Mediterranean are usually much smaller, and pressure rarely falls below 990 mbar. This does not mean that the depression is less intense, but the duration of bad weather is definitely shorter and a greater variety of conditions is experienced.

Although they are characterized by their irregularity of movement, depressions generally tend to follow a sea track. From northern Italy, they frequently pass down the Adriatic into the Ionian Sea. Here the track divides, under the influence of the land mass of Asia Minor. Many depressions continue eastwards into the Levant and Iraq, whilst others move northwards into the Aegean and Black Seas, ultimately reaching the Caspian. A second route lies in the south of the Mediterranean basin. In this case an uninterrupted sea track brings rain to the Levant coastlands, and depressions often reach the Persian–Arabian Gulf, or even the interior of Iran.

In summer and very early autumn, cyclonic disturbances rarely affect the Middle East. A more northerly track takes them across central Europe and the northern Balkans to the Black Sea, where convergence with the northerly air streams, together with the effects of relief, intensifies the rainfall of the north-east coast of Asia Minor. This is exemplified by conditions at Batum (Table 3.1).

Table 3.1 Rainfall at Batum (mm)

J	*F*	*M*	*A*	*M*	*J*	*J*	*A*	*S*	*O*	*N*	*D*	*Total*
258	153	156	128	71	150	154	209	305	226	309	254	2,473

Air masses

The most appropriate approach to understanding the weather of the Middle East is by considering the nature of the various air masses that move successively over the region, some on a quasi-seasonal periodicity, some much more sporadic and irregular. Here the special location of the area is of great importance. A link between Africa and Asia, the Middle East lies close to two of the hottest regions of the world, the Sahara and north-west India. Yet, at the same time, it forms a part of the continent of Asia, which in winter develops the lowest temperatures occurring on the globe. Intermediate between these regions of extremes, the Middle East can easily fall, for a short time, under the influence of one or the other and the relative closeness of such reservoirs of heat and cold means that little modification can take place as air currents make their way outwards from their regions of origin. By reason of its scorching heat and dust-laden appearance, African air can be felt and seen to be a 'breath of the desert', whereas, at other times, cold spells of near-Siberian intensity may freeze rivers in the north and east.

Air masses are divided, according to their area of origin, into either polar or tropical. Further division is then made on the basis of humidity. Continental air is generally drier than that originating over oceans. This gives four main types of air mass: polar maritime, polar continental, tropical maritime and tropical continental. However, division on these lines is not entirely satisfactory for the Middle East, since maritime influences are less important and continental origins cannot always be simply defined as tropical or polar.

For present purposes, it would seem useful to consider the Middle Eastern air masses as follows:

1   summer conditions;
2   maritime air from the Atlantic, reaching the Middle East via the Mediterranean; this can be either tropical or polar in origin;
3   tropical continental air; and
4   polar continental air.

Summer conditions

Air of somewhat varied origin is drawn over the Middle East during the summer months (Figure 3.1). Some of this air can be described as dried-out monsoonal air related to the lower-level monsoonal system of India. It has a long land track over the north of the subcontinent, and then crosses the Suleiman and Hindu Kush mountains, as the result of which it undergoes an adiabatic warming on descent over Iran. Air of this nature is quite dry but as a result of its previous history, it is capable of absorbing moisture if it follows a sea track. This is especially apparent in the lower layer, where the air blows from sea to land, as in the eastern Mediterranean, Persian–Arabian Gulf and the Red Sea. Humidities may thus become extremely high, though hardly any rain falls. In the north, much continental air is also drawn in from Russia. This is initially less stable, and likely to be affected by the passage of low-pressure systems or frontal conditions associated with Europe to the west and, where it meets the types of air over sea areas, as in the southern Black Sea and Caspian, mixing takes place. Given the further factor of high topography, there can develop a major convergent movement tending to uplift; and so, in the two angles fronting a major water surface, where mountain ranges meet, a zone of marked summer precipitation occurs. The two instances are the western Caspian shorelands between the Elburz and eastern Caucasus ranges, and north-eastern Turkey.

Otherwise, the predominant air masses are dry, giving almost no rainfall. Towards the west, Libya, Egypt and north-west Sudan, stability associated with subsidence tends to increase, under the influence of the subtropical high pressure zone which is a marked feature over most of northern Africa.

For the Middle East as a whole, the general result in the extreme south is to produce several months of clear skies. Near the coasts there is a regular diurnal variation. Totally clear skies occur at night and a small amount of fair-weather cumulus with sometimes slightly more over the sea during the day. However, inland cloud amounts are very small and many days may pass without any cloud whatsoever, allowing uninterrupted insolation by day, with consistent high temperatures, but considerable night radiation, providing relatively cool nights.

Towards mid-September the first signs of change become apparent. Bursts of maritime air from the Mediterranean increasingly disturb the summer flow, pressure patterns in the lower atmosphere begin to change and by late October other air masses are dominant.

Maritime air

Somewhat higher general humidity but also a generally lower temperature are characteristic of maritime air. This air usually originates over the Atlantic, passing into the Mediterranean either by way of north-western Europe, Spain or north-west Africa. There may thus be considerable differences in the air mass itself, some parts being polar and others tropical in origin. However, temperature characteristics are modified during the long passage over land and sea, whereas humidity is largely unaltered as the track of the air has lain over sea areas. On uplift, or on intermixture with other masses, considerable condensation takes place, with consequent heavy rainfall. These maritime currents blowing generally from a westerly direction penetrate most of the Middle East, with the exception of the far south between October and May. Closely influenced by the STJ, which intensifies at this season, vortices form from time to time, with an intermixture of air of differing types. Therefore, strings of small surface 'lows' can be a feature of the maritime regime (Figure 3.2), their intensity depending upon the contrast in temperature and moisture of the air masses that form them.

Maritime air exerts its greatest influence on the western margins of the Middle East with depressions penetrating as far east as Iran. The westerlies of the Middle East become progressively drier as they advance into the continental interior, and the weather disturbances to which they give rise become increasingly feeble.

The inflow of maritime air, although predominant throughout much of the period between October and May, is interrupted from time to time by outbursts of tropical continental or polar continental air.

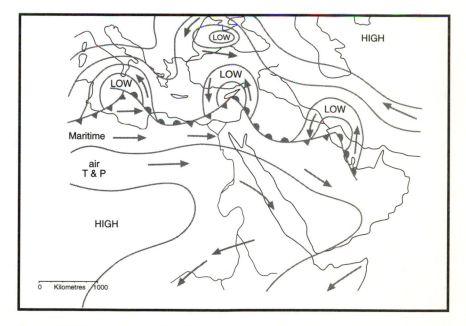

Figure 3.2 Air masses: maritime air, tropical and polar.

Tropical continental air

A feature of the Middle East is the proximity of wide expanses of desert, from which intensely hot and dry air may be drawn by the passage of depressions. If a southerly gradient of pressure develops, air from north Africa and Arabia moves northwards on a large scale, producing very special weather conditions (Figure 3.3).

A portion of the great quantity of energy in the form of heat is converted providing major variations in pressure and giving rise to strong winds, which frequently reach gale force. Temperatures rise, sometimes by 16–20°C in a few hours, and relative humidity falls to figures of less than 10 per cent. Crops may be withered in a day but the most prominent effects are driving sand and dust, which can cover roads and penetrate into houses. The effect is so marked that local names have been given to these southerly winds. The term *khamsin* is used in Egypt, *ghibli* in Libya and *shlouq* in the Levant. In Iran, *simoom* (poison wind) is a good description. The definition for a *khamsin* is now taken as a rise of at least 6°C within 6 hours, but much higher figures are of course known.

Winds of *khamsin* type develop most often when tropical air is drawn in as the warm sector of a rapidly moving depression formed in maritime air. Such depressions give the most extreme weather conditions, but are usually of short duration. In desert areas, sandstorms are almost invariably produced by strong southerly winds, and these storms, often violent, may spread into settled areas.

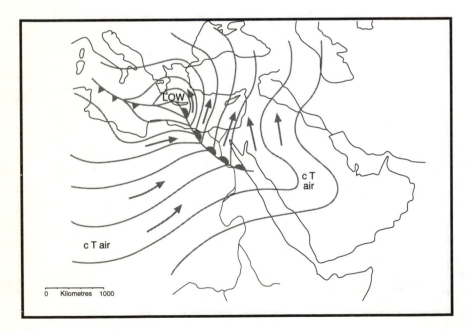

Figure 3.3 Air masses: tropical continental air.

Autumn and spring, particularly the latter, are the chief seasons at which hot winds occur. In spring, the southern deserts heat up rapidly whilst the rest of the Middle East is still cold and favourable conditions are thus established for mixture of differing air streams. Owing to the extreme dryness of the desert air, rainfall, which normally results from mixing of two air masses, is very scanty or entirely absent. However, quite often as cold air comes in at the rear, there is a most spectacular build-up of cumulo-nimbus cloud, with extremely strong convection carrying soil and dust upwards.

Polar continental air

In winter and spring, waves of cold air flow southwards and westwards from the intensely cold interior of Eurasia. Air originating in south-central Asia, overflows into Iran, and may for a short time reach the Mediterranean. The air is cold, but very stable, deriving from the Siberian anticyclone, and days are fine and clear. Over the Iranian and Anatolian plateaux very low temperatures occur but sunshine during the day mitigates the worst effects, particularly as humidity is low (Figure 3.4).

Further to the west, the air undergoes slight adiabatic heating as it descends the edges of the Zagros and the plateaux of Anatolia and Syria. Fine weather prevails, with low temperatures inland but moderately warm conditions on the coastlands. Fog, the frequent accompaniment of winter high pressure in western Europe, is absent, as the air is very dry. Anticyclonic waves are

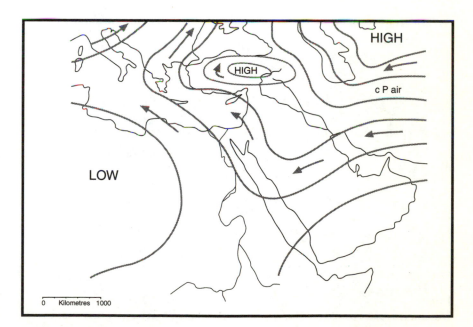

Figure 3.4 Air masses: polar continental air.

frequent in the Middle East during autumn and the early part of winter but later in the year continental air of a very different type makes its way in from the north-west (Figure 3.5).

Anticyclonic conditions frequently develop in central and eastern Europe during January, February and March. A reservoir of cold air builds up and this, unlike the air outflowing from the dry heart of Asia, is damp, since it is merely chilled and modified maritime air originating from the Atlantic. From time to time, currents of this European air are drawn into the rear of a cycle of depressions moving east along the Mediterranean and on reaching the Middle East, the lower layers of this quasi-continental air have been subjected to contact heating during the southward passage over the warm sea. Considerable quantities of moisture have also been absorbed.

Unlike the adiabatic heating of the Asiatic continental air, which affects all layers, differential heating of only the lowest layers in the European current produces much instability. This, together with high humidity, gives rise to heavy rainfall of a showery type. Practically all the snowfall and much of the rainfall of the Middle East develop in outbursts of cold damp air from central and eastern Europe. Late winter and early spring can be unpleasant seasons, liable to prolonged periods of raw, cold conditions, with frequent and heavy precipitation. If these conditions persist for any length of time, they may become intensified as much colder Arctic air ultimately becomes drawn across central Europe from the north. This can give periods of thoroughly

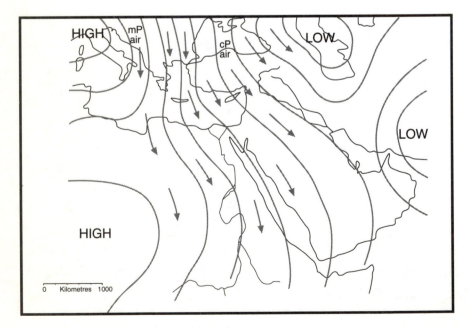

Figure 3.5 Air masses: polar maritime air.

unsettled and unseasonable weather for a few days even as far south as the Sahara, central Egypt and the upper Persian–Arabian Gulf.

Temperature

The chief features of temperatures in the Middle East are the high temperatures of summer and a wide range of temperatures, both annual and diurnal (Figure 3.6). Clear skies are the main influencing factor since during the day intense solar heating of the land surface can give rise to very high temperatures. However, during the night there is equally little check in heat radiation from the land surfaces and so away from the sea temperatures fall considerably. However, average temperatures are considerably higher than those of places nearer the Equator and, therefore, with greater insolation but relatively high cloud cover. July mean temperatures for four such places can be compared with those for four Middle Eastern cities (Table 3.2).

Another important effect is that of altitude. In this connection it should be recorded that, whilst some parts of the Middle East are very low lying, there are very extensive areas, for example most of Asia Minor and Iran, that lie at 1,000 m or more above sea level. This does not so much reduce the summer maxima, since the ground is warmed directly, as reduce winter temperatures, which can be very low indeed for the latitude. Other minor influences are the mountainous coastline in many parts which limits the

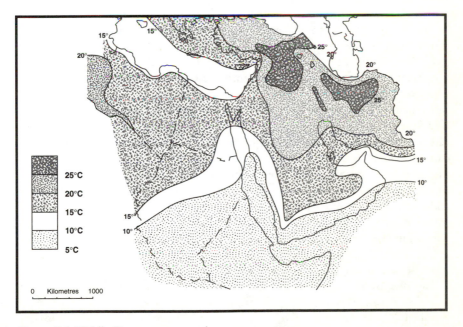

Figure 3.6 Middle East: mean annual temperature range.

Table 3.2 Average July temperatures

Location	°C	Location	°C
Cairo	28.3	Colon	26.6
Beirut	27.2	Freetown	25.5
Basra	36.1	Mombasa	25.3
Tehran	29.4	Singapore	27.2

tempering effect of the sea to a narrow strip, the basin-like character of many interiors which concentrates insolation and the absence of soil and vegetation which allows both intense surface heating and nocturnal radiation.

In view of the wide daily and seasonal ranges of temperature, averages tend to be misleading and it is more appropriate to refer to average maxima and minima temperatures in order to appreciate actual conditions. For instance, in Table 3.2, it would appear that conditions in Beirut and Cairo are very similar. In fact, Cairo may have summer day temperatures of over 40°C but with a markedly cool night whilst Beirut rarely rises above 35°C but with only 6°C difference between day and night temperatures.

A significant simplicity of rhythm prevails throughout the Middle East. With a few exceptions, July is the hottest month inland but on the coast the maximum is delayed until August because of the slower absorption of heat by the sea. Egypt demonstrates this tendency with a July maximum in the Nile valley and an August maximum on the Mediterranean and Red Sea coasts. Towards the extreme south the maximum is earlier, June in southern Arabia and Upper Egypt and even May in much of the Sudan, due to more direct solar heating towards the solstice.

In the Middle East, January is coldest month with any considerable rise in temperature often being delayed until the end of February or early March. However, once the rise in temperature begins, it is rapid. Considerable differences are apparent in winter between the north-eastern part of the Middle East and the central and southern regions. Proximity to the Eurasian interior means that continental influences are very marked in the north and east and these are intensified by the effects of topography. Therefore, many of the higher regions have severe winters, significantly colder than normal for the latitude. Extremely low temperatures occur in Asia Minor, Armenia and much of Iran, indicated by the existence of one small permanent glacier in the Zagros. Conditions at Erzerum indicate the intensity of winter cold with a January mean temperature of –11°C, an average day maximum temperature of +1°C, an average night minimum temperature of –27°C and an absolute minimum temperature of –40°C.

The Bosphorus and Black Sea sometimes become jammed by pack-ice and inland snowstorms isolate many districts, even as far south as Syria and Jordan. Small quantities of snow may fall in regions of lower elevations and, except for the lower-lying parts of southern Arabia and all of the Sudan, no part of the Middle East can be said to be entirely free from snowfall. The lower Nile Valley, the highlands of the Yemen and the hills of northern Libya experience slight falls of snow when cold air makes its way south from Europe. Exceptionally, the Jebel of Tripolitania may have up to 1 or 1.5 m of snow and there can even be snowfalls in southern Yemen and on the Jebel Akhdar of Oman.

Diurnal variation of temperature is important at all seasons but it is most developed in summer. In coastal areas, maritime influences restrict the temperature range but at a short distance inland, great heat during the day gives place at night to a most refreshing coolness. In Egypt and the Red Sea coastlands, night cooling may even bring the air below its dew-point so that early morning fog is a feature of the late spring and early summer in parts of the Red Sea coast, the lower Nile Valley and even the Mediterranean coasts of Egypt and Libya. Topography is a controlling factor and regions of high altitude have a reduced night temperature although by day little difference is apparent. This altitudinal difference is especially marked in Lebanon where several places in the mountains have a day temperature of over 25–27°C whilst at night there is a drop of 10–15°C as compared with 5–6°C on the coast.

Humidity

Humidity is a measure of the water vapour content of the air and two statistics are of meteorological value: absolute humidity and relative humidity. The actual amount of humidity, expressed in terms of weight per unit volume, is designated the absolute humidity. Relative humidity is the standard climatic measurement in that it relates directly to such factors as evaporation, evapotranspiration and precipitation. The sensitivity of the human body to heat is also largely governed by relative humidity in that heat in a humid environment is far more enervating than the same temperature under dry conditions. Thus, in normal parlance and in the way it is used throughout this book, humidity is equated with relative humidity. The warmer the air, the more water vapour it can hold. Relative humidity is the measurement, expressed as a percentage, of the amount of water vapour in the air compared with what the air could hold if saturated at that particular temperature.

In the Middle East, monthly mean relative humidities range from extremely low figures such as 24 per cent for August in Tehran to a percentage as high as 80 per cent for Istanbul in January.

As with so many variables, there are great differences in humidity within the Middle East region. Whilst, in general, the average is low, due to the prevalence of open desert, some areas may have remarkably high humidities. Evaporation from major inland water surfaces, such as the river systems of the Nile, Tigris and Euphrates or the Dead Sea, may give rise to high humidities locally but the highest humidities are often found in certain coastal zones. Wherever a narrow coastal plain is backed by a mountain barrier, well-developed sea breezes bring in much moisture which remains concentrated in the coastal zone instead of spreading into the interior. High humidity together with high temperatures make living conditions extremely unpleasant during the summer season when inland transfer of moisture from the sea is greatest. The Persian–Arabian Gulf coast was for long notorious and the shores of the Red Sea and Mediterranean are also affected.

In some places, such as Beirut and parts of the coast of Oman, humidity is actually at its highest for the year during the summer months in spite of a complete absence of rainfall and dew can occur on 200–250 days per annum, producing up to a fifth of the total recorded precipitation (Table 3.3).

Heavy dew on the Red Sea coastlands is an important aid to agriculture in the Yemen. In the past, advantage was taken of this by nomads who erected cairns of stones with a receptacle at the base. During seasons of high humidity, the stones radiated heat faster than the soil or sand and so, being colder, collected dew drops which could be sufficient to produce a small trickle of drinkable water. It is reckoned that in some places dewfall provides as much as 25 per cent of effective moisture usable by plants.

In the interior, humidity is generally low but a marked increase may occur in winter with a westerly or northerly air stream. Hence winter mist and even fog are by no means uncommon, particularly in riverine areas. Also, salt marshes may experience dense mist.

On the western side of Wahiba Sands, Oman, an extensive programme of research focused on the direct measurement of dew revealed that during humid nights the equivalent of 0.5 mm of rainfall may be deposited as dew (Anderson 1988).

High temperatures may be tolerated by humans so long as humidity remains low and so the summer heat of the interior may often be more easily borne than the muggy conditions of the coastlands. Beirut, with summer average maxima of 30–32°C and 70–75 per cent relative humidity may feel more exhausting than Damascus with 37–40°C and 30–40 per cent relative humidity. There are some localities where the wet bulb temperature may rise for a time above blood heat, 38°C, and with these conditions sustained human activity becomes very difficult. However, with the development of air conditioning this traditional handicap to life in the Middle East has been reduced. In most cities, especially those of the Gulf states, peak demand for electricity is now experienced in summer due to extensive air conditioning.

Table 3.3 Number of days with dew at Haifa (15-year mean)

J	F	M	A	M	J	J	A	S	O	N	D	Total
0	3	2	11	18	25	19	19	23	18	0	0	138

Rainfall

Except for the two small coastal regions of northern Iran and north-east Asia Minor, the Yemen uplands and adjacent areas in southern Arabia and the southern Sudan, which are climatically special cases, the whole of the rest of the Middle East has a strongly marked 'Mediterranean' rhythm of summer drought and winter rain. During summer, higher pressures over the eastern Mediterranean and north Africa act as a buffer between the low pressures of the north Atlantic, the Cyprus low and the monsoonal lows of the lower Persian Gulf and east-central Africa, thus shutting out oceanic influences from the west. For the western Mediterranean as far as and including western Libya, the northward migration of subtropical high pressure can be regarded as directly responsible for the development of dry conditions. However, it should be remembered that pressures in the eastern Mediterranean and further east are actually lower in summer than in winter (Table 3.4).

Rainfall occurs first in early autumn when the dry summer air masses are displaced by damper and more unstable currents from the west. A few short showers only occur during September but towards the end of October heavier and more prolonged falls, often with spectacular thunderstorms, announce the end of summer.

These thunderstorms usually clear up after a few days and a relatively fine period ensues until December. The real rainy season does not begin until the latter half of December and may even be delayed until the New Year. Over the western half of the Middle East January is the rainiest month with a slight tendency to a December maximum noticeable in a few areas. Towards the east, the maximum is increasingly delayed. Syria exhibits a January maximum in the extreme west and a February maximum in the remainder of the country. In eastern Iraq, Iran and parts of Asia Minor, March is often the wettest month. In these areas, the influence of interior winter high pressure generated by intense cold may deflect the rain-bearing lows elsewhere and it is not until the high-pressure system collapses in spring that maritime air can reach the interior.

Partly but not entirely for this reason, the shores of the Caspian and southern Black Sea coast have a double rhythm, with a minor maximum in spring, a major maximum in autumn and no month without rain.

By the middle of June, rain has ceased over most of the Middle East, except for the extreme north and the south, and in many parts no rain normally falls for a period of 10–15 weeks.

Table 3.4 Atmospheric pressure (monthly average in millibars)

	Benghazi	Istanbul	Limassol	Basra	Muscat
January	1,010	1,019	1,017	1,019	1,019
July	1,005	1,012	1,007	997	998

The distribution of rainfall in the Middle East is largely controlled by two factors: topography and the disposition of land and sea in relation to rain-bearing winds (Figure 3.7). It must be remembered that the Middle East is predominantly a continental area, influenced only in certain regions by proximity of relatively small areas of sea. Hence air masses reaching the Middle East from the west, even though of oceanic origin, have lost some of their moisture and it is only where a sea track has allowed partial rejuvenation that considerable rainfall can develop.

Therefore, it might be said that in most regions rainfall tends to occur in proportion to the length of coastline, or even in proportion to the length of westward-facing coastline. Regional contrasts are striking. The westward-facing shore of the Gulf of Sirte has a marked effect on the rainfall of Cyrenaica and the absence of such configuration in Egypt condemns the country to a scanty rainfall. At Benghazi, at the western end of the Cyrenaican Jebel, annual rainfall amounts to over 270 mm as compared with 77 mm at Port Said. The narrowness of the Red Sea is reflected in the lower rainfall of Arabia, whilst the influence of the broader Mediterranean is clearly shown in Asia Minor and the Levant. The proximity of the Persian–Arabian Gulf and even the extensive water surfaces of lower Mesopotamia have a favourable effect on the rainfall in the Zagros Mountains.

With such a delicate balance between dampness and aridity, it is inevitable that topography should exercise a control equally as important, if not more so, than that of physical configuration. It has already been stated that the greater part of the Middle Eastern rainfall develops under conditions of instability, to which the uplift of air currents as they are forced to rise over mountain ranges is a very powerful contributing factor. Warm frontal rainfall develops over a vertical distance of as much as 7,500 m. A few thousand metres of uplift in the lowest layers have no undue effect as ascent of air has begun independently of conditions near the ground and precipitation continues after high land is past. On the other hand, a cold front may not develop any precipitation whatever until the 'trigger action' of sudden uplift over high ground is first applied, with very considerable dynamical uplift following. Adiabatic heating on rapid descent from coastal mountain ranges to interior plains is another important influencing factor. Thus, control by topography can be so great that isohyets tend to follow contour lines with westward-facing mountain ranges or plateau edges experiencing heavier rainfall at the expense of eastward-facing slopes and lowlands. The swing of the isohyets in response to the south-eastward curve of the Turkish highlands towards the Zagros system gives rise to the beautifully developed 'Fertile Crescent' of steppeland linking the east and west of the Old World. Control by topography has resulted in the elaboration of striking regional contrasts, none more so than in the Holy Land.

The common occurrence of Middle Eastern rainfall under conditions of instability means that much of it may be heavy but short in duration and extremely capricious both in period of onset and in distribution. The

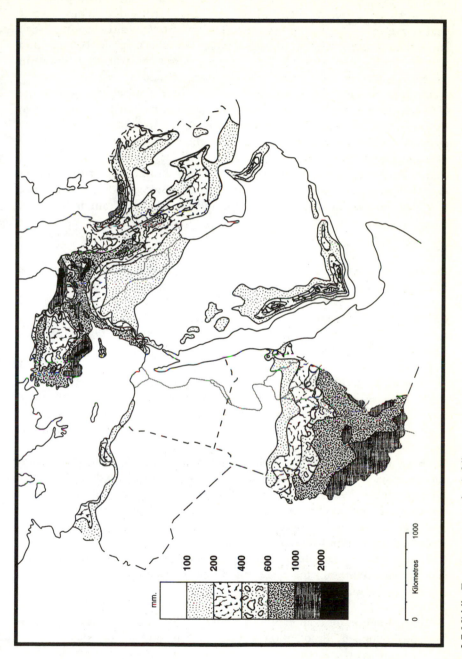

mm.

100
200
400
600
1000
2000

0   Kilometres   1000

Figure 3.7 Middle East: mean annual rainfall.

coastlands of the Levant, with annual falls of over 700 mm, receive more rain than parts of Britain but this is crowded into little more than 6 months and even so, the wettest months have only 14–18 rainy days. Thus, a great deal of rain may fall in a short time and 25 mm/h is by no means unusual. In 1945 Damascus, with an annual average 240 mm, received 100 mm in a single morning and in 1969 400 km of a newly constructed highway in central Arabia was washed away by one night of rainfall.

Averages for rainfall tend to be misleading since few regions outside the highlands of the north can count on a really regular rainfall. Whole areas in Egypt, Arabia and the northern Sudan may go years without a fall of any kind and in Libya it is reckoned that drought will prevail for two years out of every ten. A heavy fall once in several years may thus appear in climatic statistics averaged out as an annual figure. Some, but not all, of the summer rain in Turkey and Iran is of this type. In successive years, Jerusalem once had 1,060 mm of rainfall, then 307 mm and Baghdad 432 mm and 56 mm.

Evapotranspiration

In the Middle East, the effectiveness of precipitation is severely constrained by temperature. If rates of evaporation and evapotranspiration are high, the influence of rainfall upon plant growth may be negligible. The key distinction which needs to be made is between actual rates of evapotranspiration (AE) and potential rates (PE). Potential evapotranspiration has been defined as:

> Evaporation (diffusion of water vapour into the atmosphere) from an extended surface of a short green crop, actively shading the ground, of uniform height and not short of water.
>
> (Ward 1975)

Rates of water loss by evaporation are governed primarily by meteorological variables such as energy balance, temperature, windspeed and relative humidity. Plant influences include leaf area, ground cover and aerodynamic resistance. If potential evaporation exceeds or falls below actual evapotranspiration then there is, respectively, a deficit or a surplus of soil water and this water balance provides an index of aridity.

The practical uses of evapotranspiration and water balance concepts can be important but there are significant reservations which should be made:

1   Actual figures for evaporation and transpiration are, in practice, very difficult to obtain. Evaporation is not constant and an obvious difficulty arises when the surface is to be regarded as wet or dry because of the factor of latent heat of evaporation. The dryness of the average soil and its depth are important considerations.

2 Relative humidity, the extent of turbulent air mixing and differing wind velocities at differing heights, even over a small vertical range, all affect evaporation but are extremely difficult to measure accurately.

3 Transpiration from plants is a very variable factor. The same species of plant alter their rates of transpiration according to climatic conditions, their degree of luxuriance and the density of their growth.

Thornthwaite summarized these issues, concluding that potential evapotranspiration cannot be measured directly but must be computed from a number of variables: length of day, which greatly affects insolation, mean monthly temperature and precipitation (Thornthwaite 1954). The other well-known formula is that developed by Penman (1948) which accords more closely with field measurements but demands relatively sophisticated meteorological data which are only available at a few stations in the Middle East.

Given the problems of measurement, it is not surprising that there are relatively few examples of mean annual potential evapotranspiration rates available for the Middle East. Figures as high as 1,140 mm have been recorded for areas south of latitude 30°N, the effective division between steppe and desert and between perennial flow and temporary discharge. For most of the remainder of the Middle East, particularly that part within the warm temperate climatic regime, measurements down to half that value have been recorded. The lowest rates have been found in the main highland areas, restricted to Turkey and Iran (Beaumont *et al.* 1988).

Climatic change

No study of environment in the Middle East can proceed far without encountering the problem of possible variations in climate. It is clear that major climatic fluctuation is a feature of the differing geological periods and in the single Hal Far cave of Malta there is a collection of animal remains buried in silt and clay that successively ranges from hippopotamus and rhinoceros to Arctic species such as the polar bear and arctic fox. Reference has already been made to the attenuated and misfit river systems of the present day with a geomorphology that often suggests more humid conditions. Above all, there is the striking paradox in human occupancy. Looking today at the arid areas that characterize many parts of the Middle East, it is difficult to reconcile present hostile environmental conditions to the brilliant historical past. How could man have achieved in these localities such significant and fundamental advances from Palaeolithic to Neolithic and then Classical cultures in advance of most, possibly all, other parts of the world? Thus, for almost a century, the history of climatic change in the Middle East has been of unusual interest for researchers.

Physical evidence for change is reviewed in detailed by Butzer (1975, 1978). He considers the evidence for change between the Late-Pleistocene and the Mid-Holocene within three regional contexts:

1   the highland perimeter to the north;
2   the hills and plains of the Levant and Mesopotamia; and
3   the lower Nile Basin.

In the northern highlands, remains of glaciation are widespread from Turkey to Iran and the adjacent parts of Iraq together with the higher areas of Lebanon. These indicate a colder climate during the Wurm Full Glacial with summer temperatures 6–7°C lower than at present. Pollen evidence, particularly from Lake Zeribar, suggests that, during the Full Glacial, open vegetation of steppe and mountain tundra type was characteristic of these same areas. Warming during the Allerod and particularly the Early Holocene appears to have been accompanied by an increase in precipitation and a gradual reforestation of the highlands. Over the past 50,000 years, the northern highlands did not experience pluvial conditions, the Full and Late Glacials being conspicuously cold and relatively dry. Indeed, other than as a direct result of human activity, the Holocene environment seems to have remained constant and similar to what might be expected today.

In the Levant and Mesopotamia, geomorphological changes are mainly reflected in the alluvial fill of river valleys but the interpretation of these features, now dissected, are equivocal. However, in the Damascus Basin and the Jordan Valley several moist periods can be recognized from the Late-Pleistocene to the Holocene. Cave sequences indicate that most of the early Wurm and parts of both the Full and Late Wurm were sufficiently cold and moist to produce frost weathering. The resulting angular particles are interrupted by horizons reflecting the chemical weathering and more fine-grained sedimentation expected during temperate conditions. The evidence from Glacial Age fauna suggests ecological conditions similar to today while pollen data indicate that forest diminished during the Glacial periods. This characteristic was much more in evidence in northern Syria and Lebanon where rainfall decreased markedly during the Full Glacial than in Israel, and particularly the Negev, where rainfall did not decline appreciably but where reduced evaporation favoured the expansion of existing woodlands. The paucity of evidence for greater humidity implies that there was probably little alteration of climate of ecological significance in Post-Pleistocene times.

Under the hyper-arid conditions of Egypt and Arabia, true pluvial conditions were last experienced during the first half of the Wurm. In the later Full Glacial, Egypt was as dry as it is today. From wadis in Egypt and Nubia, there is evidence of increased discharge prior to 25,000 BP. Pluvial activity was more significant than now during the period 15000–3000 BC with the exception of three major dry interludes at approximately 9500 BC, 5500 BC and 4500 BC. During this time, storms became more frequent and probably of greater duration. It appears that present-day ecological conditions were established and a long arid phase with increased dune activity and comparatively low Nile floods had begun by about 2350 BC.

Thus, during the Late Glacial three distinct palaeoclimatic provinces were discernible. The northern highlands were intensely cold for most of the time and very dry during the Full Glacial. The Levant was moderately cold and relatively dry in the north but cool and comparatively moist in the south. Egypt and possibly Sinai and Arabia experienced long periods of increased rainfall except during the maximum of the Wurm. During the Mid-Holocene period there were repeated moisture episodes in Egypt and Arabia and possibly Israel but these were not in evidence further north. All the climatic changes recorded were of relatively short time period and, according to Butzer, none exceeded the duration of the standard Late-Pleistocene sub-stages developed in European chronology. Thus direct correlation with European sequences is difficult. Other than for Egypt, there is no sound evidence for pluvial periods in the Middle East. More evidence in support of Butzer's chronology is provided by Macumber and Head (1991): they cite the sediment backfill some 60 m thick in the ancestral valley of the Wadi al-Hammeh in Jordan related to the rise of Lake Lisan, the water body that filled the Jordan Valley during the Late-Pleistocene. Archaeological evidence and carbon-14 dating indicates that this lake level rise continued until about 11,000 BP and that there were fluctuations but no significant falls in the level of Lake Lisan over the period 31,000–11,000 BP. As the lake level fell post 11,000 BP, fluvial incision occurred.

With regard to the deserts of Sinai and the Negev, Issar and Bruins (1983) indicate that unique climatological and hydrological conditions prevailed between about 1,000 and 10,000 BP. In addition to the evidence of Lake Lisan, they cite three characteristics in support of their argument:

1 higher precipitation resulted in Nubian sandstone aquifers with outcrops in Central Sinai being filled and overflowing through springs along the faults of the rift system; this water being of special isotopic composition could be identified;
2 rainstorms were dust-laden causing the deposition of thick layers of loess which indicates that the rainfall was approximately double the amount received today; and
3 the drainage system of many western and northern wadis of Sinai and the Negev could not drain out all the silt-loaded water during the wet season and shallow lakes and marshes developed.

However, Roberts (1982) indicates the problems of attempting to correlate lake levels and palaeoclimatic patterns. He suggests that there have been significant spatial and temporal variations in regional water budgets over the period post 30,000 BP. In particular, he concludes that circulation patterns may have been controlled as much by the deepening of the Eurasian winter high-pressure zone as by the southward displacement of the westerly wind systems. Fluctuations in the glacial climate of the northern Middle East may thus be largely explained by the relative influence of these two weather systems and thus the existence of pluvial lakes may not imply so much increased precipitation as increased continentality.

If an analysis of past changes remains in many ways controversial, an examination of possible future variations is highly conjectural. Evidence for the Middle East is scanty and highly speculative but an indication of possible changes can be provided with regard to the Mediterranean Basin. Wigley (1992) states that transient global-mean warming between 1990 and 2030 is expected to lie in the range of 0.5–1.4°C, a rate between two and seven times faster than the warming that has occurred over the past century. Using General Circulation Models, he indicates that for the Mediterranean, the only result which can be predicted confidently is a general warming. The regional transient-response warming rate is likely to be similar to the global mean rate and Greenhouse-related changes may well be masked by natural climatic variability for several decades in the future. With regard to precipitation, there is some evidence of increased autumn precipitation occurring in the northern part of the basin and of decreases in the southern part but it is not possible to estimate the timing or magnitude of these changes.

As Wigley concludes, these suggestions are possibilities rather than probabilities and more definitive evidence must await the improvement of General Circulation Models. However, it can be stated that the Greenhouse effect is man's most pressing environmental problem and that developing projections of future climate is a major challenge.

4 Soils and vegetation

The most fundamental product of the inter-relationship of climate and the surface of the earth is soil which provides a thin surficial layer upon which animals and plants ultimately depend. Soils, vegetation and animals are the subject matter of biogeography but the development of the zoological aspects has been largely limited to highly specialized studies. One general characteristic of animals is their mobility and their tolerance of a range of environments. Indeed, many of the fauna of the Middle East have been forced by urbanization and industrialization and particularly hunting from their original environments. Therefore, while there are endemic species, such as the Arabian oryx, which can sustain themselves under the harsh conditions of much of the region, they cannot be related to environmental conditions in the same way as soils or vegetation.

This can be shown by considering the factors which control the development of soils and vegetation (Jenny 1941, Bunting 1965, FitzPatrick 1986). Soils result from the interplay over time of climate and parent material modified by geomorphology and a number of organic variables. The effects of climate and particularly the amount and movement of water within the soil determine the broad categories. Commonly, parent material is the rock on which the soil is located but soils also develop on transported material such as sand and loess. Apart from the effect of agents such as water and wind eroding and depositing soils, the major geomorphological influence is through slope. The changes in soil characteristics from the crest to the foot of a slope are together known as a catena. Organic factors include: (a) the incorporation of dead, primarily plant, material into the soil as humus; (b) the effects of fauna and flora within the soil itself; and above all, (c) the influence of human activity.

Soils also change over time from the initial development on freshly exposed parent material to soil which is in balance with the environment, and therefore analogous to climax vegetation. As with plants, the question arises as to whether conditions have, over the geologically recent past, been sufficiently stable for such an equilibrium to be achieved.

Vegetation development is controlled by the same factors except that since soils provide an interface between parent material and plants, edaphic or soil variables can be largely substituted for parent material. In the case of vegetation, organic considerations include: (a) competition between plant species; (b) the effects of grazing animals; but, most importantly, (c) the major and often catastrophic influence of human activity through wholesale vegetation clearance by anything ranging from primitive implements to fire and the most elaborate landscaping machines. Thus, both in their different ways, soils and vegetation provide a summary of the environment and geologically recent changes.

In the Middle East it has already been shown that there is a variety of climatic type while geology and geomorphology are complex. Therefore, particularly in the light of the evidence for climatic change, simple patterns of soils or vegetation would not be expected. On the one hand, the distinctive controlling factor is aridity and much of the area is relatively untouched by humans. On the other hand, the Middle East is a cradle of civilization and certain areas have been more intensively used by humanity than anywhere else in the world.

Soil classification

The unit of measurement of soil is the profile: a two-dimensional vertical section from the surface to the parent material. Each profile is, to a degree, unique but a common pattern within the profile allows the basic mappable unit, the soil series, to be identified. Common characteristics among soil series allow the establishment of associations which can be combined to produce increasingly broadly based groupings: families, groups and orders. The orders are the globally comparable and mappable units.

Soil profiles are divided horizontally into layers or horizons, which represent the sum of differences in characteristics and process. An *eluvial* horizon is one from which soluble compounds and finely divided insoluble material are removed by erosive agents, chiefly soil moisture derived from rainfall: and an *illuvial* horizon, one into which these materials are carried and redeposited. In order from the surface downwards, the 'A' and 'B' horizons are summary ways of restating this variable process and the 'C' horizon is that representing the parent material.

In humid climates, the transfer of soil water is generally downwards and the top layers of soil therefore tend to form the eluvial horizon, with the illuvial layer below. In arid climates, scanty rainfall may penetrate the soil, but as the result of capillary attraction due to strong surface heating, any soil water is soon drawn back to the surface, so that the final transfer of water is upwards leading to an accumulation of mineral salts at or near the surface.

When downward transfer of soil moisture is strongly predominant, the upper layers of soil are said to be leached, or podsolized, and they tend to be acidic, with a low pH (under 7). This is characteristic of cool temperate and cold regions with low evapotranspiration and a moderate or heavy rainfall. In the Middle East, such conditions occur only in restricted localities of considerable altitude, such as the mountains of the north. If soil exists, it may then be podsolic. However, over much of the Middle East low rainfall and high average temperatures give rise to the opposite conditions resulting in a concentration of salts near the surface and a pronounced soil alkalinity or high pH (over 7).

Besides these basic types, there are heavy alluvial soils of the riverine valleys such as that of the River Nile. These are the soils upon which early civilization was based. They may be contrasted with the most commonly occurring soils of the Middle East, the skeletal soils which consist of rock debris incompletely weathered into true soil. Rock, gravel and sand desert are all characterized by these soils which have a wide variability.

Soils in which the predominant movement of water is downwards, leaching material from the A to the B horizon and resulting in, depending upon the temperatures, podsolization or laterization, are known as pedalfers. Soils in which the movement of water tends to be upwards with salt deposition or calcification, are known as pedocals. This basic genetic distinction depends upon climate and related to it is the concept that the key determinate of soil characteristics is climate. The concept is that, with certain exceptions, a given climate will produce a particular type of soil, categorized as a zonal soil.

This broad pattern is interrupted by certain factors which outweigh climatic controls producing intrazonal soils. The main controls are the presence of limestone as parent material, waterlogging and salinization. Since calcium carbonate is dissolved during weathering and therefore lost in solution, the remaining soil comprises merely the impurities. For reasonably pure limestone therefore, the soils tend to be thin and as a result limestone soils have more in common with each other than with the zonal soils among which they occur. Waterlogging can lead to anaerobic conditions and a build-up of organic matter in what are known as gley soils. Salinization can occur naturally in some pedocals but, world-wide, is generally a product of unregulated irrigation. If land is inundated and drainage is slow, evaporation occurs leaving salt concentrations in the soil and producing solonchaks. In the Middle East, saline soils occur naturally associated with *sebkha* but also with many of the major irrigation projects, for example in Iran.

The other category is immature or azonal soils. As any soil develops on a fresh exposure of parent material, it will be in the initial stages skeletal or azonal. It will comprise broken rock fragments of the parent material and little or no organic matter. In the Middle East, with the general lack of organic sources owing to the sparse vegetation over much of the region, azonal soils are commonly encountered. This threefold classification is descriptive and genetic and particularly well suited to geography. However, over recent years, the concept has been increasingly questioned. In particular, maps of the zonal soils do not appear to correlate well with climatic maps. Theoretically, whatever the conditions, apart from those indicated for intrazonal soils, over time within one climatic zone, soils should closely resemble each other. Nevertheless, given the extremely slow rate of soil development, it is possible that insufficient time has elapsed for the production of soils which are in equilibrium and can be recognized as climate-related. Russian pedologists still favour the idea of zonal soils.

In Western Europe and North America, rather than the predominance of climate, the emphasis is upon the range of soil-controlling variables. This allows greater freedom in that the classification can be shaped to particular requirements. For measures of land capability, the emphasis might be on texture and drainage whereas for civil engineering bulk density and shear test results would be more significant. With individual properties, soils may be classified by co-ordinates in that the properties can be plotted on axes and soil types identified by their position on a multi-dimensional axial model. This is basically the scheme adopted in several European national surveys and the procedure, of course, facilitates computer analysis.

Although a number of different approaches have been adopted within the Middle East, mostly based on major soil groups, they must be viewed in the context of three more broadly based systems:

1   the US Seventh Approximation and its Supplement;
2   the FAO/UNESCO classification; and
3   land capability surveys.

The purpose of the Seventh Approximation is to group major soils in such a way that relations among them and between soils and their environment can be seen. Most of the characteristics used to classify the soils are grouped by the recognition of diagnostic horizons, but classification in the lower categories tends to be dependent upon environmental characteristics. For instance, Typic calciorthids, Ustollic calciorthids and Xerollic calciorthids are distinguished from each other by soil temperature but most significantly by the number of days that the depth 18–50 cm in the profile is moist. The Seventh Approximation has many advantages, mainly due to its comprehensive character, but for effective distinction between soils in the lower categories a considerable amount of environmental data are required. This is as yet largely unavailable in most Middle East countries.

Despite this lack, the Seventh Approximation has proved to be extremely valuable when used in conjunction with land capability studies. Land capability studies, primarily developed for agricultural purposes, are based on a much broader appreciation of the environment. Furthermore, as they are usually carried out for a specific purpose, they lack the comprehensive nature of a soil classificatory system. However, the environmental information makes them particularly suitable for use with the Seventh Approximation in prediction. In the absence of local data, soil behaviour can be anticipated by applying knowledge from areas in which the behaviour of similar soils is known.

The FAO/UNESCO classification attempts to clarify the confused situation resulting from the use of different systems as a result of which one soil has been known under a variety of names. However, this does mean the introduction of a new soil nomenclature. With the production of maps on a scale of 1 : 5,000,000 it is hoped that patterns of soil emerge.

As might be expected from the problems already identified in the context of geology and geomorphology, there are relatively few detailed soil surveys of areas of the Middle East and the use of terminology varies considerably. Therefore, to provide a basic typology, the broad classification adopted in *The Middle East* (Fisher 1978) is quoted (Figure 4.1).

Desert zones [lithosols]

In these areas, the soil-forming processes are at a very early stage, and the main feature is physical breakdown by weathering through temperature changes and moisture and by erosion through the action of wind and the rare but powerful flash floods. Coarse sand and gravel is interspersed with cobbles of various sizes and the whole is frequently permeated with desert dust. In lower horizons there may be patches of sandy loam, and deposits of salts. Many areas are dominated by loose, mobile sand which may vary considerably in colour and may include *nebkhas* (small dune-like forms developed round plants or small obstacles), *barchans* (moving crescent-shaped dunes), *seif* dunes (linear dunes with mobile surfaces) or other sand assemblages both static and mobile.

Salt efflorescences are common in desert soils and the surface may harden into a light crust. Another feature is the presence on rock masses of a hard coating resembling enamel. This varnish is formed by the interaction of desert weathering processes and compounds of a variety of minerals including iron. A patina due to the mechanical abrasion of the rock surface may also be found. Two terms have been introduced by UNESCO: (a) Ergosol, refers to sand dune formations, 'dynamic' if the dunes are shifting, 'semistatic' or 'static' if they are partially or fully stabilized; and (b) Ermolithosol, describes the hard pavement often to be found in deserts, and includes subdivisions of lithic (stony), gravelly or argillic.

Other features associated with deserts are the accumulation of salt deposits in naturally formed basins or near the coast (*kavir*, *sebkha*). In some desert areas, there may be the development of loessic-type soils as the result of wind transport. This deposit formed by the accumulation of fine sand, silt and clay has a low organic content. Loess can build up fairly rapidly, and in parts of western Libya, the Negev and Sinai, the margins of the desert are becoming covered by such deposits. Archaeological evidence suggests that near Gaza approximately 3 m of loess have accumulated within the last 4,000 years.

Arid steppes

In the arid steppes several types of soil occur. Where aridity is particularly pronounced, the processes by which mineral compounds are altered and organic matter is converted to humus are limited, so that soils are immature.

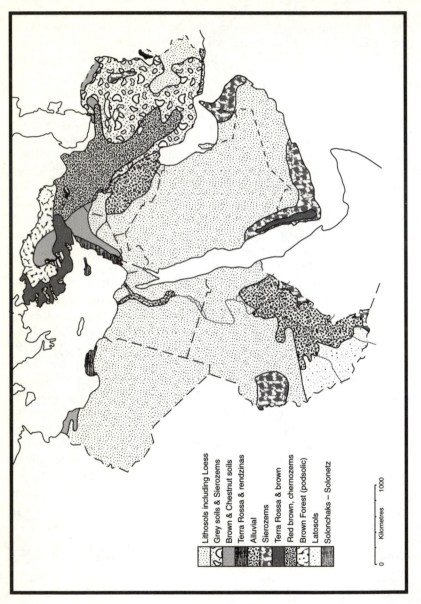

Lithosols including Loess
Grey soils & Sierozems
Brown & Chestnut soils
Terra Rossa & rendzinas
Alluvial
Sierozems
Terra Rossa & brown
Red brown, chernozems
Brown Forest (podsolic)
Latosols
Solonchaks – Solonetz

0   Kilometres   1000

Figure 4.1 Middle East: generalized soil pattern.

Of this type are the steppe soils which are basically sandy in texture, and of grey, grey-brown or grey-red colour, sometimes loamy, sometimes with clay, and often with carbonate or gypsum concentrations. Some authorities attribute any reddish colour to damper climatic phases. Cementation due to lime or gypsum may also be present.

Where aridity is somewhat less pronounced, the process of soil formation may advance further with the production of *sierozem* soils. These are rather more mature than the steppe soils with a somewhat higher, though still small, organic content, and a higher lime fraction, both being fairly evenly distributed throughout the profile. Salinity and the development of gypsum layers can be a prominent feature, but the maximum concentration of salts in a *sierozem* tends to be at a shallow depth, rather than at or near the surface.

Semi-arid to humid areas

As aridity becomes less pronounced, soils of considerably different qualities are found. A synopsis of the major characteristics is given in the context of basic soil type and location.

Brown or yellow-brown soils are associated with a wide range of parent rocks and may range from silty clays to clay loams. A particular feature is the frequent stoniness, both at the surface and at depth. Colour may vary from dull brown to browny-red or brown-yellow, silt content is from 25 to 45 per cent, and on the Jordan plateau, where these soils occur frequently over large tracts, pH values from 6.9 to 7.4 have been measured.

Terra Rossa soils are distinguishable by their characteristic red colour and also by their heavier texture, due to a relatively high clay content of 50–70 per cent, with a correspondingly lower sand content. Enriched in sesquioxides of iron and in silica, there is also evidence of decalcification, with little or no free lime, despite frequent association with a limestone parent rock. Values for pH vary from 7.0 to 8.0, and show the alkaline nature of the soil. Another important feature of Terra Rossa is its high moisture-holding capacity, which allows it to store considerable amounts of rainfall. This, together with its relatively high fertility, makes it an extremely useful soil for agriculture, though it is liable to erosion.

A highly important factor in many soils of the Middle East is the occurrence of a 'hard pan' or calcrete. Much debate has occurred concerning these formations, which may appear as a layer of hard, concrete-like material varying in thickness from a few centimetres to several metres. It is now accepted that these crusts are formed by a complex and prolonged procedure involving three major processes:

1   leaching of soils rich in calcium gypsum, carbonates and sulphates;
2   capillarity from the lower soil layers, by which moisture is drawn towards the surface. This soil moisture is thought to originate from lateral

infiltration or the accumulation of temporary water supplies following heavy but highly episodic rainfall; and

3   the effect of algae.

Topography may often produce areas where small expanses of soils of various types develop, derived from accumulation and downwash. Depending in part upon the source material and parent rock, these infill soils that accumulate in depressions or on 'flats' tend to provide patches of better arable land. Often reddish or brown in colour, they have a fairly high silt or clay content, with some humus. Typical examples of this type occur in the hollow of the low plateau of Qatar. In many of the more rugged areas of the Middle East, the occurrence of these small, scattered basins of soil is the sole basis for human settlement.

Alluvial soils are those in which temporary waterlogging occurs, due to seasonal flooding from large rivers, such as the Nile. The main effect is to give rise to a higher organic content, which produces a very dark colour although the soils are still fundamentally low in humus. Sometimes there are reddish-brown streaks due to iron oxides, with grey or purple mottlings that indicate gleying. Such soils are characteristic of the Nile Valley, especially in Sudan, and the periodic deposition of fertile clay laid down above the dark soil base gives excellent conditions for agriculture, provided that salinity problems can be avoided. Heavy-textural soils of this group are known as *grumosols* and they are characteristically plastic and sticky when wet, and hard and liable to crack when dry.

The lack of humus is widespread in most Middle East soils as basically there is very limited organic material available. Furthermore, many trees do not shed their leaves annually, hence there is less natural surface litter. Also, high summer temperatures in effect burn out much organic material.

As a result of these temperatures and the high capillary movement of soil water, fertilizing by organic or by artificial compounds is less effective than the agricultural practices in cooler, temperate latitudes. However, research by Moalla and Pulford (1995) on the mobility of metals in Egyptian desert soils subject to inundation by Lake Nasser concluded that the soil is not able to support long-term agriculture without enrichment.

Among the various specifically Middle Eastern soil classifications (for example, Dan *et al.* 1976) there is general agreement on basic types which follows broadly the pattern described by Fisher in *The Middle East* (1978). From the steppes of the Mediterranean coastlands with rainfall totals of between 250 and 350 mm to the hyper-arid interiors, there is a sequence from brown rendzinas to sierozems and desert lithosols. Desert lithosols are basically formed on hammada or rock desert surfaces but, where the surface is covered with pebbles and gravel, reg soils occur. Whereas the desert lithosols have no profile development, regs contain clear horizons. Bare sand can be considered parent material or, if it bears vegetation, a regosol. Saline soils could include, on the one hand, desert lithosols and reg soils or, on the other hand, hydromorphic salines resulting from high water tables or coastal inundation.

In his book on the flora of Iran, Parsa (1978) identified five broad soil regions:

1   the west coast of the Caspian Sea with brunisolics;
2   the east coast of the Caspian Sea with Mediterranean mountain-cinnamonic soils;
3   the extensive desert of central and southern Iran;
4   the dry Mediterranean mountains of the west and north of Iran; and
5   less dry mountains that border the Caspian Sea.

Within each region, there is a wide variety of soils. In the brunisolic region, the main soils are alluvial, organic rich but not peaty, saline and very saline (solonetz). In the Mediterranean mountains and dry Mediterranean mountains that border the Caspian Sea the main soils are recent brown and lithosols. The most important soils in the desert region are alluvial, recent brown, reddish brown, arid brown and other raw soils, lithosols, saline soils, solonetz and planosols. In the dry Mediterranean mountains of western Iran the dominant soils are alluvial, recent brown, reddish brown, arid brown and raw soils, lithosols, salines, solonetz and planosols. All these soils are formed on geologically recent parent material and have suffered little leaching.

This classification is reflected, although in more dated terminology, by Zohary (1962) to accompany his geobotanical soil map of Palestine. He identifies the following soil series, many with sub-series:

1   Terra Rossa;
2   rendzina;
3   basalt;
4   sandy-calcareous;
5   alluvial;
6   grey calcareous steppe;
7   lissan marls (related to the former Lake Lissan);
8   loess;
9   hammadas and other desert soils; and
10   saline.

These examples suffice to show that, in detail, the classification of soils in the Middle East is complex but there is a basic underlying pattern. In all, a fundamental factor is aridity and it is this characteristic which renders soils unstable. In a natural environment, undisturbed by human activity, soil erosion occurs but would appear to be generally compensated by soil formation. However, since the earliest civilizations there has been human interference and, particularly where this is associated with steep slopes, soil erosion becomes a major problem. The influence of man has been seen through the removal of vegetation and particularly deforestation, over-grazing and monoculture resulting in soil exhaustion. These are already common features of much of the region. However, if there is a temperature increase of between 3 and 4°C (Imeson and Emmer 1992), there will be an increase in aridity which will result in:

1   an increase in soil degradation resulting from the effect of increased temperature and potential evapotranspiration on organic matter, soluble salts and the moisture balance of the soil;
2   an increase in soil erodibility resulting from lower rates of water acceptance and a decrease in aggregate stability;
3   higher rates of erosion on slopes; and
4   higher sediment concentrations in slope run-off.

Vegetation

While the factors governing soil and vegetation are similar, control by climate is very clearly apparent in both the character and distribution of vegetation within the Middle East. As with soils, moisture content is the main determinant. However, for vegetation values are more clear-cut and it can be shown that for annual rainfall there are two threshold values:

1   the upper threshold of 300–350 mm; and
2   the lower threshold of 80–100 mm.

Below the lower threshold rainfall-dependent vegetation can barely exist. There are two basic ways in which vegetation can survive a prolonged period of heat and drought: (a) by completing the cycle of growth during the cooler rainy season; and (b) by special structural adaptation to resist deleterious conditions.

Plants of the first group usually germinate in late autumn, with the first onset of heavy rains, and grow rapidly throughout the winter, reaching maturity in late spring or early summer. As summer draws on, the plant itself is shrivelled and dies; but the seeds survive, to repeat the annual cycle during the next season. To this group belong grasses and cereals: wheat, barley and millet, some of which are indigenous to the Middle East and have later spread into other lands.

Structural adaptation to counter lack of rainfall may take the form of very deep or extensive roots, as for example in the vine. In some instances roots spread out just below the surface, and are hence able to absorb quantities of night dew, which is a feature of summer in coastal regions. Other plants lie dormant during the dry season, losing much or all of their portion above ground, and maintaining a stock of nutriment in bulbs, tubers or rhizomes. To this group belong the anemone, asphodel, iris, lily, tulip and narcissus. All these plants are characteristic of the Mediterranean and flower in spring but die down with the onset of summer.

Another group of plants remains in more or less active growth during the hot season, but shows special structural adaptation with the object of reducing water loss. Certain species develop a thick outer layer, on the stem or trunk as in the case of the cork oak, or on the leaves, as in the case of laurel, evergreen oak and box. In other instances, leaf surface and size are reduced; the olive being of this type, and the process may be carried further by the shrinkage of leaves to scales or spines, as in tamarisk and thorn bushes. Leaves may even be dropped at the onset of summer, the stem then performing the normal function of leaves. This occurs amongst certain species of broom and asparagus. The other water-saving device is a thick hairy coating by which the inner fleshy parts of the plant are protected from the heat. An example of this adaptation is hyssop.

As with so many other aspects of geography, the Middle East is a phyto-geographical meeting place. Four major plant-geographical regions are represented in the Middle East:

1 the Mediterranean;
2 the Irano-Turanean;
3 the Saharo-Arabian; and
4 the Sudanean.

Additionally, the vegetation of most of Yemen is classified as North-East African (Figure 4.2).

Climatically, the Mediterranean region is characterized by hot, dry summers and warm, moist winters. The Irano-Turanean region is distinguished by continentality, with high summer temperatures, low winter temperatures and low rainfall throughout. The Saharo-Arabian region is similar except that there are high temperatures in both winter and summer and rainfall is, if anything, lower, in some years approaching almost nothing. The climate of the Sudanean region is completely different, being characterized by summer rains and high winter temperatures.

The region displays not only a varied but also a rich plant life. In the desert and on its margins over 2,000 species of plants occur, many indigenous to the Middle East, whilst 10,000 species have been recorded in Iran alone. The corridor aspect of the Middle East is particularly important, for in addition to a type of flora associated specifically with the Mediterranean, there are plants belonging to two other botanogeographic provinces, one predominantly Asiatic, and the other African. Moreover, the southern Sudan, the Black Sea coastlands of Asia Minor and the southern Caspian lowlands each possess a special and distinctive flora quite different from that of the rest of the Middle East. In some localities, these last especially, the present-day flora has subsisted without major change since Pliocene or even Late-Miocene times. In north-ern Asia Minor, this would frequently seem to be the case and, despite changes in climate in the Quaternary, the resulting drop in snow-line would have been insufficient to produce major alteration at moderately low altitudes.

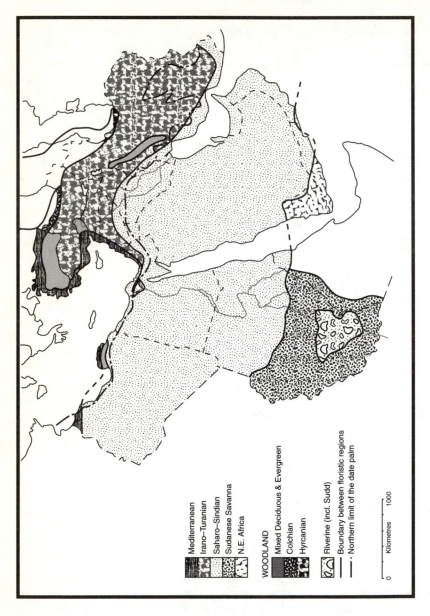

Mediterranean
Irano–Turanian
Saharo–Sindian
Sudanese Savanna
N.E. Africa

WOODLAND
Mixed Deciduous & Evergreen
Colchian
Hyrcanian

Riverine (incl. Sudd)
Boundary between floristic regions
Northern limit of the date palm

0   Kilometres   1000

Figure 4.2 Middle East: natural vegetation.

Thus, where it exists, the vegetation presents a complex mosaic although, as with soils, a number of basic types can be recognized. Fisher, in *The Middle East* (1978), recognized six main vegetation types.

Mediterranean vegetation

Mediterranean vegetation has a relatively restricted distribution, being mainly confined to the wetter parts of the Mediterranean coastal area: the narrow coastal plains of Cyprus, Israel, Lebanon, Syria and Turkey, together with the lower flanks of the mountain ranges immediately inland, including the northern slopes of the Jebel Akhdar of Cyrenaica and parts of the Jefara and Jebel of Tripolitania. In addition to the 'classic' plants of the Mediterranean regime – vines, wheat, olive and fruit trees – a large number of shrubs and herbs, many evergreen, flourish in the regions of thinner soil. Walnut and poplar trees flourish in the damper places. Cactus, introduced from the Americas and used to form hedges round fields or houses, has encroached widely in the Levant, where it has 'escaped' from its earlier, limited use.

Besides the plants mentioned above, the Mediterranean flora includes several highly characteristic plant complexes or groupings. Best known is the *maquis* or *macchia*, which is widespread throughout the Mediterranean basin, particularly in association with siliceous soils. Densely set evergreen oaks, myrtles and broom, with a thick undergrowth of thorn bushes and shrubs, form a vegetative covering that is sometimes sufficiently extensive to provide wartime cover. Through this connection, the term *maquis*, once of botanical significance only, has taken on a wider meaning involving human and even political relationships. *Maquis* is not especially widespread in the Middle East as a whole. More characteristic are degenerate types of *maquis*, in particular *garrigue*.

Garrigue is associated with the thinner soils of calcareous outcrops. Evergreen oaks, which are tolerant of greater aridity, persist, although at wider intervals than in true *maquis*, but tall shrubs are much less common, and a low scrub of dwarf bushes and thorns takes their place. Because of the more open nature of this vegetation, which rarely exceeds 2 m in height, perennial plants can develop, and for some weeks after the spring rains, a carpet of flowers and aromatic herbs appears. *Garrigue* is often discontinuous, with bare patches of soil or rock interspersed with plants. If *maquis* is cleared, it is frequently found that *garrigue* takes its place, especially if animal grazing prevents the regrowth of taller trees and shrubs. For this reason, *garrigue* is more characteristic of the Mediterranean zone of the Middle East than is true *maquis*.

Steppe vegetation

A special type of vegetation has evolved under the influence of steppe climate, of which the chief features are wide seasonal variation of temperature and generally lower rainfall. A botanical province, the Irano-Turanean has been recognized, corresponding to the geographical distribution of steppe conditions. Irano-Turanean vegetation is best developed in central Asia, but a westward extension has occurred through Asia Minor and Iraq into central Syria.

On the lower slopes of the mountains flanking the steppe, a park-like vegetation is found, with scattered carob, juniper and terebinth trees, and bushes of Christ thorn, wild plum and wormwood, separated by expanses of smaller shrubs such as sage, thyme and thorn cushions, or creepers. In regions of true steppe trees are absent, and various species of grass appear, although these have sharply restricted seasonal growth. Many grasses show adaptation to semi-arid conditions such as a hygroscopic seed casing.

More than half of the plants of the steppe region disappear in summer. There is hence a considerable difference in aspect between late winter and early spring, when numerous species of flowers and grass are in rapid growth, and the rest of the year, when most plants are shrivelled or have not yet germinated. For a few weeks each year the steppe presents a picture of luxuriant, almost lush vegetation, but with the approach of summer, only hardier bushes and thorns remain above the ground and vast expanses of bare earth appear, upon which are strewn the withered remains of earlier plant growth.

Desert vegetation

Complete adaptation by a species to the desert environment so that life continues throughout the year, even under the most arid conditions, is extremely rare. Newbigin (1948) cites only three examples of such plants. Others display an extreme degree of adaptation, evade the conditions by completing their growing cycle within a few weeks following the winter rains or are found only in desert margins. The name Saharo-Sindian has been given to the major botanical grouping characteristic of arid conditions.

Mountain vegetation

Four distinctive types of vegetation occur on the higher mountains of the Middle East. Of these, three are forest growths and the fourth is Alpine pasture or heath.

The most widespread type of woodland is of mixed evergreen, coniferous and deciduous trees. Evergreen oaks generally grow on the lower hill slopes, up to about 1,000 m, and associated with these are the carob and species of pine. At higher levels are found cedars, maple, juniper, firs and two important species: the Valonia oak and the Aleppo pine. The cedars of Lebanon, forever associated with that country, now exist only in scattered clumps as the result of centuries of exploitation. The largest of these clumps, numbering 400 trees, occurs at a height of over 200 m between Tripoli and Baalbek and, like many isolated groves in the Middle East, has acquired a semi-sacred character. By far the greatest extent of this type of forest exists in Asia Minor, with limited extensions southwards along the crests of the Zagros, and in western Syria and Lebanon. About one-eighth of the state of Turkey is classified as forested.

Above 1,500–2,000 m, or exceptionally, 2,200–2,500 m in the Elburz mountains of Iran, forests die out and are replaced by scrub or dwarf specimens of true forest trees. In eastern Anatolia and Azerbaijan, temperatures are lower than elsewhere in the Middle East, and a kind of Alpine vegetation appears. In the wetter parts, small areas of grassland may occur but, more often, conditions are too dry and vegetation is limited to bushes, creepers and 'cushion' plants.

A second type of forest is found on the northern slopes of the Elburz mountains, towards the Caspian Sea. Oak, hazel, alder, maple, hornbeam, hawthorn, wild plum and wild pear are the characteristic trees, and these are festooned by a dense growth of brambles, ivy and other creepers. Occasionally there are openings occupied by box, thorn bushes, pomegranate and medlar trees. To this luxuriant growth, peculiar to the region, and developing under conditions of abundant rainfall throughout the year with high or moderate temperatures, the name Hyrcanian forest can be given. One interesting feature is the absence of conifers.

A third type of woodland, the Colchian or Pontic forest, has developed in response to the warm humid climate of the southern Caucasus and eastern Black Sea. An indigenous species of beech, with oak, hazel, walnut, maple and hornbeam, are the chief trees, and a dense undergrowth of climbing plants is again found, with the widespread occurrence of the rhododendron. Best developed in the southern Caucasus, the Colchian forest extends westwards in an increasingly attenuated form along the north-eastern edge of the Anatolian plateau as far west as Sinop. It has been suggested that, like the Hyrcanian forest further to the east, the Colchian forest is the remnant of a flora that was characteristic of much larger areas of the Middle East during Tertiary and early Quaternary times.

Savanna

In Sudan, broadly south of latitude 15°N, in response to the markedly different climatic regime and soil types, savanna-type vegetation becomes dominant. With increasing rainfall, grassland becomes interspersed with trees. Various types of acacia occur in areas with under 450 mm annual rainfall, and then with increased rainfall, an inter-mixture with broader-leaved deciduous species. As rainfall becomes more abundant, the height and luxuriance of the trees tends to increase. Grasses continue to occur, even when, as is common, the branches of individual trees may meet and there is a complete canopy.

The intermingling of grasses and tall trees in some parts of Sudan and the absence of trees in others, leaving a grass savanna only, has been explained

in different ways. One view is that grassland without trees is a climax vegetation controlled by soil type. Another is that the practice by pastoralists of
regularly burning the savanna grass results in the extinction of trees and the
perpetuation of the more rapidly growing grasses. These grasses can regen-
erate rapidly, whereas trees take longer to re-establish themselves, and the
saplings are highly sensitive both to burning and to grazing. In the extreme
south of Sudan, the heavily wooded savanna passes in places into full tropical rain forest.

Riverine vegetation

The extensive alluvial lowlands of the great rivers have a special type of vegetation. Aquatic grasses, papyrus, lotus and reeds that sometimes attain a
height of 8 m make up a thick undergrowth in shallow, braided reaches of
the Upper Nile system and in the lower courses of the Tigris and Euphrates.
The vegetation is so dense in southern Sudan as to form a real barrier to
the movement of the water. There, it is known as the Sudd. Elsewhere, scattered willow, poplar and alder trees may occur, but the most common tree
is the date palm, which is extremely tolerant of excessive water, provided
that the temperature remains high. Cultivated palms are therefore a feature
of riverine lowlands in Egypt and lower Mesopotamia.

As with soil classification, there have been several suggested floral classifications of the Middle East and varying terminology has been used. The most
detailed is probably on the flora of the Arabian Peninsula and Socotra by
Miller and Cope (1996). They recognize three main phytogeographical areas
in Arabia:

1   the Saharo-Sindian zone, characterized by dwarf shrubland;
2   the Somalia-Masai regional centre of endemism characterized by *Acacia
    commiphora* bushland and thicket and at higher altitudes evergreen bush-
    land and thicket; and
3   the Afromontaine, archipelago-like region of endemism characterized by
    Juniperous procera, semi-evergreen bushland and thicket.

For their detailed classification of the vegetation, they identify three major
habitat types:

1   deserts and semi-deserts, including *sebkha*;
2   mountains and plains of the south and south-west of the Peninsula,
    including Socotra; and
3   coastal vegetation.

For his flora of Iran, Parsa (1978) proposed ten major vegetation types:
coastal strand, desert scrub, desert woodland, desert shrub, alkaline sink,
forests, valley grassland, mountains and cliffs, marsh and aquatic plants, and

mineral water plants. Although varying in detail, these show a similar pattern to the classifications already discussed.

If anything, the effects of human activity have been greater on vegetation than on soil. Indeed, in some important areas, it has been so extensive as to justify the view that there is now no really natural vegetation left. Long occupation by man has meant that much of the original plant cover has been removed. The most striking change is the replacement in many parts of forest and woodland by scrub and heath, for example by the practice of annual burning of the grassland and scrub carried out in many parts of central and southern Sudan. Deforestation has been widespread, particularly in the Levant where the accessibility of the wooded areas, their relative absence in nearby countries such as Egypt and reckless use in the present century, have denuded vast areas. Similarly, the Elburz uplands have lost much of their former dense tree cover, particularly since the growth in use of motor vehicles allowed charcoal to be transported for sale in the towns, particularly Tehran.

Once destroyed, woodland may not easily be renewed, since in some areas it is marginal, with natural conditions only barely suitable for its continued existence. In one sense, some, though not all, forests of the Middle East can be considered as legacies from an earlier, wetter past. Once removed, the balance turns against renewed growth. Soil is quickly eroded, the water-table may fall, and tree seedlings are not sufficiently robust to thrive in competition with scrub vegetation that springs up on deforested sites. However, for some areas it remains possible to re-afforest, though the seedlings require physical and legal protection.

Most harmful to natural or artificial regeneration is the practice of unrestrained grazing, chiefly by goats, which destroys the seedlings as they develop. In the opinion of many, unrestricted grazing by goats is one of the fundamental causes of agricultural backwardness in the Middle East. Gradual loss of forest cover leads to uncontrolled water run-off, with resultant soil erosion, and this in turn lowers the water-table in the subsoil, making it difficult to obtain an adequate water supply for cultivation.

In the future, the effects of man's activity superimposed upon probable climatic change are likely to cause alterations in the vegetation pattern of the Middle East of a far-reaching and non-reversible order. Le Houérou (1992) concludes in his study of vegetation and land use in the Mediterranean Basin that, by the year 2050 for the Afro-Asian Mediterranean countries, there will be an urban density incompatible with the maintenance of any natural vegetation. Over the past 30 years, forest and shrubland vegetation have been receding by 1–2 per cent per annum while natural steppe vegetation in the rangeland areas has been destroyed at such a rate that desert encroachment has occurred at 2 per cent per annum. By the year 2050, only Turkey may have forests remaining in its non-Mediterranean areas, Iran in its Hyrcania region and Morocco in the high and middle Atlas Mountains. Conditions are likely to be, if anything, more extreme in other areas of the Middle East and therefore, like soil, the fragile vegetation cover of the region is under immense and increasing stress.

5   Two key resources

Physically, economically, socially and particularly geopolitically, the Middle East is dominated by two liquids: water and oil. Both are extracted from specific geological structures and both in different ways are vital to the development of human society in much of the region. They therefore provide an appropriate link between the physical and the human geography of the Middle East. Water is highly significant as a result of scarcity within the region whereas with oil the major deficiency is outside the region. Both are therefore considered strategic resources, the supplies of which may be disrupted by geopolitical actions. Therefore, they provide a key connection between the systematic and the geopolitical sections of this book.

WATER

Aridity has already been highlighted as the dominant climatic characteristic exercising a major influence on geomorphology, soils and vegetation. It provides a common thread which, in so many ways, gives the Middle East its distinctiveness. However, although the effects may appear obvious, the definition of aridity remains illusive. Furthermore, there is no general agreement over the related terms drought and desert. Climate, as shown in Chapter 3, is the fundamental characteristic which embraces all three terms but, if the areas concerned are to be mapped, aridity, desert and drought need to be related to physical variables on the ground such as landforms, soils or vegetation.

The landscape features characteristic of arid zone landscapes have been described in Chapter 2 but it must be concluded that they are not sufficiently sensitive to serve as boundary markers either for arid zones or deserts. There remains the startling paradox that the major landforms of drylands are mostly the products of the action of water (Graf 1988). In the light of the processes indicated in Chapter 4, it might appear that the arid zone soil profile would provide a significant feature for mapping. However, in reality vast areas are covered by thin immature infertile soils which show little if any signs of pedological development. Entisols, immature soils, and aridisols, red-brown desert soils suitable only for rough grazing, cover according to Dregne (1976) 41.5 and 35.9 per cent, respectively, of the arid zone. Vegetation, as shown in Chapter 4, can demonstrate a wide range of adaptation to aridity but, although climate may be predominant, plant distribution depends upon a number of other variables. Therefore, it is difficult to produce any distinctive classification of vegetation as a result of the diversity of arid environments. In very general terms, Dregne (1976) recognizes three broadly based categories: semi-arid vegetation (7 million km²), arid vegetation (33.4 million km²) and hyper-arid areas in which there is little or no vegetation (6.3 million km²).

Thus, although there are in the Middle East landforms, soils and plant communities which possess agreed arid zone characteristics, no universally agreed basis either for the definition of aridity or for the mapping of arid areas has been produced. Similar problems obviously occur in attempting to use the same variables to define deserts. Therefore, the basic definition set out by Goudie (1985) states only that deserts are areas of low or absent vegetation cover with an exposed ground surface.

On the face of it, drought would appear to offer more scope for precise definition but drought is multi-faceted. If drought implies at its most basic the lack of water then clearly both the supply side and the demand side need to be examined. Water shortages can result not only from low rainfall or declining ground water supplies but also through population growth or agricultural changes. Meteorologically, drought may be defined by the number of consecutive days without rainfall but the effects will be mitigated by such factors as the season. Furthermore, throughout the arid zone including most of the Middle East, there are few data recording stations and therefore precipitation deviations from the mean must for large areas be conjectural. For the farmer, drought implies a lack of moisture in the root zone of crops whereas for the politician or economist, a drought is only of significant shortage when it affects the economy or the stability of a country. As Copan (1983) concludes: 'A drought becomes a drought when someone wants it to be so'.

Nonetheless, one fundamental demarcation which can be made is between regions where rivers have a perennial flow, and regions where river flow is only seasonal or intermittent. The latter area is extensive, including central Iran, most of the Arabian Peninsula, the plateau of central Syria, the Sinai Peninsula, and apart from the Nile valley and delta, the whole of north-east Africa.

The intermittent flow zone is characterized by wadis, defined as stream courses which are normally dry but sometimes subjected to large flows of water and sediment (Goudie and Wilkinson 1977). Wadis vary from deeply eroded ravines to stream courses across plains discernible only by variations in the texture of deposits. Although there is no standard morphology of wadis, there is a characteristic process, the flash flood. This is normally produced by a localized convectional storm of high intensity which results in a rapid rise in water level and maximum discharge, frequently in a matter of minutes. The typical flood hydrograph has a steep rapidly rising limb, a sharp peak and an equally steep falling limb. Discharge may be hundreds or even thousands of cubic metres in one hour. Such rates are likely to cause significant erosion, transport and deposition. Flash floods are infrequent and between the events, weathering produces depths of unconsolidated material which is easily removed owing to the general lack of constraint provided by vegetation or organic matter. As with most aspects of arid zone geomorphology, geology and hydrology, there is a dearth of reliable and accurate data on such events. However, the results can be measured and Schick (1985) cites the effect of a storm, with a one in fifty-years recurrence interval over the Sinai Peninsula, which produced a deposit measuring 300 m by 500 m at the outlet of Wadi El Arish.

Wadi processes are therefore characterized by long periods of slow rate weathering interrupted perhaps at intervals as great as fifty or a hundred years by cataclysmic flood events. The geomorphological form is produced in perhaps a matter of hours. In upland areas, the cross profiles of wadis indicate steep sides and flat floors while the long profiles normally exhibit a gentle gradient (Anderson and Curry 1987).

Apart from flash floods, other types of flood can also be distinguished in the Middle East (Graf 1988). Single peak floods lasting perhaps a few days are the product of storms over large catchments, while multiple peak floods occur with more persistent rainfall. However, the best known floods are the seasonal floods exhibited by the exogenous rivers such as the Nile. Such seasonal floods are the product of the entire catchment and, in the case of the Nile, this covers a variety of climatic regimes. Not only is there a time lag between the advent of the rainfall and the increase in discharge but the complexity of the seasonal hydrograph is enhanced because each tributary exhibits its own flood pattern.

In the Middle East, perennial surface flow is restricted to most of Turkey, western and northern Iran, central and eastern Iraq, the Levant coastal zone, Cyprus and certain limited areas of the North African coastal fringe (Figure 5.1). Only two rivers considered major on a global scale, the Euphrates and the Tigris, are wholly fed by indigenous rainfall. The third major river, the Nile, is maintained almost entirely by rainfall from outside the region in Ethiopia and East Africa. As a result of this pattern, the only Middle Eastern countries which are relatively water-rich in terms of surface flow are Egypt, Sudan, Turkey and Iraq, each of which has a mean annual flow of between 75 and 85 bcm (billion cubic metres). Iran enjoys approximately half that total and Syria about one-third. Lebanon has approximately 1.5 bcm while Israel and Jordan share the same total. Elsewhere in the region, surface flow is intermittent and, in terms of the figures given, minimal. However, these data take no account of other sources of water and represent only the supply side of the budget. For example, given the size of its population and the importance of agriculture, Iraq can be considered relatively water-rich as can Sudan but Egypt, with its burgeoning population certainly cannot be characterized in the same way.

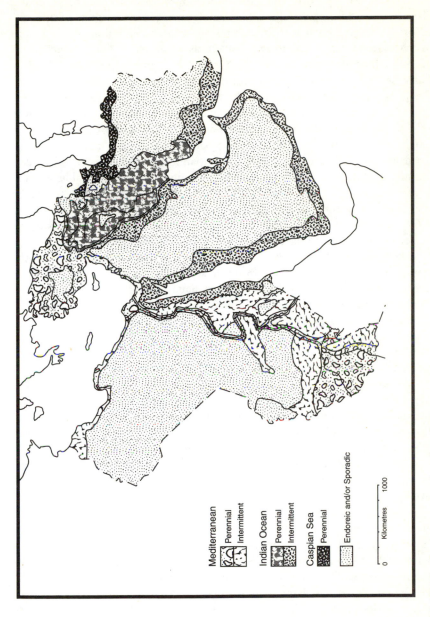

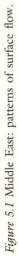

Figure 5.1 Middle East: patterns of surface flow.

Another major feature of the Middle East is the extent of endoreic or inland drainage. The main influencing factor is of course low annual rainfall, but structure and physiography have important roles. Interior basins shut off from the sea by encircling mountain ranges, lava flows across existing valleys and tectonic basins along fault-lines have all, in a region of episodic and often scanty rainfall with high rates of evapotranspiration, given rise to extensive areas of inland drainage, usually as a complex of sumps or basins.

These inland drainage basins often show a characteristic assemblage of landforms. Defined by an upland rim, the closed basins have an outwash area or 'alluvial fan', composed of rock debris, gravels and silts, immediately below the mountain slopes. At lower levels the topography becomes more level and plain-like with finer silts that may be re-deposited by aeolian action, though there can be benches or relict shorelines and plain rock pediments. At the lowest level, there is usually an expanse of water, marsh, salt desert or *kavir*, shallow and highly variable, often saline but not inevitably so. Iran has the greatest extent of inland drainage, since no major river has broken through the encircling fold ranges, and thus the interior is divided among *kavir* and endoreic basins. Only in the extreme north and west are streams sufficiently developed to have cut back deeply into the mountain chains. The Arabian Peninsula has virtually no surface drainage and, owing largely to the eastward tilt of the rock strata, *kavir* or *sebkha* formation occurs mainly in the extreme south-east, on the borders of Saudi Arabia and Oman.

Although heavier rainfall on the northern and western coastal margins has led to a great development of rivers, many of which have cut back through the coastal ranges to reach the plateau basins of the interior, there remain a substantial number of closed drainage basins in Turkey, chief of which are occupied by Lake Tuz and Lake Van.

A well-known, but relatively small basin of inland drainage is that of the River Jordan, with the Dead Sea at its lowest point. Further to the north, the rivers that occupy the tectonic trough, the Litani and the Orontes, have both broken through to the sea, but an area of endoreic drainage lies just to the east and both Damascus and Aleppo lie on the alluvial zones of small closed basins. In North Africa there are also a number of drainage sumps, but none has a feeder stream. Instead, there is, as in the Qattara Depression, upwelling of ground water.

Apart from the very few large rivers, most Middle Eastern surface flow comprises relatively short streams which descend the mountainsides and exhibit steep gradients and incised cross profiles. The Anatolian plateau is ringed by a number of such rivers that are broken by cascades and waterfalls and gorges. There are only a few more developed rivers such as the Buyk Menderes. Similar effects on a much smaller scale characterize western Lebanon. Most of all, perhaps, the western Zagros shows an elaboration of short immature streams that have cut spectacular gorges even across anticlinal ridges, and because of their erosive power carry immense quantities of sediment.

Therefore, unlike humid latitudes, the water management system depends upon aquifers rather than surface storage. There are obvious exceptions to this such as Lake Nasser and Lake Kinneret but of compelling concern for water abstraction throughout most of the region are the shallow and deep aquifers. Burdon (1982) provides a guide to the key hydrogeological conditions within the region. Since precipitation averages less than and, often much less than, 150 mm annually while potential evapotranspiration over the same period exceeds 2,250 mm, it is only under exceptional circumstances that there is any infiltration. The main condition under which infiltration occurs is as a result of floods which can result from either the intensity of individual storms or the concentration of run-off into channels. The other important factor is temperature at the time of precipitation in that during the winter, particularly in the northern part of the region, potential evapotranspiration rates will be lower. Other than that, two local factors may be of significance: the rapid infiltration of even small amounts of moisture into sand (Agnew 1988) and the infiltration of dew (Agnew and Anderson 1988).

Therefore, the sources of renewable ground water are very limited. Indeed, much of the ground water which originates within the region itself occurs in very circumscribed areas, predominantly those of higher rainfall and those of major rivers where there is continuous infiltration.

In the Middle East, two main types of aquifer can be distinguished (Beaumont *et al.* 1988): (a) shallow alluvial aquifers which are generally small and unconfined and react rapidly to the onset of precipitation; and (b) deep rock aquifers usually located in sandstone or limestone which are often confined. A further distinction is possible between the deep aquifers which are naturally recharged and the fossil ground water aquifers.

Current research indicates that between the sources of non-renewable fossil ground water and the aquifers that are replenished regularly occurs water in major faults such as the Jordan transform fault. This water has been shown to have collected from the mountains of Lebanon and southern Turkey in what can be described as a geological mega-watershed (Anderson 1992).

Ground water enters the more arid part of the Middle East from regions to the north and south which have higher precipitation. Some of this moves through extensive aquifers and some enters as snow which recharges the surface aquifers bordering the major rivers. The main examples are the Nile and the Tigris–Euphrates, most of the flow of which is stored on the ground or beneath the surface. Some major aquifers extend into areas of higher rainfall. The Rub' al Khali basin receives recharge from the highlands of Yemen and Asir and water from the mountains of northern Oman replenishes the aquifers of the UAE.

Highlands, mostly around the periphery of the Middle East, which receive higher rainfall and therefore generate ground water, range from the mountains of northern Libya to the mountains of Lebanon, Jordan, Syria and southern Turkey. More central, in that they are in the Arabian Peninsula, are the mountains of Oman and Yemen. These present a contrast since, for example, the mountains of Lebanon produce perennial rivers whereas the flow from those of Yemen and Oman is seasonal.

Good quality, usually shallow, ground water can also be found under certain conditions of lower precipitation. In Libya, the Mediterranean littoral of Egypt and the eastern Province of Saudi Arabia, freshwater lenses float on more saline regional ground water. A secondary coastal freshwater lens system has been identified which stretches along coastal Saudi Arabia, across Bahrain, Qatar and into the central UAE with segments in northern Kuwait and the eastern UAE. This zone occurs between the coast and the *sebkha* zone and results from particular climatic conditions. Sub-surface ground water flow also occurs in wadi drainage systems and the deposits on wadi floors constitute important shallow aquifers.

The other important source of water now used in certain parts of the Middle East, notably Libya and Saudi Arabia, is fossil ground water. This is defined as 'infiltrated meteoric water' which is contained in deep aquifers with little or no replenishment over the past few thousand years (Burdon 1977). Since there is no replenishment, the extraction of fossil ground water is analogous to mining. Whilst, as already indicated earlier, there is some controversy about the changing climatic oscillations of the Quaternary, there seems little doubt that water entered the deep fossil aquifers during pluvial periods. This has been shown by the use of isotope techniques as a result of which the water itself can be dated.

The most clearly identifiable fossil ground water is that found in the Palaeozoic terrestrial sandstones of the great basins although such ground water does occur in the marine aquifers. In general, the temperature of the water tends to be high, from 20–70°C, since much of it may have originally been as deep as 5,000 m. In the main sandstone aquifers most of the waters are only slightly mineralized whereas in the carbonate aquifers they are of medium chemical composition.

Within the Middle East there are seven major ground water basins (Blake *et al.* 1987) which contain mainly fossil ground water (Figure 5.2). Four occur in north Africa: Chad (14,000 km³), Fezzan (4,800 km³), Kufra (3,400 km³) and the Western Desert of Egypt (18,000 km³). In addition, it has been proposed that for the northern Kufra and Sirte Basin a figure of 2,000 km³ is appropriate. Furthermore, there is additional fossil ground water in Sudan. In the Arabian Peninsula there are three major basins: Nafud (4,000 km³), Riyadh (1,500 km³) and Rub' al Khali (14,500 km³). There are also minor basins in Sinai and Jordan. All these statistics are conservative estimates and it would seem reasonable to suppose that 65,000 km³ of fossil ground water exists in the region. These deep aquifers underlie almost 50 per cent of the land area of the Middle East and they are obviously of crucial importance as a ground water resource.

Although, for the Middle East as a whole, water management is based upon ground water and the aquifer, there are notable exceptions particularly in the basins of the major river systems. The use of surface storage to regulate

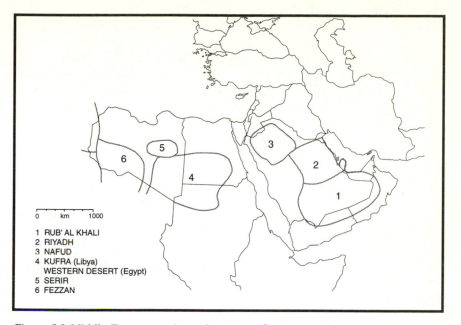

Figure 5.2 Middle East: approximate locations of major aquifers.

water distribution is limited by the general lack of perennial surface flow and, where it is possible, it is constrained by rates of evaporation. Comparison can be made between Lake Nasser which extends across the Egypt–Sudan boundary in the flat lower course of the Nile, and Lake Tana, the source of the Blue Nile located in the Ethiopian highlands. The climate of the region of the former is for most of the year hot to extremely hot with a very low relative humidity while the latter is characterized by a considerably cooler mountain climate with a generally higher relative humidity. From the viewpoint of potential evaporation losses, Lake Tana would provide a far more efficient control for the Nile system. However, hydropolitics dictated the development of the use of Lake Nasser with consequent evaporation losses of some 10 bcm annually, 10 per cent of the entire flow and twice the saving which would be achieved by a diversion through the Sudd region.

Other than the obvious climatic considerations, the key factor governing the rate of water loss by evaporation is the relationship between the surface area and the volume stored in the lake or reservoir. The volume to area (V/A) ratio offers a useful measure of potential for water loss. Kolars (1994) shows that the lake behind the Ataturk Dam, the major control of the Euphrates system, has a volume of 48.7 bcm and an area of 817 million square metres giving a V/A ratio of 59.6. This can be compared with the far less conservative shape of Lake Nasser with a volume of 78.5 bcm and a surface area of 3.5 billion square metres giving a V/A ratio of 22.4.

Apart from the potential for water loss associated with surface storage, the requirement for dams to either develop or enhance the storage introduces another series of problems particularly obvious in the Middle Eastern environment. If the impounded lake is maintained at a reasonably high level, the dam will be unable to act as an effective flood control mechanism at times of rapid discharge increase. Middle Eastern rivers are characterized by such rapid increases and also by large-scale sediment transfer. The storage lake acts as a sediment trap and is therefore in danger of rapid infill and subsequent loss of capacity. The removal of sediment also affects the character of the water downstream which has enhanced erosive power but in general lacks the nutrients which were considered so vital in traditional irrigation practices.

Dams may also be constructed in an attempt to enhance the recharge of ground water. Two particularly notable constructions are those, each over 20 km in length, which have been erected in Oman, one at Wadi al Khawd in the Greater Capital Area and one near Sohar. Key problems for recharge from this approach include evaporation losses and the effects of sedimentation on infiltration rates. Nonetheless, there are many recharge dams both relatively large and small throughout the Arabian Peninsula. The most effective approach is probably through the use of gabions which can be employed to facilitate the temporary construction of dams. Similar structures, usually on a small scale, have been developed for water harvesting but recharge is likely to be more effective through surface spreading or the use of wells. In particular, wells constructed near areas of surface flow can be used to draw down the water-table thereby inducing the transmission of water into the aquifer.

One particularly attractive aspect of artificial recharge is that natural filtration of the water occurs. Artificial filtration or recycling can be carried out by physical, chemical or biological means. There are now numerous wastewater recycling plants in operation in the Middle East but for most of the region, the product is used for municipal purposes or irrigation. Israel in particular has focused on recycling and it is estimated that almost 80 per cent of its municipal water is now reused.

The other major source by which the supply of water can be increased is desalination, the separation of water of a required purity from saline water. It is estimated that approximately two-thirds of the global desalination capacity is located in the Middle East, almost exclusively in the oil-rich countries, especially Saudi Arabia, Kuwait, Qatar, Bahrain, the UAE and Libya. Capital costs are related particularly to the form of power while operating costs are dependent upon the technique (Agnew and Anderson 1992). To arrive at comparable figures entails reliance upon a number of assumptions and the only safe conclusion is that water from this source remains relatively expensive. Therefore, the product is likely to be devoted essentially to the maintenance of a guaranteed potable water supply. The two major approaches to desalination are: (a) through increasingly sophisticated forms of distillation and (b) through the employment of membranes. Modern techniques of distillation are known as Multiple Stage Flash (MSF) and the basic membrane technique is reverse osmosis. Other techniques include vapour compression and electrodialysis.

For completeness, it must be noted that supply-side contributions may be increased through the natural use or deliberate harvesting of occult precipitation, specifically dew, mist and fog. None of these sources is likely to increase water availability significantly but each may be important for flora and fauna. For example, the southern Dhofar region of Oman enjoys a period of relatively constant mist during the Kharif season coinciding with the Indian Monsoon. Modern appliances are being used to intercept and capture this moisture but there is evidence of more traditional techniques. In many cases, containers measuring as much as 60 m in height and more than that in diameter have been constructed abutting on to trees to capture the stem flow which results from the mist. Precipitation may also be increased by cloud seeding using, commonly, silver iodide or dry ice. The results of such practice are usually difficult to assess in that the natural series of meteorological events can only be conjectured. However, work on cloud seeding began in Israel in 1948 and some impressive results have been reported (Goldreich 1988).

Despite this variety of possible water sources, the Middle East remains a region, the greater part of which is distinguished by water scarcity. In particular, Egypt, Cyprus, Jordan, Palestine, Israel and the countries of the Arabian Peninsula are water deficient at present rates of usage. However, in the region as a whole, some 70 per cent of water is devoted to agriculture through which it provides a return far smaller than that available from industry. As a result, there has been a focus upon water conservation and improved irrigation techniques. As with any other scarce resource, a cost–benefit analysis can be made and the resource allocated either directly by price or by government policy. In the case of agriculture, there is a further possibility in that the import of food can be perceived as the enhancement of supplies through the addition of 'virtual' water.

Irrigation is the process by which water is applied to the soil in a controlled manner with the aim of maintaining soil moisture within the range required for optimum plant growth. However, there are so many techniques and purposes that precise definition is difficult. Traditional methods through inundation have been practised in the Middle East for millennia and Heathcote (1983) records sites as old as 5500 BC in Iran and 2300 BC in Mesopotamia. Indeed, Egypt claims a structure built some 5,000 years ago as the oldest irrigation dam. Before the advent of twentieth-century machinery, water requirements not obtainable as a result of natural floods were lifted from channels or wells by human or draft animal power using a variety of devices from a rope and a bucket to the Archimedes Screw. The water was applied, and still is throughout a large part of the Middle East, through the construction of basins or furrows.

Water for irrigation may be transported from its source to its point of application through surface or sub-surface channels. For surface channels, conservation can be improved through channel linings but sub-surface channels have the added advantage of limiting evaporation. Such under-ground water courses, known as *qanats, aflaj, foggara* or *karez* are found in many areas of the Middle East and their use in southern Arabia has been documented in detail by Wilkinson (1977). Such subterranean channels may be many kilometres in length and in Iran alone over 20,000 alone remain operational. From the point where the *qanat* intersects the surface, the water is first put to human use and then used for what is frequently an elaborate system of surface irrigation.

Surface irrigation accounts for some 95 per cent of world irrigation (Kay 1986) but more conservative methods of application are increasingly found particularly in the oil-rich Middle Eastern countries. The most prevalent is undoubtedly through the use of sprays and the centre pivot aerial sprays are commonly encountered in the Arabian Peninsula. Indeed, wheat production in Saudi Arabia is almost exclusively dependent upon them. However, the use of sprays entails obvious evaporation losses and various forms of drip irrigation have been installed in the region particularly for high-grade crops. Israel, with computerized systems of water application, is at the forefront of Middle Eastern technology. While irrigation has resulted in some cases in a threefold increase in cereal yields, there are problems. In particular, drainage systems must be effective if salinization is not to occur. The influence of irri-gation on the demand side of the water budget throughout the Middle East has already been mentioned but there are also interesting questions concerning the optimum size of schemes bearing in mind environmental and health hazards (Agnew and Anderson 1992).

The potential for climatic change in the region and the possible effects of this on precipitation have already been discussed. However, the other major consideration which would affect aquifers is sea-level change. As Lindh (1992) confirms, the societal consequences of climatic change, those affecting the hydrological cycle and water resources, are expected to be particularly serious. Also, salt water intrusion as a result of sea level rise will cause increasing salinity in aquifers. Since large areas of the Middle East are dependent upon water from aquifers, this would be a particularly disastrous change.

PETROLEUM

The Middle East is the richest petroleum province in the world and is unique among all the provinces in terms of the magnitudes of its fields (Shannon and Naylor 1989). For the producing or proven fields, twenty-five are recognized as super giant fields, with in excess of 5 billion recoverable barrels, while sixty-nine qualify as giant fields, with between 500 million and 5 billion recoverable barrels. The largest field is Ghawar in Saudi Arabia with conservatively estimated recoverable reserves of 83 billion barrels. The most important fields and their dates of discovery are given in Table 5.1. No other hydrocarbon province globally has such a concentration of super giant and giant fields.

Table 5.1 The key oilfields of the Middle East

Location	Year of discovery
Iran	
Agha Jari	1938
Gach Saran	1928
Marun	1964
Bibi Hakimeh	1961
Ahwaz	1958
Iraq	
Kirkuk	1929
Rumaila	1953
Kuwait	
Burgan	1931
Saudi Arabia	
Ghawar	1948
Safaniya	1951
Abqaiq	1940
Libya	
Sarir	1961

The majority of the reservoirs are of Mid-Jurassic to Upper-Cretaceous age and the next productive group of reservoirs occurs in the rocks of the Oligocene to Miocene age. The oil shows great physical and chemical similarity which suggests a common source, the most likely being Lower- to Mid-Cretaceous dolomites, limestones and shales. A number of general features are evident aereally and stratigraphically. The oilfields are typically ellipsoidal to elongate in shape and show two distinct trends: north–south and north-west to south-east. The fields of Saudi Arabia, Iraq and, to an extent the UAE, display the former alignment while those of Iran exhibit the latter. Clearly the vast continuity of the Arabian platform can be contrasted with the dramatic structural form of the Zagros domain. Another characteristic is that the principal reservoirs become progressively younger from south-west to north-east.

Origin of oil

The exact processes by which petroleum is formed are complex and in their finer detail some controversy remains. Suffice it to say that in broad terms, petroleum results from the partial decomposition of organic matter under anaerobic conditions. The organic matter comprises animal and plant micro-organisms, the remains of which accumulated on the floors of seas. They were covered by deposits of thick fine sediment which excluded air and light thereby producing anaerobic conditions. This was the basic environment but, given the relative scarcity of oil, the specifics with regard to the exact nature of the organic matter, the sequence of events and the rates of sedimentation must have been far more exacting.

With continued sedimentation, compression occurred and rock strata were formed. The pressure squeezed the oil globules, water and natural gas from the original consolidated silts and clays in which they were formed to nearby permeable rock measures, chiefly limestones and sandstones. Once located in these reservoir rocks, the oil, natural gas and water separated according to specific gravity so that in a rock containing all three there would be a sequence downwards of natural gas, oil and saline water. The final part of the process appears to have required some disturbance or irregularity of the strata allowing the oil to migrate into a trap or basin. Without this stage, the oil would have remained so dispersed throughout the rock that commercial exploitation would have been precluded. However, excessive disturbance would have broken up the retaining structures allowing the oil to escape. Petroleum deposits are therefore normally associated with the outer margins of fold structures where disturbance of the rock was restricted.

An examination of the various accumulations indicates three types of structure in which oil is trapped (Shannon and Naylor 1989):

1   salt-controlled structures;
2   basement-controlled structures; and
3   fold structures.

It can be concluded that oil accumulation depends upon a conjunction of factors:

1   the original richness of the oil-forming material;
2   the occurrence of porous strata;
3   tilting or other changes in the rock succession to allow separation of oil from water, and its concentration in workable quantities; and
4   impermeable cap-rocks to prevent leakage of oil to the surface.

The main contributory causes of the prolific petroleum province of the Middle East are (Shannon and Naylor 1989):

1   the presence of a thick and extensive sequence of salt in the eastern Saudi Arabian shelf region;
2   the long and tectonically quiet history of almost continuous sedimentation;
3   the significant volume of marine sediment including extensive carbonate reservoirs;
4   the close association of reservoirs and source rocks as the result of a series of transgressive and regressive cycles in evolution of the region;

5 the presence of efficient seals including the regionally extensive evaporites, especially in the Late-Mesozoic and Cenozoic;
6 the presence of very broad and simple anticlinal structures; and
7 the presence of a simple system of basement lineaments reactivated only slightly during the evolution of the region.

Of these, the first is regarded as of fundamental importance and it is likely that many of the other variables derive from the presence, location and movement of the salt.

Oil-producing areas

In the geographical pattern of Middle Eastern oil and gas production, three locations can be discerned:

1 under and around the Persian–Arabian Gulf, with its geosynclinal extension into Iraq as far as north-east Syria and south-east Turkey;
2 along the Red Sea–Aqaba rift zone, as far north as Israel; and
3 the northern rim of the shield or basement of north-east Africa, with deposits in the interior deserts of Libya and western Egypt (Figure 5.3).

In the Persian–Arabian Gulf area, the main fields are in Iran, Iraq, Saudi Arabia, Kuwait, the UAE and Qatar. In Iran and Iraq, the oil is found along the western flanks of the Zagros mountains on either side of the Pushti-i-Kuh range. This range descends steeply to the plain of Mesopotamia but to the south-east and to the north-west there are wide lowland embayments where the plain rises more gradually through a series of foothills to the Zagros proper. The southern of these embayments, drained by the Karun River, contains the Khuzistan oilfields of Iran and the northern embayment, the fields of north and north-west Iraq. In Khuzistan, the oilfields are related to the Asmari limestone of Oligocene–Miocene age which is some 300 m in thickness and acts as the reservoir. Above the Asmari series is a complex of shale, salt and anhydrite beds of Miocene age known as the lower Fars series. These beds together act as the impermeable cap-rock except where they are broken by faults. The oil occurs in dome-like anticlines each measuring some 30–30 km in length.

Oil was first produced in 1908, from the Masjid-i-Suleiman field. After 1918, output steadily increased, and in 1928, the Haft Kel field was developed. The remaining fields came into production during or following the Second World War. Iran also has offshore producing fields in the Persian–Arabian Gulf, the most important being Darius, Cyrus, Esfandiar and Feridun.

The oilfields of Iraq, in the embayment to the north of the Pusht-i-Kuh range, follow a similar geological pattern to those to the south. The reservoir rock termed the Main Limestone, is of Eocene–Miocene age and is similar to the Asmari series but of greater thickness. The major field is at Kirkuk

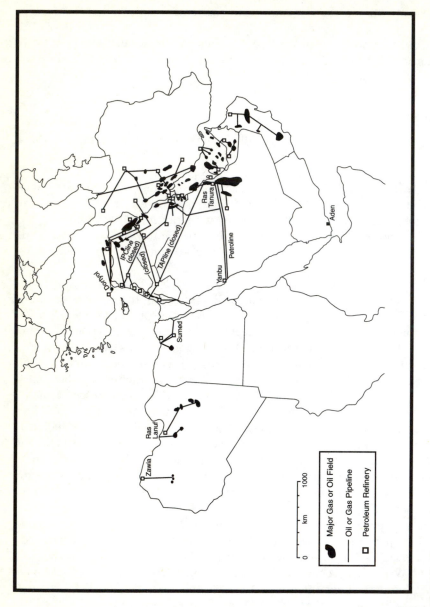

Figure 5.3 Middle East: main elements of the petroleum industry infrastructure.

and occupies a single narrow anticline between 80 and 90 km in length. The field was developed in 1927 and the oil exported by pipeline through terminals on the eastern Mediterranean coast. However, owing to the various bouts of political turmoil in the Levant coastal area, the Haifa branch was closed in 1947 and the other branches in 1976. In 1976–77, two replacement pipelines, one the strategic pipeline to Basra and the other to Dortyol in Turkey came into operation. During the Gulf War (1991), the Dortyol pipeline was closed.

In 1952, oil production began in the Ain Zaleh field north of Mosul where the oil is contained in Cretaceous limestone. The other key fields of Iraq are known as Zubair and Rumaila and are located in Cretaceous sandstone to the north of the boundary with Kuwait. In these fields, the geological conditions are completely different from those in the Zagros and resemble those of Kuwait and Saudi Arabia. Rumaila is one of the world's major oilfields and featured in the Iraq–Kuwait boundary dispute. The UN-controlled boundary settlement has produced a line cutting across the southern tip of the field so that a few kilometres remain within Kuwait.

In Saudi Arabia, along the western side of the Persian–Arabian Gulf, conditions are entirely distinct from those in the Zagros. Oil structures are broad open anticlines and the reservoir rocks range from Jurassic to Cretaceous in age. Furthermore, the oil of Arabia is more frequently held in sandstones than limestones.

In Saudi Arabia, the first oil strike was made in 1948 at Ain Dar and six months later at Haradh. It was later realized that these constituted part of the Ghawar structure which measures 200 km from north to south and 25–30 km from east to west. This is the greatest single oil-producing field in the world. To the north is a complex of producing areas, some on land and some offshore. The greatest is Safaniya, some 5 km offshore, which is the world's largest offshore field. Some production in the Neutral Zone between the two countries was shared with Kuwait but now the land boundary between the two states has been demarcated. The maritime boundary is in the process of delimitation. However, some sharing of oil production offshore with Bahrain remains. The main export terminal is at Ras Tanura and pipelines connect the region with the Mediterranean (TAP Line) and Yanbu on the Red Sea. Like those to the north, the TAP Line was closed as a result of conflict in the eastern Mediterranean.

In Kuwait, there are two reservoir layers of Cretaceous sandstone, each approximately 300 m thick at a depth of about 1,200 m. Production began in 1946 at Burgan, a major world oilfield. The large size of the deposits in Kuwait and their relatively shallow depth have resulted in considerable economies of scale.

In Qatar, production began at Dukhan on the western side of the peninsula in 1947. The reservoir rock is Jurassic limestone which occurs at some 2,000 m below the surface. The current dispute between Bahrain and Qatar over the Hawar Islands which are located adjacent to this field has an obvious hydrocarbon component. Qatar is, however, better known for its offshore gasfields to the north of the peninsula which have made it a major producer on a global scale.

The key producer in the UAE is Abu Dhabi where production began in 1962. The first field was that at Murban and subsequently other onshore and offshore fields have been discovered. The preponderance of offshore fields has drawn attention to the continuing geopolitical dispute over Abu Musa Island between the UAE (Sharjah) and Iran.

In contrast to the Persian–Arabian Gulf region, the geology of the Suez–Sinai fields is extremely complex with a wide variation of rock types ranging from pre-Cambrian to Recent in age. The Red Sea, together with the Gulfs of Suez and Aqaba, is a fault system and major tectonic movements have produced severe dislocations so that environments suitable for large-scale oil accumulation are infrequent. Discoveries have been made since 1913 in various locations on either side of the Gulf of Suez but the more important exploitation only began in 1974 when Egypt became a producer of some significance. The fields occur both offshore and onshore and the reservoir rock is Nubian sandstone at a depth of 2,000–3,500 m. Small fields occur in Israel, Syria and Turkey but these three countries remain very modest producers.

Completely different are the oil deposits of the interior deserts of Libya which ranks among the most important oil-producing countries of the world. The major discoveries have been made in the hinterland of the Gulf of Sirte where a broad lowland embayment extends southwards and eastwards. The largest reserves occur in the Sarir field, another of the world's major production areas. Zelten was the first major field to be developed (1954), producing both oil and gas. It is linked by oil pipeline (1961) and gas pipeline (1969) to the terminal at Marsa el Brega.

With the development of liquid natural gas carriers, the export of gas, both associated and unassociated with oilfields, has made rapid progress. Iran is the second largest producer in the world after the Former Soviet Union (FSU). Major unassociated fields are located near Bushire and Sarrakhs while there is also considerable export of associated gas from the southern oilfields.

Conclusions, problems and prospects

The Middle East is pre-eminent globally for oil resources and a key producer of natural gas. Furthermore, given the highly advantageous geological environment, it is likely that other sources, although almost certainly on a scale smaller than those already discovered, can be identified. The location of the region, central in the World Island and penetrated by seas which link it to the global ocean, facilitates distribution to international markets. Transport from the region has been enhanced by the development of a pipeline network which also provides insurance against some of the geopolitical vulnerabilities such as the narrow straits or choke points. The construction of refineries and gas-processing plants near the sources of supply has added considerably to the value of the resources for the producer states. The basic pattern of pipelines and refineries is shown in Figure 5.3. For economic and geopolitical reasons, both discussed in later chapters, additional pipelines and refineries are planned.

With regard to the petroleum industry, there remain three basic problems for Middle Eastern states:

1   to ensure that the benefits of the resulting gross domestic products (GDP) for Saudi Arabia, the UAE and Kuwait, among the highest per head of population in the world, reach as widely as possible among their own people;
2   to make provision by development of other activities, for the time when oil wealth will decline; and
3   to devise a means whereby the economic power of the oil states is used globally and particularly for the Arab and Islamic worlds in a positive and constructive way.

All this can only be achieved given stability within the Middle East.

With regard to the future, exploration is likely to concentrate on such areas as western Iraq, the Rub' al Khali basin, north and south Yemen, the overthrust belt of the UAE and Oman, and offshore northern UAE (Shannon and Naylor 1989). The focus will be on progressively smaller structures and traps. The huge potential of the region should guarantee that, whatever happens with the Caspian Basin exploration and exploitation, the Middle East will continue as the dominant oil-producing region of the world well into the twenty-first century.

6   Historical geography

The change from nomadic wandering as hunters or gatherers to settled occupation of the land with the development of plant cultivation is recognized as the first real stage in growth of civilization. This critical event is generally thought to have taken place first in some part of the Middle East, though opinions differ as to the precise locality and the date.

Certainly, the Natufian peoples of western Palestine were among the first known to have developed a way of life based on cultivation (with some herding of animals), as a secure and fixed, rather than nomadic, base. The time of this development is put very soon after the last glacial period, 12,000–10,500 BP, with other developments *c.* 8,000 BP at Catal Huyuk (Asia Minor) and Jericho (Palestine). Rapidly improving ecological conditions produced first a warm, temperate, lightly wooded environment that later became drier and still warmer.

However, the development of communities other than on a purely local basis involves extra factors, since cultivation must be on a scale large enough to produce the surplus of wealth necessary to support specialist crafts and services. It was in the two major river basins of the Middle East, the Tigris–Euphrates and the Nile, that the organization of political units on a wider scale first took place. In both instances, there were several key factors. First, and most important, was the existence of a large alluvial tract, watered annually by river flooding that also renewed fertility by deposition. Second was spatial location, allowing contacts with other groups. At this early period there would seem to have been three major ethnic groups within our area: Semitic related to the Arabian peninsula; Hamitic in north-east Africa; and various groups of Indo-European cultures in the northern mountain zone. This last group comprised: Aryans who spread into Iran by 4000 BC, Hittites who entered Asia Minor by 2000 BC, and various Caucasian groups.

Another factor, the ability first to absorb other influences and develop, and then to demonstrate the capacity for social and political co-operation at wider levels, is also involved in the process of civilization. With the development of the city which offers a better environment for all these, the process gained even greater momentum.

There has been controversy as to the origin of this 'urban' level, which can be termed 'civilization', of recognizably Neolithic character and which involved integration of rural territory and a number of organized urban communities. Mesopotamia and Egypt are generally recognized as the earliest such civilizations in the world and the oldest major culture is thought to be that of Sumer, *c.* 3500 BC, with a parallel but slightly later development in Egypt *c.* 3100 BC. Both had an advanced technical level of cultivation which supported a highly evolved social life, with organized civil and religious bodies. Copper and bronze were used, though on a limited scale, and trading contacts had begun, especially with Syria, which, located between the two prime areas and endowed with natural resources not present in the other two, participated in the general material and cultural advances initiated from Egypt and Sumer.

The second millennium BC saw much invasion and disturbance with a relative decline in the influence of Mesopotamia and Egypt. From the more arid areas of Arabia, the Hyksos entered Egypt and Armenian peoples, among them Canaanites, Israelites, Philistines and Phoenicians, established themselves in the better-watered regions of the Levant, whilst the Kassites entered lower Mesopotamia. In the north, originally pastoral peoples, the Mitanni and the Hittites, invaded and introduced the horse and the chariot. The Hittites would have appeared to have invented iron smelting around about 1300 BC. At the same time, because of extensive commercial relations developed by the Phoenicians, whose chief cities were Tyre, Sidon and Arad (Ruad), Aramaic languages began to spread over much of the Middle East, as far as Europe and Iran. With the Phoenician alphabet these were a considerable advance on the hieroglyphics of Egypt and the cuneiform of Mesopotamia.

After a relatively short period that saw the ascendancy of an Assyrian Empire (1200–1000 BC) based on an area of rain-fed cultivation on the plains of the upper Tigris, came the rise of Persia. Settlement had developed in parts of the inner plateau of Iran as early as 5000 BC and in *c.* 1000 BC there occurred eruptions of Indo-Aryan peoples, among whom were the Medes who settled in the north-west and the Persians (Farsi) who settled further south, around Fars. Unification by conquest was achieved by the Achaemenid dynasty, the outstanding monarch of which was Cyrus. His conquests, beginning in 539 BC with Mesopotamia, brought together under a single unified administration, a territory that ultimately extended from the Hindu Kush to the Aegean and North Africa (Figure 6.1).

A system of national trunk roads, with posts and rest houses, a metal coinage and a centralized administration with Aramaic as the official language, led to a major change. Iran was drawn towards the West and brought more into Western political life, a factor of great geopolitical importance from then until the present time. The rise of a specifically Aryan culture alongside the older non-Aryan civilizations of Egypt, Mesopotamia and Syria was of very profound significance, and its effects are felt to this day.

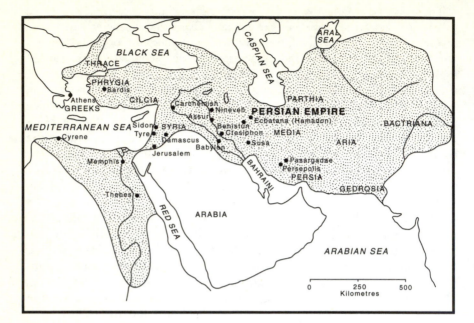

Figure 6.1 The Persian Empire at about 500 BC.

Classical period

Another influence of lasting importance to the Middle East was conquest by Alexander the Great of Macedon who in 331 BC overcame the Iranian Darius at Arbela (Erbil). Although Alexander himself died very shortly afterwards, Hellenic influence in the Middle East was maintained by his generals, certain of whom established themselves as successors of Alexander in various areas. To one general, Ptolemy, fell control of Egypt, with at first Syria and Palestine, and the Ptolemic dynasty ruled Egypt until the Roman conquest of that country in 30 BC. Another general, Seleucus, established a dynasty in Iran, Mesopotamia and, at a later date, in Syria. The Macedonians who shared with the Iranians a strong military tradition with extensive use of the horse, intermarried to form a Perso-Macedonian ruling class. Both groups found that as a minority their influence could best be exercised amongst urban populations, rather than rural cultivators and herdsmen. Thus within the towns, Hellenism was often dominant, while in the countryside, older Semitic traditions persisted. In time Seleucid rule declined, to be replaced by a new, native Parthian state in Iran and the growing Roman Empire in the West. For several centuries this division between Parthian and Roman spheres was the basic geopolitical fact within the Middle East. Despite the power of Rome, the centres of the Parthian Empire, doubly insulated by the Syrian desert and the Zagros mountains, were too remote to be attacked in strength.

Conversely, the Parthians could never conquer the western provinces of Syria, which were ultimately sustained by a powerful Roman fleet.

Under Roman rule, western Anatolia, Syria, Palestine, North Africa and Egypt entered upon a period of prosperity which has probably never since been bettered. Tranquillity led to a great expansion in agriculture and trade and irrigation was practised on a scale greater in places than at any time until the present day. As Italian agriculture declined in the later Roman period, the greater became the demand for Middle Eastern wheat. Increasing material prosperity stimulated a demand for manufactured goods and the textile industry, first woollen and linen, then silk, had the advantage of abundant raw materials and natural dyes. Glass-blowing and metal-working also reached a high level of skill and at a number of centres on the coast easily accessible woodlands provided material for shipbuilding.

The Middle East as a whole derived much profit from the expansion in world relations which took place under the Romans, for besides the two empires of Rome and Parthia, there were now equally highly developed civilizations in India and China. Imports from the East included silk, jewels, spices, drugs and sandalwood.

Land and sea routes were both used, but land transport, although more costly, was preferred as safer. A northern route passed via the central Asian oases of Khokand, Bokhara and Merv to Hamadan in Iran, where it was joined by another route from India via southern Iran. From Hamadan, the route continued to Babylon, which was still a focus of communications, and then turned northwards by way of the Middle Euphrates, whence a choice of routes skirting the Syrian desert led to Palmyra, Aleppo and Antioch, or Damascus. Access was then gained to the numerous Syrian ports. A more southerly route utilized the Indian Ocean and Red Sea as far as the Gulfs of Suez and Aqaba. From the latter point shipment then took place by way of Petra, Jerash and Philadelphia (Amman) to the south Syrian ports, while from Suez, a route led overland through Egypt to Alexandria.

Similarly, further to the east, the Parthians derived advantage from trade routes, maintaining active trading relations with China, India, Arabia and Siberia, in addition to those with Rome. Although they left almost no written records, it is clear that the Parthians had a relatively high level of cultural development. However, the Hellenistic influence of Seleucid times gradually gave way to an increasingly local and Oriental outlook.

In AD 224, the Sassanids gained power in Parthia and though most of the existing Parthian way of life continued, the later Iranian Empire showed itself increasingly Oriental in its internal life and in external affairs more aggressive in the west, against the slowly weakening power of Rome. A long period of intermittent fighting between Rome and Parthia, with border skirmishes and occasional major expeditions, continued until the downfall of both Empires in the seventh century AD. The Roman Empire had at some time previously fallen into an eastern and western portion and of these, the former, now spoken of from AD 373 as the Byzantine Empire, took up the Iranian challenge. The Middle East from the Black Sea to the deserts of Arabia became a battleground and the inconclusive results of this warfare, prolonged over many years, led to the ultimate exhaustion of both sides. The resulting material devastation, ideological intolerance and crippling taxation of the native populations were important contributing factors in the rise of Islam during the seventh century AD.

Arab period

The Arab conquest of the Middle East between AD 630 and 640 brought an active, virile, but essentially uncultured desert community into contact with the rich and highly developed civilizations of Rome and Iran. The early followers of Muhammad were largely desert Bedouin, without any tradition except that of a hard, patriarchal society.

The rapidity with which these semi-nomads became absorbed into the existing life in the Middle East and the extraordinary cultural development which resulted from the fusion of the two groups must be considered one of the outstanding events in the development of human society. Politically, the Islamic state at first enjoyed unbroken success. With the capture of Mecca in AD 630 as a beginning, there followed in 636 the seizure of Syria, of Egypt in 641 and of Libya in 642. Within a century, the whole of north Africa and most of Spain had become Muslim, and France and Italy were threatened. In the east, Islam had reached the oases of Turkestan and was stretching towards India and China. Only in the north was expansion arrested, and that only temporarily, by a reduced but revivified Byzantine Empire (Figure 6.2).

Materially, the impressive level of achievement of the Roman era was maintained. Egypt and Syria retained a high agricultural productivity and taxation was lighter than in Byzantine times. A degree of religious toleration allowed collaboration between various sects, with a fruitful interchange of ideas. In Mesopotamia, the development of irrigation reached its highest point with the construction of numerous canals. Manufactures, including metal-work, leather and textiles, gave additional prosperity and ancient trade routes between Europe and the Far East continued in full use.

Intellectually, at a period when learning was practically extinguished in Europe, the Arabs took over and expanded classical philosophy, particularly in the fields of medicine and science, adding a considerable body of new thought, partly developed by themselves and partly derived from Iran and even Hindu sources. They introduced paper from China, where it was first invented, together with the present system of numbers and the use of the zero sign. It may therefore be said that, almost until the Renaissance, the main stream of traditional classical culture deriving from the Ancient World was to be found in Arab lands rather than in Europe, and Arab commentaries and expositions on ancient authors were used as textbooks in European universities until the seventeenth and eighteenth centuries.

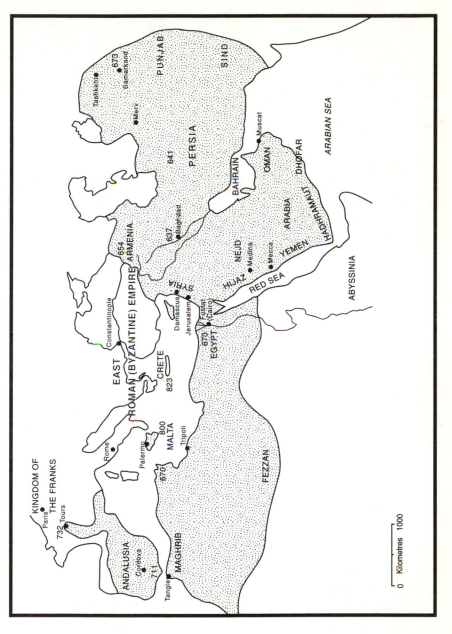

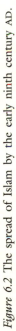

Figure 6.2 The spread of Islam by the early ninth century AD.

The Caliphate, or political leadership of Islam, was on the death of Muhammad first established in the family of the Omeyyads, who reigned from Damascus. After a century, a rival family, the Abbasids, supplanted the Omeyyads and the political centre of Islam shifted to Baghdad. To this period, during which material prosperity in Mesopotamia probably reached its greatest peak, belongs the Caliph Haroun al Rashid (AD 786–809).

Later, towards the end of the ninth century, the political unity of Islam broke down, partly due to the great cleavage between Sunni and Shi'a. Ever since the original rise of Islam the Byzantines had succeeded in holding Asia Minor and the existence of this advanced base of Christianity greatly favoured the operations of the first European Crusaders, who in AD 1079 landed in Asia Minor, and proceeded to an invasion of Syria.

Crusader period

Within a short time, the coastal area of the Levant, from Cilicia to central Palestine, had fallen under Crusader domination. Feudal principalities were established on the European model and Jerusalem was held from 1099 until its recapture by Saladin in 1187. The Crusaders did not, however, penetrate far inland. Except for somewhat more extensive, but short-lived kingdoms in Cilicia (Kingdom of Armenia) and Judaea (Kingdom of Jerusalem), their influence ceased at the crest of the mountain ranges backing the coast and the important cities of Aleppo, Hama and Damascus remained in Muslim hands.

The Crusaders were not wholly inspired by religious motives. Trading activities had attraction, especially to the Genoese and Venetians, who supplied sea transport for crusading armies in return for substantial commercial concessions. Throughout most of the hundred years of occupation by the Crusaders, a considerable volume of trade in precious stones, spices, silks and other luxuries was carried on with the Muslim states, part being a transit traffic from China and India. The Crusaders learned much from Islam, an indication of the advances made in the Middle East during the period of the Dark Ages in Europe.

During the thirteenth century Arab power revived and the Crusaders were gradually driven from Syria. By 1299, Acre, the last Christian stronghold, had been taken. A remnant of Crusader refugees fled from the mainland to the Levant and established themselves on the island of Cyprus where European control, first under a Crusader dynasty and later under the rule of the Republic of Venice, lasted until 1571. The Lusignan period was one of great brilliance in the history of Cyprus, for the island became an entrepôt between Europe and the East.

Trading posts on the mainland of Asia, first developed under the Crusaders, continued to function after the Muslim re-conquest and European traders, chiefly Italian and French, had depots at Latakia, Tripoli, Beirut and Alexandria. Much commerce passed through Cyprus, which, secure from attack because of the lack of a seafaring tradition amongst the earlier Muslims, took its place as a natural centre of the eastern Mediterranean. It was not until the sixteenth century that the Ottoman Turks, having developed a maritime tradition, invaded the island which passed to Muslim control.

From the Crusader period onwards, repeated incursions of Mongols from the steppes of central Asia caused much destruction and devastation. The brunt of Mongol attacks fell on Iran, where between 1220 and 1227 the 'hordes' of Genghis Khan sacked and almost completely destroyed many cities. Many of the irrigation works, maintained for centuries, now fell into disrepair and vast stretches of land returned to steppe or desert. By 1258, the terror had reached Mesopotamia, Baghdad was destroyed and the Abbasid Caliphate was extinguished.

A century later, fresh Mongol invasions occurred. The armies of Timurlane reached Syria, where Aleppo, Homs and Damascus were burned and looted, while in Iran the Nestorian Christian community, which had played a great part in the intellectual life of the country, was reduced to a tiny minority. Mongol invasions were, however, arrested in Asia Minor, where the Ottoman Turks had risen to importance.

Ottoman period 1517–1923

In the thirteenth century AD, a small group of Mongols known as the Ottoman or Osmanli Turks invaded Asia Minor and received a grant of territory in north-west Anatolia from a somewhat insecure Sultan. By 1400, the Ottomans had extended their domains to include central and many parts of western Anatolia, together with a considerable area of Balkan Europe. In 1453, after a number of previous attempts, the city of Constantinople was taken and the Byzantine Empire, of which it had been the capital, finally destroyed. From then onwards, the expansion of the Ottoman state was rapid. By 1566, the entire north coast of Africa as far as Algiers had been occupied, together with Egypt, Syria, Palestine, Anatolia and Iraq. In addition, the whole of south-east Europe between Croatia and the lower Don fell into Ottoman hands and Austria was menaced. A number of campaigns had led to a stalemate in the east, where the region of eastern Armenia, the Caucasus and the Zagros mountains separated the Ottoman state from Iran, which retained its independence.

The early expansion of Ottoman power was accompanied by a process of administrative reorganization, the form of government that resulted lasting without serious change from the fifteenth to the twentieth centuries. Certain features of Turkish rule are extremely interesting, showing to a marked degree the influence of the social background of the early Ottoman tribesmen, who became the ruling class of the new Empire.

The Ottomans had entered Anatolia as pastoral nomads from the steppes of Asia and they brought to the task of administering their Empire much of the technique that had served them in the handling of animals. In the first place, as a herdsman keeps separate his sheep and goats, the Ottomans made no attempt to develop a single, unified state, but rather were content to allow existing differences of race, religion and outlook to continue. At first, this tolerance of division amongst peoples of the Ottoman Empire was probably mere indifference, but at a later period it became an important basis of policy. By the 'divide and rule' principle, Muslim was turned against Christian, Shi'a against Sunni, Kurd against Armenian, and Orthodox Greek against Roman Catholic. Sectarian feeling was provoked and increased by the creation of the millet system, and after the seventeenth century the grant of Capitulations to various non-Turkish nationals was a further recognition of the separate status of certain communities within the Empire. Provided that the subjects of the Sultan showed themselves amenable to rule and willing to pay the taxes demanded and, in some cases, support military conscription, many communities could exist largely in their own way.

The Ottomans created a special corps of picked subjects, known as Janissaries, to police the Empire. Bought as slaves from neighbouring states, or obtained as tribute, usually from Christian families, within the Empire, the Janissaries were separated from their parents and grew up without family ties or local sympathies. Obedient, ruthless and fanatically Muslim by reason of their early training, the Janissaries were the instrument by which the Turkish government maintained a hold upon its mixed population.

As pastoralists, the Ottomans had had no tradition of industry and commerce and in their Empire both tended to be despised or left to Armenians, Greeks, Jews or foreigners. Trade activities were tolerated, but little positive encouragement was afforded by the government. Much business was therefore carried out half secretly and this long tradition of dissimulation in business has had a most harmful effect. Even in some modern Middle Eastern states, methods inherited from the not-too-remote past still persist amongst small-scale traders. Accounting systems are rare, banks are often mistrusted, 'presents' are frequent accompaniments to commercial transactions and fixed prices, even in retail shops, are by no means universal. A problem of many modern governments has been to apply taxation to traders, often the wealthier section of the community, but also the more adept at concealing their resources.

It must also be recognized that the Ottomans achieved power after a long series of wars, mainly against Christians. Unlike the original followers of Muhammad, who practised some toleration of Christians and Jews, the Turks were fanatical Muslims and destruction of infidels became a highly meritorious action. Intolerance and fanaticism led to a narrowing of outlook among Muslims themselves and Islam became for several centuries a closed and rigid system which ceased to develop further.

The discovery in 1498 by the Portuguese of the Cape route to India had a profound effect on the Middle East. Transit traffic between Europe and Asia, which from time immemorial had enriched the countries of the Middle East, was now diverted to the sea and the flow of goods through Iran, Syria and Egypt ceased. The establishment in 1514 of a Portuguese fort at Ormuz at the head of the Persian Gulf excluded the Ottomans from any possibility of interference with Portuguese communications with India and, as a result, the Middle East entered upon a period of decline. It was not until the cutting of the Suez Canal in 1869 and the development of motor and air transport later still, that the region recovered some of its ancient importance. Between the sixteenth and late nineteenth centuries, the Middle East was increasingly remote from the main commercial currents of the world.

In a political sphere, the Ottoman Empire reached its maximum power during the seventeenth century, when Austria was threatened and Vienna besieged (Figure 6.3). From this high-water mark, with a turning-point at the repulse of the Ottomans outside Vienna in 1683, the Empire entered on a period of slow decline. One reason for the decline was pressure by newly developing states such as Russia and Austro-Hungary, but the principal factor was unrest amongst non-Turkish subjects of the Empire, especially amongst Christian communities. A symptom of the growing weakness of the Ottomans was the rise of autonomy in outlying provinces, where the governors tended to follow an increasingly independent policy and ultimately to become hereditary rulers. The Sultan could do no more than issue a formal confirmation on the succession of a new provincial ruler. Such a development occurred in Algiers, Tunis, Egypt and the centre and south of Arabia, all of which by the nineteenth century had evolved into more or less independent states.

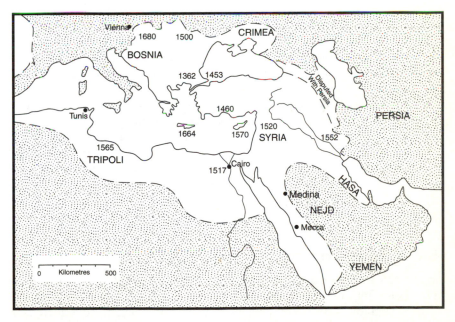

Figure 6.3 The Ottoman Empire by the seventeenth century.

Commercial decline in the Asiatic and African provinces of the Empire, together with the rise of political autonomy in certain regions, caused the Ottoman rulers to devote their greatest attention to the European part of their dominions. By the later nineteenth century, however, the rise of national feeling amongst Balkan peoples and the aggressive action of outside powers were rapidly reducing Ottoman territory in Europe and the Sultan felt it necessary to revise the system that had been followed in non-European districts. Accordingly, a 'forward' policy, designed to reassert Turkish supremacy, was undertaken in the Middle Eastern provinces. Its exponent, Sultan Abdul Hamid II, was aided by the recent construction of the Suez Canal, which allowed the easier movements of troops from Turkey south-wards. Garrisons of Ottoman soldiers were established in the Hijaz (1869), Yemen (1872) and Hasa (1871).

Further steps in the strengthening of Turkish control in the Middle East were taken by the construction of railway routes that were designed to bring outlying provinces into closer contact with Constantinople. Lines were built with the aid of foreign capital, chiefly German, from Anatolia south-eastwards to Baghdad via Aleppo, and southwards from Aleppo through Damascus and Ma'an as far as Medina.

The efforts of Abdul Hamid delayed but did not prevent the break-up of Ottoman power. Nationalist feeling continued to develop, and although at first restricted to the Christian peoples of the Balkans, it later became a disruptive factor amongst the Islamic populations of the Ottoman state. Arab as distinct from Turkish national feeling was fostered by the rise of semi-independent tribal leaders in Arabia, and by the organization of secret political societies of anti-Turkish complexion, chiefly in Syria. By the twentieth century, nationalism had begun to affect the Turkish people themselves. In 1908 a revolution occurred, as the result of which Abdul Hamid was deposed and a more strongly nationalist impulse was given to the policy of the Empire.

Six years later, Turkey entered the First World War on the side of Germany and following defeat in 1918, the complete extinction of Turkish power seemed at hand. However, a revival under Mustapha Kemal and his Nationalist Party altered the situation and a new, purely Turkish state, came into being. For a short time, the Ottoman Sultanate continued, although without effective power, but in 1923 Turkey was declared a Republic. A year later, the religious title of Caliph, which had been borne by the Sultan, was also extinguished and since that date Turkey has functioned as a nationalist, secular state.

The nineteenth century and the First World War

So far, this survey of the historical geography of the Middle East has identified dominant powers and their geopolitical legacy. Both the historical and the geographical records are incomplete in that for long periods there was no dominant power and few powers encompassed the whole area, which is defined in this book as the Middle East. This simple picture of a basically unipolar Middle Eastern world was shattered with the onset of the First World War and the period 1914–18 therefore reflects a major historical cleavage. The final demise of the Ottoman Empire can be thought to herald the modern era in the Middle East, but the death throes of that Empire extended over most of the nineteenth century.

One element in the long-continued decline of Ottoman power was the control gained by foreigners over much of the economic life of the Turkish state. By the end of the nineteenth century, a considerable body of rights, concessions and privileges, known collectively as Capitulations, had been granted to foreign traders. Capitulations gave to nationals of Austro-Hungary, France, Germany, Great Britain, Italy and Russia virtual exemption from most of the internal taxation of the Turkish state, together with immunity from police search of foreign-owned premises. Foreigners were also immune from trial in a Turkish court of law, special tribunals being appointed for each nationality.

Concessionaires gained control of resources which were then exploited, without regard to the general interests of the country, and by lending money at usurious rates, foreigners gained a stranglehold on much of the economic life of the whole Ottoman Empire. Most gas, electricity and water companies, where these existed, together with the greater part of the railway system, were foreign-owned. Profits from economic activity were drained away from the country and, at the same time, the Turkish state was confronted with numerous non-Turkish communities over which it had no legal jurisdiction. In effect, 'states within a state' had been created.

Further emphasizing the economic dominance of foreign powers, the Ottoman government had been forced to agree that its national debt should be administered by an international committee, the 'Council of Administration of the Ottoman Public Debt', which, presided over alternately by a Frenchman and an Englishman, was composed of nationals from Austro-Hungary, Germany, Holland, Italy and Turkey, together with a representative of the Imperial Ottoman Bank which was itself largely foreign-owned. The Council repaid interest and capital from the ordinary revenues of the Turkish state and hence had a certain control of taxation within Ottoman dominions.

Economic penetration of the Ottoman Empire was accompanied by equally active political pressure, but the actual break-up of Turkish dominions was delayed because of mutual jealousies among interested powers. For example, in 1854 Britain and France went to war to thwart Russian schemes of expansion at Turkey's expense in the Black Sea and eastern Anatolia.

In Ottoman affairs, the nineteenth century was a period of accelerated decline as the result both of internal weakness and outside pressure. The more outlying provinces of the Empire, those of north-west Africa, achieved virtual independence, only to fall later under French, Spanish and Italian dominance and by the opening of the twentieth century, even the position of Egypt as an Ottoman province had become increasingly doubtful.

Egypt and Sudan

Since, during the nineteenth century, the effects of the Ottoman decline largely dictated the course of Middle Eastern affairs, some prominence needs to be given to events elsewhere in the region. For this synopsis and also for the historical geography of the post First World War period, the original framework developed by Fisher in *The Middle East* (1978), including such terms as the now dated 'Asia Minor' and 'Levant' will be retained. Fisher further refined these terms to provide the basis for Part III of his book, Regional Geography of the Middle East. However, such terminology can also be used to illustrate the evolution of the current geopolitical pattern.

In Egypt, one highly influential ruler, Mohammed Ali, was thwarted in his attempts to overthrow the Ottomans in Syria and Crete. However, his dynasty ruled until 1953 and his legacy can be seen in the continuing French influence in Egypt. This influence was further strengthened by the construction by de Lesseps of the Suez Canal, which opened in 1869. Nevertheless, the geopolitical situation changed radically in 1874 when a successor of Mohammed Ali, the Khedive Ismail, decided to sell his holding in the Canal Company. The British government came forward as a buyer and thereby found itself the principal holder in the Suez Canal Company and the possessor of a vital stake in Egyptian affairs. Subsequently, in the reorganization of Egyptian finances, France withdrew and Great Britain became effectively the controller of Egyptian affairs.

Mohammed Ali was also influential in Sudan which, before 1800, had been under a vague form of Egyptian suzerainty. Following his re-conquest of the country, Mohammed Ali used Sudanese troops in his campaigns against the Ottomans and maintained Egyptian garrisons in Sudan. As the central government weakened, these garrisons became less effective and in 1881 an outbreak of religious fanaticism, headed by the self-proclaimed Mahdi, resulted in initial successes against both the Egyptian and the British forces. The military power of the Sudanese dervishes was finally ended in 1898 at the battle of Omdurman by combined British and Egyptian forces, under Kitchener, and a Condominium by Great Britain and Egypt was established.

Although Britain dominated the Condominium, Sudan retained a special place in Egyptian affairs. Politically, Sudan had long been regarded as an integral part of Egyptian national territory, while control of the Nile, a key factor in economic prosperity, remains a vital element in Egyptian–Sudanese relations.

On the declaration of war by Turkey, in 1914, Egypt under Ottoman suzerainty was declared independent. In fact, internal control remained largely in British hands and British forces were stationed in the country. At the same time, Cyprus, also technically an Ottoman territory, came under British rule as a Crown Colony.

Asia Minor and the Levant

Throughout the nineteenth century Asia Minor and the Levant, the 'heartland' of the Ottoman Empire, remained outside foreign political control. Penetration of these regions was limited to the economic and cultural spheres, and in the latter instance sectarian differences provided the main opening for intervention. Russia remained in close contact with Orthodox Christian minorities, France championed the Uniates and Britain supported Muhammadanism, sometimes orthodox, sometimes heretical, as in the case of the Druzes in the 1840s. In 1860, the massacre of Maronites by the Druzes led to intervention, chiefly by France, and the creation of limited autonomy for Lebanese Christians in the form of an 'Organic State of Mount Lebanon', which, restricted to the higher parts of the northern Lebanon range, was cut off from the sea and contained no important towns.

By 1914, chiefly in Damascus and Beirut, a number of secret societies designed to advance Arab nationalist aims were in existence. When war broke out between Turkey and the Western powers, these societies attracted the attention of Britain and encouragement of Arab nationalism both in Syria, which in 1914 included Palestine and Transjordan, and Arabia proper came to be appreciated as a useful political weapon in the struggle against Turkey. A small group, including T. E. Lawrence, within the Allied armies in the Middle East was allotted the task of fostering Arab revolt against Ottoman overlords by means of monetary subsidies and small-scale military assistance.

Of greater significance was the fact that France and Russia were now committed to the break-up of Ottoman power, although Italy presented a complicating factor. Having seized the Dodecanese Islands and the coastlands of Tripolitania and Cyrenaica during the Turkish–Balkan wars of 1910–12, Italy was party to the Triple Alliance, on the side of Germany and Austria. It became Franco-British policy to detach Italy and territory was offered in the Sykes-Picot Treaty (1916) which delimited the future intended territorial allocations within the Middle East (Figure 6.4). However, the agreement never came into effect since Tsarist Russia collapsed in 1917, the year before the final defeat of Turkey.

In enlisting Arab help against the Ottomans during the 1914–18 war, the promise was made by the British that Arabs should achieve independence and sovereignty in Arab lands captured from the Turks. The one reservation made concerned Palestine, about which promises had also been made to the Jews. Following the entry of the Arabs into Damascus in 1918, the Arab leader Feisal took over control of the government in Syria, with the acquiescence of Great Britain. At the same time, France, which had long entertained the ambition to rule in Damascus and Aleppo and was not bound by negotiations between Britain and the Arabs, adopted a different policy.

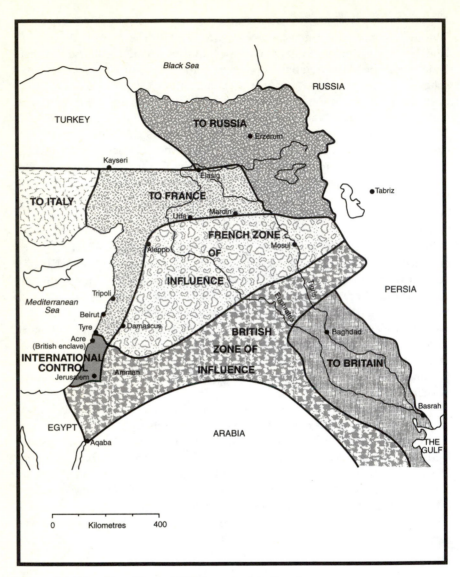

Figure 6.4 Sykes-Picot Treaty (1916), territorial allocation.

Iran

Since reaching a high point in the eighteenth century, the power of Iran had declined and it was coming under increasing threat, particularly from Russia. In 1814, a defensive alliance was concluded between Britain and Iran, by which Iran was to assist if needed in the defence of Afghanistan, in return for a money subsidy.

Despite this alliance, Iran lost much territory between 1804 and 1828. By the Treaty of Turcomanchai, which closed the war in 1828, Iran was forced to agree to the loss not only of Georgia, but also of the districts of Erivan and Lenkoran; to the payment of indemnity to Russia; and to the grant of capitulatory rights within Iran to Russian citizens. It was this treaty that brought into existence the present Russo-Iranian frontier between the Black Sea and the Caspian. Thus, reduced in influence and unable to resist either Russian or British inroads on its sovereignty, Iranian diplomacy aimed at playing off one country against the other.

This situation was the fundamental feature in Iranian affairs down to the present, and although one side has at times gained a temporary advantage, the balance has in the main been preserved, and still continues its uneasy existence, with, however, the USA replacing Britain as the principal opponent of first the Soviet Union and then Russia. Iran was currently of prime importance as a political unit in the struggle between East and West, and having gained markedly over the past few years in economic strength, was able to derive much political advantage from this 'uncommitted' position. With the demise of the Soviet Union and the present weakness of Russia, the situation has now changed.

With the western shores of the Caspian occupied, the Russians were free to turn their attention to the east, and, after 1840, a series of aggressive expeditions brought Turcoman, the Uzbeg tribesmen living in the Aral region, under Russian rule. In 1864 Tashkent was taken, in 1886 Samarkhand, in 1873 Khiva, and in 1876 Khokhand. The final state was reached between 1880 and 1893, when the northern slopes of the Kopet Dagh ranges were occupied, and Merv, loosely held by Iran, passed to Russian control. A series of boundary commissions fixed the final Russo-Iranian frontier east of the Caspian Sea, as the result of which the fertile lower slopes of the Kopet Dagh were allotted as part of the Russian Empire, to the disadvantage of Iran.

During the same period, the Iranian frontier with Afghanistan was demarcated by a number of British officials, but Iran remained dissatisfied and the question was finally settled by a Turkish arbitrator during 1934–35. On the west, the ancient indefinite frontier zone between the Ottoman and Iranian states was precisely delimited by an Anglo-Russian commission. The relative weakness of Iran was illustrated by the allocation of the entire Shatt al Arab to Turkey (later Iraq), with the international boundary running along the eastern bank of the waterway. Thus, Iran controlled only the immediate approaches to Abadan, Khorramshahr and Khorzabad and had no possibility of developing other ports on the Shatt al Arab.

Although closely enveloped in the north, Iran was strong enough in the south to maintain or even extend control over a number of Gulf islands (Qishm, Hormuz, Larak and Henjam). However, with sea power based in India, Great Britain had a strong position in the Persian Gulf, on the southern shores of which a number of Arab sheikdoms: Bahrain, Kuwait and Muscat were induced to accept British protection. Nonetheless, Iran remained involved along the southern Gulf coast, having ruled various territories, particularly Bahrain. The other influence in the region towards the end of the nineteenth century was that of Russia, with its quest for a warm water port. By the Curzon Declaration (1902), Russia was warned by Great Britain that attempts to increase its influence in the Gulf would be forcibly resisted.

To the north, the situation was far more favourable for Russia which surrounded Iran on three sides. The long-continued expansion of Russian power was, however, checked by the Russo-Japanese war of 1904 and in 1907 a compromise was reached over Iran by which the country was partitioned into three zones, one Russian, one British and one neutral. The Russian zone included most of the towns and the fertile land in Iran, but the subsequent discovery of oilfields in the neutral zone and their development by British capital, made the entire neutral area, in effect, a British preserve.

Although neutral from the outbreak of the First World War, Iran was immediately involved as the two sides fought on Iranian soil. Both Russia and Turkey breached the boundaries, while there was guerrilla warfare in the south-west, where Britain was attempting to protect the oilfields. After the Russian collapse in 1917, the British force was sent to north-west Iran to resist German and Turkish pressure in the Caucasus. By the end of the war, therefore, British influence had increased considerably.

Arabia

As a result of the extreme physical conditions and the lack of resources, the Arabian Peninsula remained largely outside the sphere of imperialist influence. It was only in the coastal areas that European occupation occurred. Most importantly, Britain secured the Aden district in 1839 as a base for the sea route to India. On the Gulf coast of the peninsula, several treaties were negotiated with the local sheikhs by which British protection was accepted, but the extent of British influence inland was never fully defined. In the interior, the Turkish hold was precarious and away from the garrison towns of Mecca, Medina and Sana'a, tribal government and inter-tribal rivalry persisted.

At the beginning of the twentieth century, the most important of the non-Turkish rulers in Arabia was King Hussein, Sherif of Mecca, Hereditary Guardian of the Holy Places of Islam and direct descendant of the Prophet Muhammad. At the start of the First World War, Hussein was persuaded to take up the cause of Arab nationalism and his son, Feisal, assumed command of the Arabian army that conducted irregular operations against the Turks in Arabia and Syria. However, in the early years of the twentieth century, a second more significant Arab leader arose, Abdul Aziz Ibn Saud. From a family which had earlier exercised influence, Ibn Saud was a natural leader, a great military tactician and a skilful practitioner of the art of geopolitics.

By his diligence and skill, Ibn Saud achieved a commanding position in the affairs of the Arabian peninsula and after the First World War was able to dispute with Sherif Hussein the leadership of Arabia.

Libya

At the beginning of the twentieth century, Libya was under Ottoman rule with Turkish officials and military garrisons, but economically, the region was extremely backward. In 1911–12, during the Turkish–Balkan war, Italy occupied Tripoli and at the start of the First World War was in the process of extending its hold on the country. The main opposition to Italian penetration came from the Sanussi, under their leader, the Grand Sanussi, and the presence of a Sanussi–Turkey army for some time posed a threat to the British position in Egypt. Nevertheless, after 1918, Italian forces gradually subdued the coast and, later, the interior.

With the decline and eventual end of Turkish hegemony in the region, the influence of other European states, particularly Great Britain, increased, and a number of significant local rulers, such as Mohammed Ali and Ibn Saud, arose. Never again would the Middle East be governed predominantly by one power.

From the First World War to the present day

Egypt

In 1924, a treaty was signed by which Britain handed over the bulk of its affairs to Egypt, retaining the right to maintain garrisons of British troops in Egypt for the defence of the Suez Canal. Later, changes in the political situation, following the Second World War, including the eclipse of Italy and the enhanced wealth of Egypt, together with a decline in British influence, led to more and more insistent demands for complete independence. By slow stages British garrisons were withdrawn and bases evacuated until, following the 1956 Suez episode, the last remaining groups were withdrawn. Egypt was then not only completely independent and in control of all its territories, but the acknowledged leader of Arab nationalism in the Middle East.

Nasser, who became President in 1954, benefited greatly from the Suez war of 1956 and the cult of Nasserism developed. This came to symbolize, above all, pan-Arabism, but also non-alignment and anti-imperialism. Pan-Arabism probably reached its zenith with the union between Egypt and Syria in the United Arab Republic which lasted from 1958 until 1961. However, defeat by Israel in 1967 was on such a scale that Nasser's role was shattered. The Sinai was lost and the Suez was closed, leaving an extremely difficult legacy for the next President, Sadat. Nevertheless, he enhanced his authority by expelling Soviet military advisors in 1972 and at the same time releasing members of the Muslim Brotherhood from prison.

Sadat was therefore in a good position to lead Egypt into the October War (1973) against Israel, which was intended to erase memories of the 1967 defeat and also break the diplomatic impasse in the region. Although the results did not totally fulfil his expectations, Sadat was able to address the Israeli Knesset in 1977 and in 1979, to sign an American-brokered peace treaty with Israel, the Camp David Accords. While this treaty provided a measure of security for both Israel and Egypt, it attracted disapproval throughout the Arab world and Egypt was expelled from the Arab League.

After the assassination of Sadat in 1981, Mubarak succeeded with an increasingly authoritarian stance. The elections of 1990 and 1995 were both won with great majorities, but despite the introduction of draconian anti-terrorist laws, terrorism remains a major problem.

Sudan

With Egyptian independence after 1924, Sudan passed first from direct rule, in effect by the British, into a period of Indirect Rule or Native Administration (1924–43), with a parallel 'Southern Policy' applied to the non-Islamic peoples of the south, who were regarded as possible adherents to a federation that might in the future be based in East Africa. Muslims were compelled to move out of the southern areas which were declared 'Closed Districts', with the aim of discouraging migration out of them by their non-Islamic inhabitants.

Nationalist feeling developed in strength, but was faced with the dilemma of either total independence or connection with Egypt. Under Nasser, Egypt believed that the Sudan Union political party would opt in favour of permanent union with Egypt and supported it, so that its leader was able to assume power in 1955 as head of an independent state of Sudan. However, he declared for total independence rather than any form of union with Egypt. During the 1960s, there was much insurrection and separatism in the non-Islamic areas of the south, due partly to heavy-handed, uncomprehending rule from Khartoum, but mostly reflecting the ferment for national freedom and aspiration that swept much of central Africa at this time. After a period of unsuccessful military action by the north against the south, which led to a mass flight of southerners as refugees in adjacent territories, an amnesty was declared in 1965, and conditions improved, particularly following the declaration of regional autonomy for the south by President Nimeiri.

However, following the introduction of Shari'a law and the re-division of the south, rebellion by the Sudan People's Liberation Movement (SPLM) and the Sudan People's Liberation Army (SPLA) broke out in 1983. By 1985, drought, famine, bankruptcy, conflict and large-scale economic collapse brought Nimeiri down. Subsequent elections brought the Umma Party to power under Sadiq al-Mahdi with an Islamic constitution. Further political drift ensued, the war in the south was effectively lost and, in 1989, the army overthrew the regime. The coup provided the opportunity for the National

Islamic Front, the political wing of the Muslim Brotherhood, to come to power and to bring an increasingly Islamic orientation.

By invoking a jihad in the south, the regime attracted support from other Muslim states, but the war continues and to the present day Sudan has been in an almost constant state of crisis. Aid has declined in protest at Sudan's economic policies and human rights record, the economy has been marked by high inflation and there has been large-scale emigration. Not only has the country grown increasingly poorer, but support for Iraq and Iran and alleged involvement in the World Trade Centre bombing (1994) have led to its designation by the USA as a pariah state.

Asia Minor

With the defeat of the Ottomans in 1918, it seemed as though the end of Turkey as a state was at hand. The Treaty of Sèvres, proposed as a political resettlement of Asia Minor in 1920, would have resulted in a virtual extinction of Turkish power. In effect, the position of Britain, France, Italy and Greece as previously defined in 1914–16, each with its own share of the country, was to continue unaltered. New elements were that the zone of the Straits, formerly allocated to Russia, was to pass under international control, and American interests led to proposals for an Armenian state to be created in the former Russian zone of eastern Anatolia. The Treaty of Sèvres can be regarded as the high water mark of European intervention in the Middle East since, had it been implemented, practically the whole of the Middle East region, except for Iran and inner Arabia, would have fallen under foreign domination or influence.

Faced with the possibility of near extinction as a political unit, the Turks rallied, and, owing partly to division between the Allies, partly to general war-weariness and partly to their own courage, an unlooked-for revival of Turkish power occurred under the leadership of Mustapha Kemal Pasha. After some months of fighting, the French withdrew claims to the part of Anatolia allotted to them under the Treaty of Sèvres. A year later, the Greek-held city of Smyrna (Izmir) was taken by storm and a large Greek population, numbering over half a million, fled or was forcibly expelled to Greece. Thus, a great measure of ethnic and political unity was achieved in western Asia Minor, though at a heavy price, both in human suffering and in economic disruption.

In the east, the Armenian question had been to some extent resolved by the massacre of large numbers of Armenians by the Kurds and outside military support from one of the great nations, which was not forthcoming, would have been necessary to establish a separate Armenian state. Hence, the new Turkish state was able to negotiate a more advantageous peace settlement, the Treaty of Lausanne (1923), by which control of Asia Minor and the adjacent area of European Turkey remained firmly in Turkish hands. Since that time, Turkey has become more powerful, as a result of fundamental secular, social and economic reforms undertaken by Kemal Pasha, who took the name Ataturk.

Despite its power in terms of area, population and economic potential, the threat from the Soviet Union drove Turkey to seek alliances in the West and in 1952 to join the North Atlantic Treaty Organization (NATO). Notwithstanding its economic potential, domestic politics in Turkey remained unstable until 1983. A succession of elected governments, several without an overall majority, was interspersed with periods of military rule. In 1974, following a failed attempt by the then junta in Greece to take over the island, Turkey invaded Cyprus. Population movements ensued and the island was divided between Turkish Cypriots and Greek Cypriots.

From 1983 with the election of Ozal and the Motherland Party, there was almost a decade of stability with military disengagement from politics. The Turkish economy gained in strength and national confidence was evident in the initiation of the vast South-East Anatolian Project (GAP). Since the death of Ozal, then President, in 1993, the major political event has been the election of an Islamic government in 1996. As a result, secular Turkey has seen a growth of more radical Islamic thinking, which opposes closer ties with both Europe and the USA. A balance is currently sought between the secularists and the Islamists. The situation has been complicated throughout the 1990s by the continuing struggle in the east with the Kurdish Workers Party (PKK) which received support from Syria and serves as a constant reminder of Turkish human rights issues.

The Levant

In the Levant in 1920, France took over the administration under mandate of Syria as a federation of territorial units. However, this proved an onerous task and, as British colonial power was thrown off in first Egypt and then Iraq, tensions in Syria grew. Nevertheless, by 1939 Syria was still an integral part of French overseas territories, but this situation effectively ended in 1941 when, to thwart German operation, British and Free French forces invaded and occupied the country. After the Second World War, France attempted to regain control, but when large-scale uprisings in Syria and Lebanon seemed imminent, the British government again intervened. The result was that within a year both British and French forces had been withdrawn and complete independence had been granted, with the creation of the sovereign republics of Syria and Lebanon. For some years the two countries maintained an economic union, but in 1950 this was dissolved. In 1958, Syria joined with Egypt to form the United Arab Republic, a union that ended with the withdrawal of Syria in 1961.

Following a similar event in Iraq, the Ba'ath coup occurred in Syria in 1963 and resulted in a transformation of the Syrian state structure. Between then and 1970, power struggles continued, particularly within the armed forces where the initiative lay. The key posts were assumed by Alawites,

Druzes and Ism'ilis, all non-Sunni minorities who seized the opportunities offered in a sectarian state. Further coups followed in 1966 and 1967 when, as a result, Al-Asad came to power. He introduced a degree of economic liberalization which resulted in an economic boom lasting until the late 1970s.

Since the advent of Al-Asad, the power invested in the presidency has drastically increased, supported by the Ba'ath Party, the armed forces and the bureaucracy. In 1973, Syria fought alongside Egypt against Israel, but the Golan Heights remain in Israeli hands. More significantly geopolitically, the Syrian army entered Lebanon in 1976 and since then Syria has been even more intimately involved in Lebanese affairs. Following the demise of the Soviet Union, its chief supporter, economic problems were offset by financial rewards from the USA and the Gulf states, following Al-Asad's decision to join the coalition against Iraq in the Gulf War. However, Syria remains crucial as the most powerful front-line state deployed against Israel and therefore a key factor in any Middle Eastern peace talks.

Palestine

At the same time that France received a mandate for Syria, Britain was also allotted a similar mandate for Palestine and the region lying east of the Jordan river. In the former area Britain was committed to what was obviously an impossible task of creating a 'National Home for the Jews' without prejudicing the rights of Arabs living there. In the latter, a purely Arab area, the Emir Abdullah, an elder brother of the Emir Feisal, was proclaimed ruler of a state of Transjordan, under British Mandate. Though small and poor the country of Transjordan proved for several decades to be a stable political unit. After some twenty-five years as mandated territory, during which a slow but consistent development took place, the full independence of Transjordan was recognized in 1946. Since 1949 the country has been known as the Hashemite Kingdom of Jordan.

In Palestine itself from the end of the First World War, increasingly large-scale immigration of Jews has continued to arouse apprehension among the Arabs who were, at that time, vastly in the majority. Later the Jewish call for a 'refuge' and a 'home' within Palestine gradually turned into far stronger demands for a sovereign Jewish state.

Jewish colonization of Palestine/Israel has occurred in a series of waves. In the latter 1880s, alongside an Arab population of about 440,000, there grew up some 25,000 Jews, chiefly refugees from Tsarist Russian pogroms. A second immigration took place in the first years of the twentieth century, again composed of Russians, this time of more pronounced socialistic and communistic leanings. It was these immigrants who started the kibbutz idea. A third wave of immigrants after 1918 was made up of dispossessed refugees from the devastated areas of central Europe. Another considerable increase in immigration, this time of Jews affected by anti-Semitism in Poland, took place between 1924 and 1928. Persecution in Germany under Hitler then led to a fifth wave of immigration, on this occasion by German Jews.

Such mounting entry of Jewish immigrants led to serious riots and civil disobedience by the Arabs in 1929 and again in 1936–39. This disorder and rioting achieved a certain success in that a limit of 75,000 persons within five years was placed on Jewish immigration. During the Second World War, arms and ammunition were accumulated by both Arabs and Jews and with the end of warfare in 1945 the extremist section of the Zionists opened a campaign of terrorism and intimidation to effect a return to unrestricted Jewish immigration, as the first step on the final objective of a Jewish Palestine.

In 1947, the question of settlement in Palestine was referred to the United Nations (UN) and a Commission, the nineteenth in twenty-five years, once more surveyed conditions in the country. The report of the Commission was in favour of partitioning the country between Arabs and Jews, with Jerusalem under international control. However, mounting terrorist action, chiefly by Zionists, prevented implementation of the report and in May 1948 British Mandatory rule was officially declared at an end. Three months later the last British forces were withdrawn and fighting immediately broke out between Zionists and the states of the Arab League. Between 700,000 and one million Arabs fled from Palestine into nearby Arab countries, where a high proportion remain as refugees.

Eventually an armistice was declared, with existing battle positions stabilized to form a frontier. A small strip near Gaza, occupied by the Egyptian army, was incorporated into Egypt, and portions of the Judaean highlands and Jordan Valley held by the forces of King Abdullah formed an enlarged state of Transjordan, then renamed Jordan. The remainder of Palestine, including the New City of Jerusalem, but not the Old City, became the Jewish state of Israel.

Between 1949 and 1951, the population of Israel doubled, mainly due to massive immigration, this time mainly from Arab countries, where Jewish groups, often of long standing, were no longer welcome. Late in 1956, following a secret agreement with France and Britain, Israel occupied the Gaza Strip, the entire Sinai as far as Suez, and islands in the Gulf of Aqaba. UN action led to later Israeli evacuation of these territories, but in 1967 a demand by Egypt for the withdrawal of the UN peacekeeping force led to the 'Six Day War' during which Israel, after a brilliant campaign, was able to re-occupy the entire Sinai region as far as the Suez Canal, overrun all the Jordania West Bank, including the Old City of Jerusalem and push the frontier with Syria over beyond the Golan Heights. This brought a further 1.1 million Arabs under Israeli rule in the so-called Occupied Territories (450,000 in the Gaza Strip and 650,000 in the West Bank). At first, with the exception of the Old City of Jerusalem which was fully absorbed into a single Jewish municipality, Jewish settlement in the Occupied Area was not permitted, but gradually clusters of Jewish settlements of various kinds have developed. The illegality of these settlements is now a major issue both as regards external politics and also within Israel. Jerusalem has been rapidly colonized by Jews.

The wars in 1948 and 1967 produced some 1.5 million Arab refugees who fled from Palestine in 1948 and from Jordan and Syria in 1967. The majority still live in camps, but those who have managed to obtain an education have been able to move to various Arab countries, particularly in the Gulf and secured, in many cases, professional positions. The remainder are cared for by the United Nations Refugee and Works Agency (UNRWA) in camps or suburbs. From the Arab viewpoint, they left their homes under the threat of imminent massacre by Israeli soldiers. The official Israeli view is that they fled either of their own accord, or were deliberately moved away by Arab leaders in order to embarrass the Israelis. The government of Israel is disinclined to have them back, since they would constitute a large dissident minority and would also present a serious long-term demographic threat, owing to their relatively high birth rate. While the camps have become effectively permanent, local employment for the refugees has been restricted by the need to conserve local economies.

The various Arab states are vehemently unwilling to absorb more than a very few refugees, on the grounds that:

1 to do so would be tantamount to acquiescing in the permanent loss of Arab Palestine;
2 there would be enormous social and economic difficulties in trying to absorb them; and
3 so long as the refugees remain as obvious victims, there is still hope they might one day return or receive some reparation.

It is from the desperate, embittered and forcibly idle menfolk in these camps that the Arab guerrilla leaders have drawn their recruits.

When all of Palestine was under Israeli control until the present, the entire history of the region has been dominated by Palestinian efforts to obtain firstly recognition and secondly the lost territories. In contrast, Israel has striven to avoid the recognition of a legitimate Palestinian territorial entity. The overriding reason is said to have been security but another significant factor has also been water supplies.

The Palestine Liberation Organization (PLO), originally founded in 1964, was taken over by Fatah, led by Yasser Arafat, in reality a guerrilla organization, but transformed into a national liberation movement. From 1967 until 1970, operating from Jordan, the focus was upon the perpetration of revolutionary tactics against Israel and Zionism. This ultimately threatened the rule of King Hussein in Jordan and in September 1970, 'Black September', his forces smashed the PLO infrastructure and the organization moved its headquarters to Lebanon. There was to be one further move in exile in 1982, following the Israeli invasion of Lebanon, when the PLO headquarters moved to Tunis.

While other Palestinian-supported organizations were variously considered to comprise terrorist or freedom fighters, the PLO was seen to become less radical and more nationalistic with its decision in 1974 to aim at the establishment of a Palestinian national authority in areas liberated from Israeli control. This of course implied a two-state solution with Palestinian recognition of Israel. Following this, Arafat was invited to address the UN General Assembly.

The overall result of these changes was that the solution to their problem was being sought simultaneously by Palestinians within Palestine and by Palestinian bodies based outside the area. This raised the question of legitimacy which continues until today. Inside Palestine, the major events which shaped the struggle were the large-scale confiscation of land in the West Bank and Gaza Strip by the Israelis and the dependence of as much as 40 per cent of the Palestinian labour force upon Israel for employment. While these factors increase Palestinian vulnerability, the development of an indigenous university system provided a new and positive dimension for the Palestinians. The resultant educated 'élite' was able to construct internal organizations, both to parallel the existing imposed structure and to oppose Israeli occupation.

A major turning point occurred in 1979 with the Camp David Accords, which effectively neutralized Egypt as a partner in opposition to Israel. This provided a realization that the solution needed to be long term and generated by the Palestinians themselves. The effect was that Palestinians in the West Bank and Gaza Strip became more aggressive towards the occupation. The situation was further exacerbated when in 1982 Israel invaded Lebanon for the express purpose of destroying the PLO. Following the move of the PLO headquarters to Tunis, it was even more apparent that solutions must come from political power inside the Occupied Territories, rather than from that exerted by externally based organizations. Nevertheless, the PLO presence remained as many of its factions, including Fatah, the Marxist Popular Front for the Liberation of Palestine (PFLP), the Democratic Front for the Liberation of Palestine (DFLP) and the Palestine Communist Party were active within the new élite in the Occupied Territories.

In December 1987, a spontaneous popular uprising against the Israeli occupation, known as the *intifada*, occurred. It was not particularly violent and was aimed at mass disengagement from Israel. It was, of course, possible because of the earlier activities of the newly educated élite and the structures they had created, which allowed the boycott of Israeli goods and the refusal to pay taxes. Israel, as had become its custom, responded harshly and over 1,000 Palestinians were killed by the Israeli forces. Stones were ineffective weapons against guns. The *intifada* had two other far-reaching effects. International coverage, particularly on television in the USA, provided clear evidence of Israeli brutality and of the overwhelming power which Israel was prepared to deploy. It also allowed Islamic groups such as Hamas and Islamic Jihad to gain an increasingly strong foothold.

However, much of the goodwill generated, particularly in the Arab world, was dissipated during the Gulf War of 1990–91, when the PLO sided with Iraq. The actions of Saddam Hussein, the first Arab leader for a very long time to be in a position to challenge Israel and its major protector, the USA, gained huge support among the Palestinians, particularly when he linked the withdrawal of Iraqi forces from Kuwait with a similar movement of the Israeli military from the Occupied Territories. This may have been cynical manoeuvring on the part of the Iraqi leader, but both Iraq and Israel had been called upon to comply with Security Council Resolutions of which Iraq had been in breach for weeks and Israel for over twenty years. Nevertheless, the overall result was that Kuwait and Saudi Arabia cut off financial support to the Palestinians and the viability of the PLO itself was threatened.

One positive outcome was that the US Administration realized that, to achieve its long-term policies in the Middle East, a solution to the Palestine problem must be found. The result was the US-sponsored Madrid Conference of October 1991, which represented a major diplomatic breakthrough in that it was the first time that Palestinians had been in attendance at a major peace conference. However, no concessions were won and the real change occurred as a result of the Oslo Accords of 1993, which resulted from secret meetings between Israel, now with a Labour government under Rabin, and the PLO in Norway. In October of that year, the PLO and Israel signed a Declaration of Principles, designed to ease the two countries through the final stages of conflict. There were two stages envisaged, an interim period leading to a final settlement in five years. During the interim time, there would be partial Israeli withdrawal, Palestinian self-rule in some sectors and Palestinian elections. The most controversial issues, those of Jerusalem and the Israeli settlements constructed in the West Bank were left on one side.

Initially, deadlines were ignored, but in May 1994 there was the Gaza–Jericho agreement by which self-rule in these areas was agreed, although the level of autonomy was rather less than that envisaged by the Oslo Accords. For some Palestinians these terms were totally insufficient, but for others the key point was that the Palestinians had become a nation with territory. In September 1995, Israel and the PLO signed the full interim agreement which called for the redeployment of Israeli Defence Forces away from most Palestinian towns.

Nevertheless, progress was slow as a result of Israeli intransigence and problems within the Palestinian National Authority (PNA), which relied for its authority mainly upon members from outside the West Bank and Gaza Strip, together with loyal cadres of Fatah. The problems were greatly compounded later in 1995 with the assassination of Rabin and election victory of Netanyahu, at the head of a hard-line right-wing Likud government. Israeli Defence Force activity promptly increased, further Israeli settlement construction began and the result was escalating violence from the Palestinians.

However, the policy of Likud led to clear splits in Israeli opinion and generated support for the Palestinians from groups such as Hamas. At the same time, international criticism of Israel has increased, particularly as its activities in southern Lebanon have become more obviously inept. Arafat continues to have problems of legitimacy and, in attempting to reach some agreements, has been forced to clamp down on his own people, thereby creating further splits in the Palestinian population. The new Labour government under Barak elected in 1999 may provide the catalyst needed to get the peace negotiations completed.

Iraq

The present political unit came into existence after the Versailles settlement of 1919 when it was to a great extent inevitable that the lower valleys of the Tigris and Euphrates should fall to Britain. River navigation and rail communications had for long been in British hands, the development of oil in Iran had enhanced the value of the river ports of the Shatt al Arab and British influence was paramount in the Persian Gulf.

In 1919, Britain was first allotted a Class A mandate for Iraq south of latitude 35°N, and shortly afterwards the Emir Feisal was elected king by almost unanimous plebiscite. Thus Britain could claim that promises made to Arab nationalists were to a great extent fulfilled in that states ruled by the leaders of the revolt, Feisal and Abdullah, had come into existence, though subordinate for a limited term of years to a mandatory power.

Problems were experienced over boundaries. In the north, the Mosul Vilayet, with its future oil possibilities was in contention between Britain, Turkey and France, with some interest expressed from the USA. In 1926, the Vilayet was finally awarded to Iraq. In the south-west, most of the population was nomadic, with pastoralists moving between Saudi Arabia, Iraq and Kuwait. Eventually, 'neutral territories' were created to facilitate movement, but these were partitioned in the 1970s, in view of their value as oilfields. The boundary between Iraq and Kuwait has long been controversial with major incidents in 1961, 1973 and 1976. An international peace-keeping force from other Arab countries was interposed for some months in 1961.

The British Mandate ended in 1932 and a treaty was negotiated by which Iraq took over full powers of government, but the political evolution of the country has been erratic. Rifts between the then governing groups, largely Sunni, and the mass of Shi'a peasantry, grew and in 1958 a revolution overthrew the Hashemite dynasty. Since then Iraq has pursued political aims similar to those of Syria, although relations with Syria itself have frequently been strained. A long-continued source of internal strife has been the demands of Iraqi Kurds for autonomy in the area of the Zagros, but as this includes the Kirkuk oilfield zone, the demands were resisted by Baghdad.

The monarchy was overthrown by military officers, led by Qasim, who immediately faced the problem of whether or not to join the United Arab Republic formed by Egypt and Syria. The Ba'ath Party supported the move which was opposed by the Communists and the decision not to join eventually alienated all support and resulted in the 1963 coup which brought the Ba'athists to power. A second Ba'athist coup in 1968 instituted economic reforms and, in foreign affairs, strident anti-imperialist and particularly anti-Zionist policies. The deputy in this regime was Saddam Hussein who, throughout the 1970s, centralized power in his own hands and was able to succeed in 1979.

In the same year, the Iranian Revolution, occurred and the Algiers Agreement of 1975 was abrogated. The powerful Shi'a theocratic government in Iran was a great concern and, Saddam, underestimating the remaining strength of the Islamic state, attacked in September 1980. The immediate causes for the war were given as disagreements over the Shatt al Arab and the status of Khuzistan.

In contrast to the quick victory which Saddam had anticipated, there was effectively a long drawn-out stalemate. The Kurdish Democratic Party (KDP) was able to take a great deal of the Iraqi Zagros region and Iraq became heavily reliant for funding upon the Gulf sheikhdoms. By 1987, the war had spread to the Gulf with each side targeting the oil infrastructure of the opposition, but throughout, Iraq could count upon the support of Saudi Arabia, Kuwait and the USA. To conserve oil supplies, the USA took an active role in what became known as the 'tanker war' in which it firmly supported Iraq.

In spring 1988, Iran, with the assistance of the two main Kurdish groups, captured Halabja. In retaking it, Iraq used chemical weapons. Eventually, a ceasefire was called on 20 August 1988 and UN Security Council Resolution 598 ended the war. The war had cost Iraq over $450 billion, 400,000 dead and 750,000 wounded and there was a foreign debt of between $60 and $80 billion. The economies of both Iraq and Iran were devastated and there were few tangible gains. Furthermore, Iraq was left with an army which was approximately one million strong and an extensive armaments industry, which was to prove a major source of anxiety not only to its neighbours, but also the wider world.

The aftermath coincided with the end of the Cold War and the fall of many oppressive regimes and this may have influenced the Iraqi leadership which became increasingly aggressive. The main concerns were over oil pricing agreements with the Organization of Petroleum Exporting Countries (OPEC). Kuwait and the UAE were acknowledged to be producing more than their quota, thereby lowering prices and Iraq desperately needed a price increase. Kuwait was specifically targeted as Iraq added a further episode in what had become a long saga in the search for guaranteed access to the Gulf through the use of Bubayan and Warbah Islands, both part of Kuwait. There was a further issue of the Rumaila mega-oilfield which, though predominantly in Iraq, does underlie the Iraq–Kuwait boundary.

On 2 August 1990, Iraqi forces crossed the boundary into Kuwait with the support, in some cases only tacit, of Jordan, Yemen, Algeria, Sudan, Tunisia and Libya, together with the Palestinians. Five days later, US troops were ordered to Saudi Arabia and Operation Desert Shield began. During the build-up which lasted almost six months, there were feverish negotiations and President Bush of the USA was able to establish world-wide consensus on behalf of what became known as the alliance. UN Security Council Resolution 678 enabled the alliance, primarily the USA, to use force and on 17 January 1991 Operation Desert Storm began with a massive and prolonged aerial bombardment. This was followed by the ground offensive on 23 February and a ceasefire 100 hours later. Again, Iraqi losses were huge with more than 100,000 killed, 300,000 wounded and 2.5 million people displaced.

UN Security Council Resolution 688 placed sanctions upon Iraq and, with enemies on all sides with the notable exception of Jordan, Iraq was effectively under siege. Sanctions will remain until UN arms inspectors certify that all Iraq's weapons of mass destruction have been destroyed. In the meantime, UN Security Council Resolution 986 and a number of successors allowed Iraq to sell oil for food and other vital necessities.

Increasingly, as the problems of the Iraqi population have become better known, Arab sympathy has swung behind Iraq, if not Saddam, personally. The situation has been exacerbated by the devious nature of US foreign policy towards Iraq and also the grossly disproportionate treatment of Iraq compared with Israel since both have been in breach of Security Council Resolutions.

Iran

Following the First World War, revival of Iran, like that of Turkey, was masterminded by a former army officer, Reza Khan. A treaty, remarkably favourable to Iran, was negotiated with the Russians, who relinquished all their commercial and economic rights in Iran, with the exception of control of the Caspian fisheries. Strengthened in this way, Reza Khan was able first to undertake a reduction of British influence, and later, a pacification of the tribal areas, which had for long been almost independent of Tehran. Kurds, Lurs, Qashqai, Bakhtiari, Baluch and Khuzistan tribesmen were in turn subdued by a military force. As master of Iran by 1925, Reza Khan, was able to secure the deposition of the reigning Shah and to assume the title himself. He immediately embarked upon a policy of extensive modernization, including the development of modern industry, the emancipation of women and the forceful settlement of nomads. A degree of prosperity returned to Iran, but in 1941 the Shah refused to expel German agents at the request of Great Britain and the Soviet Union and a joint invasion by the two former rivals followed. The Shah was deposed, to be succeeded by his son.

During the Second World War, Iran became an important supply base for the Middle East, particularly with regard to the transfer of materials to the Soviet Union, to which communications were built from the Gulf. After the war, for a number of years the Iranian government was subjected to tensions from within and pressures from without the country. There were local demands in the north-west for regional autonomy and the Soviet Union remained in occupation of the Tabriz area for several years after the end of the war. For some years the left-wing Tudeh party engaged in opposition to the Shah's government, with tacit support at least from the Soviet Union. Internal unrest from tribal elements, chiefly the Bakhtiari and Qashqai, produced a situation of near civil war in some of the more distant areas of the mountains.

However, in the late 1950s the economic situation began to improve and the Shah found himself able to undertake the 'white revolution', forcing land reform on reluctant landowners. This was successful and economic prospects were totally transformed by the rapidly rising level of oil revenues. A series of Five-Year Plans greatly expanded the infrastructure and allowed a wide range of economic activity, particularly in the industrial sector. This made Iran one of the leading countries in the Middle East and also an influential member of OPEC.

By the 1970s, the Shah had greatly increased his own power, but the economy had become highly inflationary. Growing social unrest and political dissent culminated in revolution when, in demonstrations in February 1979, the army declared itself neutral. The Ayatollah Khomeini assumed power and a Islamic theocracy was created. From the start, there were major disagreements over the constitution and in particular the roles of the religious and the political leadership. A number of constituent bodies were set up, but the overall policy appeared to be based on terror.

Then, in less than eighteen months after its establishment, the Islamic Republic was faced with the attack by Iraq. In contrast to Iraq which received military supplies from both the West and the Soviet Union, Iran had to rely on less legitimate markets. To compensate for these and other shortcomings, large numbers of volunteers were incorporated into the military and, as a result, the Iran–Iraq war had a devastating effect in both human and economic costs.

At the end of the war, the government had to introduce rationing in an attempt to repair its shattered economy. Politically, Iran has in many senses withdrawn from Middle Eastern politics, particularly in the Gulf. However, it has developed an expansive foreign policy in central Asia, although in this it is rivalled by Turkey. Iranian foreign policy remains suspect in the Middle East and the crisis was deepened when the USA placed embargoes in 1995. Key problems remain the economy with repayments running at approximately half of Iran's oil revenues and internal politics, since the present state structure seems designed to produce friction between the President who is secular and the Supreme Leader of the Islamic republic who is a cleric. A *rapprochement* with Saudi Arabia in 1999 has improved prospects of Gulf security and may have returned Iran to mainstream Middle Eastern politics.

Arabia

Following the defeat of the Sherif of Mecca by Ibn Saud in 1925, the interior of Arabia was effectively united under a single ruler. Ibn Saud, after his enthronement as King Abdul Aziz pursued a strictly Arab policy, increasingly

aligned after 1945 to collaboration with the USA, rather than British interests. For a time in the 1950s and 1960s, Saudi Arabia, like other oil-rich
and monarchical states, was criticized and opposed by radical Middle Eastern
states, led principally by Egypt under President Nasser. However, the rising
wealth of Saudi Arabia, backed by cautious, but sagacious and quietly determined policies, has brought Saudi Arabia to a position of leader, arbiter and
patron of Arab opinion and policies. The radical states have become clients
rather than opponents and Saudi Arabia has developed an increasingly influential foreign policy, affecting not merely the Middle East, but adjacent areas
of Africa.

After his death in 1953, King Abdul Aziz has been followed in succession
by his sons, the most astute of whom was King Faisal who, with the aid of
oil revenues, established the modern state. The 1970s were the period of
the oil boom and with the quadrupling of world oil prices in 1973–74, Saudi
oil revenues increased astronomically. During this period the infrastructure
of the country was constructed along ultra-modern lines and it was only in
1979 that problems occurred. In that year, the Iranian Revolution, followed
by the occupation of the Grand Mosque in Mecca, provided precursors of
the problems which were to follow. The fall of the Shah not only produced
ideological and security challenges, but resulted in the Iran–Iraq war which
for the best part of the next decade dominated the Gulf region.

Taking over from the British in the region, the USA had adopted a twin
pillar policy of reliance upon Iran and Saudi Arabia. The one remaining pillar
was soon afflicted with problems when the oil price increases of 1979–81
resulted in budget deficits. By that time, the conservation measures introduced globally after the 1973–74 price rises had begun to take effect, while
the economy suffered from reduced oil prices and loans to Iraq. The 1990s,
following Operation Desert Storm, have been characterized by increased instability. Rather than accept the overthrow of a neighbouring monarchy, Saudi
Arabia chose to activate its US security connection and allowed US forces
into the Kingdom with profound political effects. In foreign policy, Saudi
Arabia supported the US line in the Middle East Peace Process and bought
vast amounts of US military equipment.

In domestic politics, the aftermath of Operation Desert Storm was unprecedented discussion concerning internal political structures and the possibility
of reform. King Fahd responded with three decrees and produced a constitutional document, *The Basic System of Government*. The decrees set up a con-
sultative council, the *Majlis Al-Shura*, and a system of regional government.
Nonetheless, core criticism remains that as a result of its entanglements with
the USA, the Kingdom has abandoned its legitimization by Wahhabi Islamic
law. Thus, there are two key issues, both in effect new: the management of
economic retrenchment and the management of the Islamic opposition.

The sheikhdoms of what is now correctly called the Persian–Arabian Gulf
have had a varied political evolution. Kuwait, in fear of Turkish occupation,
became a British protectorate in 1869 and gained independence in 1961.

Bahrain was claimed in the 1960s as a sovereign territory by Iran, but local feeling led to a report by the UN, which proposed independence. This came into effect with Iran dropping the idea of union, during 1971. As Great Britain had previously announced the intention to leave the Gulf, in which it had acted as protector of the small states, by 1971, the small sheikdoms felt particularly vulnerable in the presence of their more powerful neighbours: Iran, Iraq and Saudi Arabia. Initially, the Trucial States, Bahrain and Qatar, formed a federation based around Abu Dhabi and Dubai. However, once Iran had dropped its claim to Bahrain, that state and Qatar decided not to join the grouping and both became independent in 1971. The six remaining sheikhdoms formed the UAE, which came into being at the end of 1971 and was joined in 1972 by the seventh, Ras al Khaymah. The UAE comprises the oil-rich state of Abu Dhabi and the less oil-rich states of Dubai and Sharjah together with Ajmam, Umm al Qawain, Fujeirah and Ras al Khaymah.

The remaining states of the Arabian Peninsula, Yemen and Oman, are both considerably larger than the original sheikhdoms. For most of the modern era both continued to be isolated and aloof from world affairs. The original Yemen, which later broadly comprised North Yemen or the Yemen Arab Republic, has a history going back into the mists of time.

There was some Italian attempt to develop influence during the 1930s, but the Yemen remained independent, first under the despotic rule of the Shi'a (Zaidi) Imam, then after 1962 divided between Royalists and Republicans, with seven years of civil war. Saudi Arabia gave support to the royalist cause, whilst Egyptian troops were sent in by President Nasser on the Republican side. Periods of *rapprochement* between Egypt and Saudi Arabia meant de-escalation of the war in the Yemen and events moved towards a Republic, with the Union of Soviet Socialist Republics (USSR) giving mate-rial support and Britain refusing recognition until Egyptian troops were withdrawn.

Earlier efforts which had been made to bring about federal union between Yemen and the United Arab Republic had come to nothing and, with the evacuation of Yemen, it was possible to negotiate a moderate settlement in 1970. Two years later, after a period of fighting against South Yemen (formerly Aden Protectorate), it was agreed that both Yemens should form a single unitary state. However, this remained an ideal until 1990.

The southern coastlands of Arabia were ruled by Britain under treaties with various local rulers. Aden city was occupied in 1839 by troops of the British East India Company, and until 1944 was administered from India. Extending some 800–1,000 km further along the coast were some 23 Arab states whose rulers had accepted British suzerainty in foreign affairs. This eventually became the South Arabian Federation organized as a loose association of units that varied in size and influence and in 1967, after a series of varied experiments in administration, none of which provided viable or successful, the British withdrew and the People's Republic of South Yemen succeeded to all the territories. Later, South Yemen embraced

Marxism and became known as the People's Democratic Republic of South Yemen. Thus, the union in 1990 of a traditional conservative state with a radical Marxist state produced something of a hybrid which remains politically unsettled.

Until the exploitation of oil in 1967 the Sultanate of Muscat and Oman, now known as Oman, remained isolated from the main economic and political currents of the Middle East. Indeed, contact and any concessions to modernity were strongly discouraged by the Sultan. The situation was further complicated by tribal dissidence, particularly that fostered by the Imam based at Nizwa and supported, after 1976, by South Yemeni troops. The Sultan was deposed by his son in 1970 and power became highly centralized, even by Gulf standards. Guerrilla fighting on the boundaries of South Yemen and Oman in the Dhofar was only put down in 1976 and since then there has been relative political stability, although Islamist activity has surfaced in the 1990s. Economically, Oman has followed a pattern of breathless material development as an oil state.

Libya

The pacification of Libya by the Italians was not complete until 1934, but eventually, under Mussolini, some 250,000 Italian settlers became established in the country. During the Second World War, the British entered into an agreement with the Great Sanussi, then living in exile, and recognized him first as ruler of Cyrenaica and later as king of a united Libya. Full independence was gained in 1951 and for a time Libya was governed as a federation of free territories: Tripolitania, Cyrenaica and the Fezzan, with the capital of each acting in turn annually as federal capital. This proved unworkable and was dropped in 1963. In 1969, a bloodless coup under Colonel Gaddafi overthrew the Sanussi monarchy in favour of a republic and since that time Libya has followed a highly idiosyncratic policy aligned strongly to the left. Strong links were developed with the Soviet Union and oil revenues used to support dissidence and revolution in various parts of the Middle East and the world in general.

At different times in the 1980s, Libyans were implicated in assassination attempts and sorties were flown by the USA against the country in 1986. After suffering heavy losses in fighting in the Aouzou Strip, the boundary dispute in that region went to the International Court of Justice (ICJ) which, in 1987, found in favour of Chad.

While the attitude of the leader, Gaddafi, seems to have moderated with age, the country was beset with UN sanctions imposed as a result of the Lockerbie crisis. A Pan-American aircraft was destroyed in mid-air over the Scottish town of Lockerbie in December 1988, with considerable loss of life. Despite some strong evidence to the contrary, the Gaddafi regime was held responsible for this outrage. In 1999, agreement was reached on the trial of the suspected bombers and sanctions were lifted.

Non-state groups

Throughout the Middle East there are minorities such as the Eris and the Baluchi, but, in the post Second World War period, only three could be considered in any sense nations without states: the Palestinians, the Armenians and the Kurds. For all its limitations, the Palestinians now have a territorial identity. The Armenians, formerly grouped inside and outside the Armenian republic of the Soviet Union now have an independent state. The Kurds, alone, have no state and little prospect of obtaining even autonomy in most of the countries in which they live.

With a population estimated conservatively at over 22.5 million, the Kurds are recognized globally as the largest nation without a state. They are divided principally between four countries: Turkey (10.8 million), Iraq (4.1 million), Iran (5.5 million) and Syria (1 million), while there are approximately 0.5 million in the FSU and almost three-quarters of a million elsewhere (Figure 6.5). In Iraq, the Kurds make up just under one-quarter and in Turkey, approximately one-fifth of the population. The three countries in which the Kurdish question has arisen most obviously are Iran, Iraq and Turkey.

In Iran, during the Iran–Iraq war, the Kurdish Democratic Party Iran (KDPI) was virtually eliminated by government forces and in the process some 25,500 people died. Following that, in 1989, the leader of KDPI was assassinated and there is now no question of mounting campaigns against the government. The best KDPI can hope for is further concessions from an increasingly liberal government of the Islamic republic.

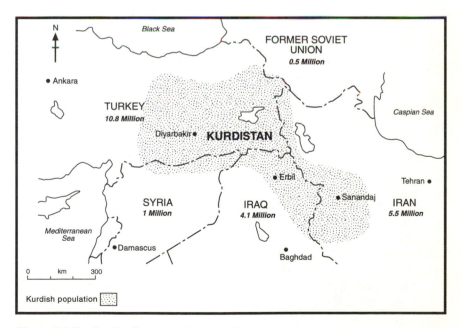

Figure 6.5 Kurds: distribution and populations.

Iraq has recognized Kurdish rights to a greater degree than either Iran or Turkey, but in its treatment of the minority, it has also been more oppressive. The second key dimension in the Kurdish issue in Iraq is that of the incessant internal conflicts within Kurdish society. These have interacted with the government to the detriment of any Kurdish national aspirations. Throughout three decades, from the 1950s, succeeding Iraqi governments have made concessions to Kurdish nationalism at times of weakness, but have never been able to accept Kurdish autonomy.

Until 1975, the Kurdish political scene was dominated by the KDP which was established by Barzani while in exile in the short-lived Mahabad Republic (1946). In 1977, the Patriotic Union of Kurdistan (PUK) was formed as an opposition to Barzani under the leadership of Talabani.

In the latter stages of the Iran–Iraq war, the first phases of Operation Anfal began. In a series of major air attacks, using chemical and high-explosive weapons, the Iraqi military devastated many of the *Peshmerga* (Kurdish freedom fighters) controlled areas in what has come to be seen as genocide. The Anfal operation accounted for between 150,000 and 200,000 lives, 4,000 villages were destroyed and at least 1.5 million people resettled. By July 1989, 45,000 of Iraqi Kurdistan's 74,000 km² had been cleared of Kurds.

The Kurds are a mountain people, partly semi-nomadic, but now mainly settled agriculturalists, who themselves claim to be the descendants of the ancient Medes. In the past, many Kurds practised semi-nomadic pastoralism, with transhumanance between summer quarters on the higher mountain slopes and winter quarters in the valleys and lowlands. Their language is clearly Indo-European rather than Semitic and is related to Farsi (Persian). Whilst about three-quarters are Sunni Muslim, the remainder (especially those within Iran) are chiefly Shi'a Muslim, Sunni Dervish, or even Yazidi, who could be said to be an offshoot from Zoroastrianism.

The Kurds have rarely been independent and their homeland is now divided between Turkey, Iraq and Iran, with enclaves of Kurds also in Syria and in the FSU. It has been said that, since the Kurds have been settled in their present home since about 2400 BC, they have claims to racial purity and to continuity of culture that are stronger than those of any European nation. Perhaps as many as three-quarters are illiterate and there is a high death-rate, general and infantile, as compared with the population generally of the three countries they inhabit. From the days of Xenophon onwards they have been renowned as a war-like people and at the present time they have remained at least partially free. The Kurds are at best, neglected and victims of central government indolence, at worst, actively oppressed.

The Treaty of Sèvres (1920) made provision for the creation of an independent Kurdish state, alongside an Armenian one, but the rise of Kemal Ataturk prevented any implementation. In 1946, in an effort to achieve territorial identity, Iranian Kurds set up the Mahabad Republic which only lasted a few months.

Authoritarian rule in Turkey and Iran has given little scope for Kurdish separatist aims, but in Iraq where the internal regime has been weaker, sustained attempts have been made since 1958 to assert some form of independence. The Kurdish–Iraqi war of 1970 led to an agreement to allow limited autonomy, but this was modified and not fully implemented, so warfare broke out again in 1974. This time the Shah of Iran, hitherto opposed to the Iraq government and a tacit supporter of the Iraqi Kurds, made an accommodation over the Shatt al Arab boundary and withdrew support from the Kurds, who were then overrun by Iraqi military forces.

Following the Iraqi invasion of Kuwait in 1990, there was a spontaneous uprising throughout Kurdistan and by 19 March 1991, virtually all of the area had fallen to the Kurds. The Iraqi government counter-attacked nine days later and UN Security Council Resolution 688 was passed, calling for a cessation of hostilities. By the end of April 1991 there were an estimated 400,000 refugees in camps on the borders with Turkey, over one million had crossed into Iran and approximately 1,000 were dying daily on each frontier. The UN solution was the establishment of a safe haven in the north which became operational on 27 April 1991.

With Kurdistan effectively under siege, an election took place in May 1992 which resulted in a dead heat between the KDP and the PUK. However, civil war erupted again in 1994 and further fighting occurred in 1996. The current issue of importance is the implementation of UN Security Council Resolution 986 which allows Iraq to sell oil, the proceeds of which are used for humanitarian aid. Thus, the problems of Iraqi Kurdistan are of great complexity, internally and externally. Furthermore, none of Iraq's neighbours is likely to support an independent Kurdish state.

During the 1970s, there were ten different governments in Turkey, not one with an overall majority, and the general political instability and economic failure provided the environment in which the PKK was formed. The Turkish state considered the PKK to be a major threat to state integrity and targeted the organization. The whole of the Kurdish area, eastern and southern Anatolia, was militarized and two of the four Turkish armies were based in the region. By 1990, the Turkish garrison in the Kurdish provinces had reached 150,000 troops.

With its left-wing ideology and its espousal of violence, the PKK creates ambivalent feelings amongst Turkish Kurds. The government has taken draconian powers to control the region, but Kurdish national feeling has not been repressed.

Regional state organizations

The League of Arab States

In 1945, the seven Arab countries which were fully or partially independent formed the League of Arab States, commonly known as the Arab League.

They were Egypt, Iraq, Lebanon, Saudi Arabia, Syria, Jordan (then Transjordan) and Yemen (North Yemen). Upon attaining political independence, further countries joined until when Djibouti became a member in 1977 there were 22 members. Subsequently, the number has been reduced by the unification of the two Yemens. In only Djibouti, Mauritania and Somalia is the spoken language not Arabic. Palestine, although then without territory, was admitted in 1976 and was represented by the PLO. Thus, all the Arab states from Mauritania in the west to Somalia in the east, are in membership.

The original objective of the organization was resistance to Zionism and foreign exploitation, but its activities have been extended to embrace economic, social and cultural issues. To carry out this role, the Arab League has developed a range of specialized agencies, including the Arab Monetary Fund, the Arab Fund for Technical Assistance and the Arab Cultural Organization.

When Egypt signed the Camp David Accords with Israel in 1979, it was expelled from the Arab League and the headquarters was moved from Cairo to Tunis. However, Egypt re-joined in 1989 and the headquarters returned to Cairo in 1991. As a result of lobbying by the League, Arabic became the sixth official and working language of the UN from 1 January 1983.

Organization of Petroleum Exporting Countries (OPEC)

Among the countries of the Middle East, only Israel, Lebanon, Jordan, Sudan and Turkey have little or no crude oil production. However, all the Arab states are linked economically or politically to oil through a range of activities, including joint venture companies. In order to exert control over their own oil resources, most of the major producers at that time joined together to form OPEC in 1960. The founder members were Venezuela, the instigator of the move, Kuwait, Iran, Iraq and Saudi Arabia, with Qatar becoming a member in 1961. These were later joined by Algeria, Ecuador, Gabon, Indonesia, Libya, Nigeria and the UAE. The notable absentees from the list of major producers have remained the USA and the Soviet Union/Russia.

OPEC has been the co-ordinating body behind the successful attempts to raise the world selling price of crude petroleum and in using embargo and joint collective action to bring about first increased taxation and then expropriation of Western ownership and operation. The power of OPEC was particularly apparent in 1973, after the October War, when the six Gulf members cut production and posted prices of Gulf crude oil which rose from $2 per barrel to $12 per barrel between 1973 and 1977.

However, co-ordination among the OPEC members has been difficult. Broadly speaking, countries with large reserves have been content to maximize their revenues by higher production, whereas those with more limited reserves have been intent on higher pricing. Eventually, during the 1980s, as a result of regional instability created by the Iran–Iraq war, a rise in non-OPEC production and improvements in fuel conservation techniques world-wide, the oil price had to be reduced.

In parallel with OPEC, the Organization of Arab Petroleum Exporting Countries (OAPEC) was founded to include most of the Middle Eastern Arab oil producers. Nevertheless, it was felt that given the regional instability created by the Iran–Iraq war, a further body, the Gulf Co-operation Council (GCC), was required to protect Arab oil interests, particularly with regard to security. The GCC was formed in 1981 by an agreement between the rulers of Saudi Arabia, Kuwait, Qatar, the UAE, Bahrain and Oman. Joint projects have included the formation of an Arabian free trade area, the abolition of many tariffs and the establishment of a joint defence force of brigade size. After the Gulf War in which all the GCC members supported the alliance, they formed with Syria and Egypt the Arab Mutual Defence Organization. Divisions within the GCC membership remain, but clear progress towards co-operation has been made.

7 People and population

Few regions in the world can surpass the Middle East in heterogeneity of population. From the earliest times, the region has attracted waves of immigrants from various parts of the Old World and fusion of diverse elements has sometimes tended to be slow, or even incomplete. Frequent references in the literatures of various peoples of the Middle East indicate an acute awareness of racial and cultural distinction, and insistence by these groups on their separate identity can be taken as an indication of the close juxtaposition of many types of racial stock, cultural patterns and social organization.

Distinctions between Jew and Christian, Believer and Infidel, Turk or Arab, Semite and Hamite, or between mountain dweller and plainsman, and Bedouin (herders) and Hadhar (cultivators), have long coloured the geography of the Middle East and are still at the present day factors of prime importance in the human relations of the region. Despite improved communications and the spread of education, it is currently an unfortunate tendency of societies the world over, and not just in the Middle East, to have become more rather than less conscious of nationality and cultural divisions. Race, colour, culture and nationality, whatever any of these terms may mean specifically, are increasingly considered significant.

It has proved difficult to establish criteria by which the peoples of the world can be subdivided. Appearance and physique, blood serum, character, language, religion and nationality have all been invoked as a basis of definition, but no single one of these can provide a wholly satisfying and acceptable point of reference. For the Middle East, this insufficiency is obvious.

Nationalism is an increasingly potent element, political consciousness having greatly developed over the period since the Second World War. It is evident within states, but within non-state national groups, even more obviously. However, within this second category smallness of numbers, intermingling with other, equally conscious groups, or external political obstacles, have tended to inhibit effective realization, so that existing international frontiers can hardly be said to allow complete political expression to all nationally conscious groups. The Kurds are the most obvious example, but there are also the southern tribes people of Sudan and, prior to 1974, the Turks of Cyprus.

Furthermore, there is a wide range of political structure from the social-
istically inclined republican states, typified in Egypt, Syria and Iraq, to
monarchical, oligarchic and theocratic forms of government. Pan-Arab,
pan-Iranian, pan-Turkish and pan-Kurdish movements are political issues of
varying intensity, but they all encounter practical difficulties which have so
far prevented their full realization. In an attempt to throw further light on
the complicated question of cultural, political and social organization, the
population of the Middle East will be considered from the standpoint in
turn of five criteria: race, language and religion, which will be addressed
in this chapter; and social organization and national consciousness, which
will be considered separately in Chapters 8 and 10, respectively.

As a preliminary, the influence of certain geographical factors should be
noted. In this respect, the region has a dual aspect. It is a broad corridor
linking several major parts of the world, either overland or by short sea
journeys, but there are also local environments which have favoured the
preservation of distinctive physical human types and cultures, that in more
accessible areas have been submerged by later arrivals.

A somewhat discontinuous belt of fertile steppe and oasis runs along the
southern slopes of the Elburz mountains of Iran and is followed by an ancient
route that in the opinion of some anthropologists was one avenue by which
certain types of early man spread into western Eurasia. Further to the west
the route divides, one branch turning south-west into south-west Iran and
Iraq, and the other continuing through Azerbaijan to Asia Minor. A zone
of steppeland in inner Anatolia completes the link between Central Asia and
Europe.

More strongly marked is the steppe area of the Fertile Crescent which
links western Iran, Iraq, south-east Anatolia and the Levant. By this route
numerous invaders from the east and north have gained the shores of the
Mediterranean and the Nile Valley, while as a return movement Egyptian
culture has also spread into Asia. The Mediterranean itself has facilitated
rather than impeded intercourse between the peoples near its shores. The
island-studded Aegean was a nursery for man's first sea adventures and it is
not surprising that some of the earliest cultural developments occurred in
Cyprus, which is adjacent to both Asia Minor and the mainland of the Levant.
In the south, close physical connection between Arabia and north Africa is
paralleled by similar cultural contact. The Sinai has formed a major route for
movement of peoples and some authorities stress the importance of the Bab
el Mandeb region as the gateway between Arabia and Sudan.

Alongside these routeways through the Middle East, there occur regions
in which difficulty of access or hostile natural conditions have restricted
penetration. The largest of these is the desert of Arabia, but another such
region comprises the highland plateaux of the Mediterranean eastern seaboard,
where Samaritans, Druzes, Maronites, Metwalis, Alawites and various sects
of Asia Minor, such as the Bakhtahis and Takhtajis, have preserved a certain
cultural autonomy.

Further east, the Kurds of the Anatolian and Zagros mountain belts have maintained a separate identity; and smaller units can be traced in the Yazidid of Jebel Sinjar of north Iraq, the Circassians (Cherkasski) and the Turcomans. Variation on a large scale is also characteristic of the Zagros.

However, it is also necessary to note that nowadays several influences are working to erode the separation and distinctiveness of these long-standing minority communities. One is the increased efficiency of communications which allow formerly isolated groups to reach and be reached by vehicle, radio and other forms of media. Another is the enhanced 'pull' from new occupations and better living conditions in the cities, due in large part to the prosperity based on oil exploitation. A third influence is the medical care frequently available only in major urban centres and finally, there is the stronger control by central government which, using modern methods of communication, is able to reach out far more effectively and consistently to impose centralized rule upon once isolated communities.

Race

Racial division has always been a difficult issue for anthropologists, but in recent times more objective systems of analysing racial diversity have been developed, based primarily on blood grouping, but also on other genetically controlled factors. Quite distinctive elements, often relatable to genetic factors, can be isolated in human blood and many individual systems of classification are therefore possible. The best known of these is the ABO blood grouping system in which an individual may have the A gene only, or the B gene only, or both A and B genes, or neither (O). These characteristics would seem to reflect the geographical location, since the proportions of A, B or O are not random, but vary recognizably from place to place.

Recognition of other blood group substances allows the construction of further separate systems. Besides being easily measurable and objectively determined, racial grouping based on this kind of genetic information has the merits of showing linkages, both geographical and historical. That is, two communities, even though separated in habitat, but of similar blood grouping, could be shown to be derived from a single common group. Such connection has been demonstrated, for example, for the gypsy communities of Europe and the Middle East. Furthermore, close relationship to genetic factors allows the tracing of outside influences such as geographical intermixture or isolation, social barriers or inbreeding.

Of special interest is the position of the Jews. For long, anthropologists have disputed whether or not it was possible to define or recognize a distinctly Jewish type, or whether this type, where it existed, was really a part of a Semitic race that also includes Arabs and Armenoids. Blood grouping has shown that many, though not all, Jews show certain similarities among their own group and fewer similarities with the surrounding communities. That is, they do have a biological basis which is to some extent distinctive.

Other blood factors, enzyme deficiencies and biological traits, would seem to have a patterned distribution among human groups, suggesting some form of differential response to environment. The importance of this would seem to lie in the fact that there appear to be recognizable regional variations between communities of human beings, with, however, variability rather than uniformity the principal factor.

There are definite differences between human groups as groups, but these tend, in the main, to occur spatially as gradations rather than as sharp divides. Moreover, the elements that vary are not confined to matters of skin colour or physical appearance. These may be part, but only a limited part, of the whole highly involved complex that makes up the biological distinction between individuals. No single 'nation' or 'race' is entirely homogeneous in its physical attributes, yet at the same time, the persistence of genetic differences, whether of skin colour, hair types or blood group does occur.

The picture is very confused and in itself, the question of race, often highly emotive, is of very little consequence. More interesting is the concept of the racial group and, in the Middle East, the racial–political group. The emphasis is on the political rather than on the racial. Furthermore, the concept of race can be more obviously and clearly understood in its relationship to the linguistic group.

Language

With regard to language, the position of the Middle East is much clearer (Figure 7.1). Although it is possible that at one time each racial group had its own language, intercourse soon breaks down linguistic differences, and one language may quickly establish a dominance, to the exclusion of others. Whereas physical characters can persist through many generations as a result of biological inheritance, a cultural feature such as language often becomes modified within a short time and may even die out, particularly if the language itself possesses no literature.

In the north, Turkish, as the language of Osmanli conquerors, became dominant in Asia Minor. Spoken only by Asiatic nomads, Turkish at first possessed no alphabet, and Arabic characters were borrowed, but these letters are not particularly well adapted to expressing the sounds of Turkish. By decree of the Turkish Republic, Roman letters replaced Arabic script in 1923.

However, Turkish is by no means universally spoken in Asia Minor. Towards the east, Armenian, a more ancient language with its own characters and literature, still persists, and further to the south, Kurdish has a wide extension. In the difficult and inaccessible hill districts near the Russian frontier many remnants of Caucasian languages occur, especially Circassian, Lazi and Mingrelian. Some of these represent ancient cultural or racial groups submerged by later arrivals, but others are Finno-Ugrian dialects brought in at comparatively recent dates by invaders from the east. At one time Greek

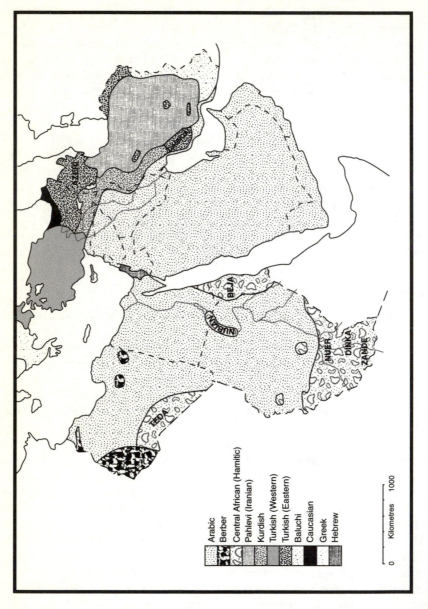

Arabic
Berber
Central African (Hamitic)
Pahlevi (Iranian)
Kurdish
Turkish (Western)
Turkish (Eastern)
Baluchi
Caucasian
Greek
Hebrew

0   Kilometres   1000

Figure 7.1 Middle East: distribution of major languages.

was spoken by a considerable minority in western Anatolia and it remains the native tongue of a majority of the inhabitants of Cyprus.

From the interior of Arabia have come two great groups of languages, North Semitic and South Semitic. To the first group belong the Aramean and the Canaanitish dialects: Aramaic proper, Hebrew, Phoenician and the now extinct Palmyrene. To the second group belongs Arabic. In the first and second millennia BC, extensive migrations spread North Semitic languages into Mesopotamia, Syria and the Levant. Aramaic, the language of a people living in the western edge of the Syrian desert, became predominant over a wide area in the Middle East, most probably because of the extensive commercial relations developed by the Aramean states. Aramaic dialect was the vernacular of Palestine during the time of Christ and at least one of the Gospels can be considered as pure Aramaic literature. Aramaic also had a considerable effect on certain languages spoken in Iran and traces are still apparent in the north-west of the country. The language itself lingers in a few villages near Damascus and Mosul and is the sacred language of a number of Christian sects in Syria and Iraq. A revived and simplified Hebrew has been adopted as the official language of the State of Israel.

Arabic, originally spoken by a small group of traders, townsfolk and desert nomads in the district of Medina and Mecca, was the language of Muhammad and his early followers and with the rise of Islam in the seventh century AD, quickly replaced existing languages in Libya, Egypt, the Levant and Iraq. As the language of the Koran, Arabic is one of the great unifying influences of the Islamic world, since although local deviation in vocabulary and pronunciation occur, a standard form of classical Arabic is understood by most literate Muslims.

It should be noted that the dominance of Arabic in the Middle East is by no means complete. Penetration of the mountain zone of the east and north did not occur and over much of Sudan languages other than Arabic are current, from Nubian in the extreme north and Beja in the Red Sea hills to the many central African languages of the south. Arabic has, however, intermingled in many areas of the country. In Asia Minor, Arabic script only was adopted and the numerous non-Semitic languages of the region remained in current use. Similarly, Iran has retained its own distinctive speech, Farsi, which can best be described as an ancient indigenous language, termed Pahlevi, much modified by extraneous influences, but using Arabic script. Extensive interchange of vocabulary and ideas between Iranian and Arabic shows that the cultural influences of Arabic have been strong, yet not sufficient to drive out the native form of speech. Again, reminiscent of Asia Minor, Iran has a number of minority languages, chiefly spoken by tribes living in the Zagros regions: Turkic, Arabic, Baluchi and Kurdish. There are other minority languages in south Arabia, particularly the Hadhramaut. One way of subdividing the Middle East linguistically is into an Arabic area in which the single language predominates to the virtual exclusion of all others and composite regions in which many languages remain current. However, it is probably more acceptable to distinguish Arabic, Farsi and Turkic core areas between which are a mixture of developments and indigenous languages.

Religion

It is true to say that religion has always played a dominant role in the Middle East. The region was the origin of four of the world's great religions, all monotheistic, which arose in the following order: Zoroastrianism, Judaism, Christianity and Islam. While the region is now dominated by Islam, the development of religion can be better understood if the four are taken in the order of their genesis.

Zoroastrianism

Zoroaster, or Zarathustra, a native of Iran, lived during the period 700–50 BC and his teaching that human life is a battleground of opposing forces, good and evil, was adopted as the official creed of the Persian Empire under Cyrus and his successors. The religion of Zoroaster, by its insistence on moral and ethical standards, represented a great advance on the older pagan and polytheistic creeds, which had frequently appealed to the basic human instincts of self-interest and sensuality. The older gods were, however, not entirely abandoned, but retained as subsidiary in influence to a single Lord of Goodness and Light, or sometimes as demons.

The association of the Lord of Goodness with light led in time to the use of fire as an important element in Zoroastrian worship, and with the spread of religion amongst peoples who found difficulty in appreciating the abstract conceptions of Zoroaster, some earlier ideas were abandoned in favour of a simpler form of belief that was greatly influenced by pre-existent pagan creeds. Fire also came to be used as the principal element of ritual. Thus, Zoroastrianism became associated with fire-worship and remained the distinctive religion of Iran until the rise of Islam. One of the largest fire temples was at Masjid i Suleman, in the Iranian Zagros, where a natural oil seepage occurred. Muslim persecution over a long period has reduced the fire-worshippers to a small remnant, who now exist as a minority in the Kerman and Yazd districts of Iran. Large numbers emigrated to India, where they formed the Parsee community.

Judaism

The Jewish people entered Palestine during the second millennium BC and successfully established a small state in the highlands of Judaea. It was subsequent to this entry into Palestine that the religion of the Jewish people, at first worship of a purely tribal deity and strongly influenced by a pastoral way of life, developed the lofty conceptions which now characterize it. Following conquest by the Assyrian king in 722 BC, the first dispersal of the

Jewish people took place, when numbers of Jews were forcibly settled in Mesopotamia. Later, however, these immigrants were allowed to return and the Jewish state revived. In AD 71, as the result of a revolt against Roman overlords, a second and more permanent Dispersal or Diaspora, took place. Jews became established in many parts of Europe and Asia, where they absorbed much of the culture and even some of the racial traits of the people amongst whom they settled. Yemenite, Persian, Turcoman, Georgian and even Abyssinian Jews are still distinguishable and by reason of their Oriental culture and outlook, these groups contrast with the Jews of Europe. Numbers of Oriental Jews settled in Palestine during the Ottoman period, and for a time in the twentieth century showed some disinclination to identify themselves with Zionism. However, since the creation of Israel in 1948 a mass exodus of Jews from Arab countries has taken place.

Of the European Jews, a minority entered the Iberian peninsula, while the remainder settled chiefly in Poland and adjacent countries and it is from this latter group that the majority of modern European immigrants to Israel have come. Today the Jews from the Middle East, Oriental Jews, are designated Sephardic and those from Europe are known as Ashkenazi.

Christianity

In AD 313 Christianity was adopted as the official religion of the Roman Empire. This adoption led to many changes in Christianity itself, which, from being the fiercely held creed of an active, but disliked and mistrusted minority, became an accepted and integral part of the Roman state. In undergoing this development, Christianity was subjected to two influences. The first, an attempt at defining a single body of dogma and ritual in order to preserve unity within the Church itself, is of little concern to the geographer, but the second, the creation of administrative provinces on a territorial basis, has much geographical interest.

Within the Roman world, four cities could claim a certain material and intellectual pre-eminence. Rome itself, for long the centre of the empire, began to decline in influence with the increasing economic and political importance of the lands of the eastern Mediterranean during the later Roman period, and a symptom of this shift of balance was the corresponding rise of Constantinople, which supplanted Rome as the capital of the Roman Empire in AD 330.

Further to the east, Alexandria and Antioch, both extremely wealthy commercial centres, had each developed a distinctive tradition based on special regional interests. Accordingly, the early Christian Church was organized into four provinces, based on these four cities, with each province headed by a patriarch. Within a relatively short time, however, instead of the unity that had been hoped for, strong regional particularism came to be manifest.

The province of Constantinople was in closest touch with the seat of government and therefore came to be identified with imperial authority.

Moreover, by reason of its location, the Church in Constantinople was greatly influenced by Classical Greek rationalist thought and, like the pagan religions of the country, tended to employ much painting, sculpture and music in its ritual.

In the West, the patriarch of Rome inherited much of the ancient authoritarian tradition of the former capital city, and in spite of the newer supremacy of Constantinople, was able to exert a dominating influence on the Christians of the West. Far less affected by Greek thought, the Church in Rome developed a separate tradition of its own, more particularly as a Western European Church.

Christianity in Alexandria and Antioch was influenced to a varying extent by Oriental mysticism and speculation. Traces of the religion of ancient Egypt were manifest in the dogma and ritual adopted by the province of Alexandria, while other differences became apparent in Antioch. Within a relatively short space of time, increasing divergence in doctrine and ritual led to a complete separation of the four provinces. From Constantinople arose the Greek Orthodox Church; from Rome the Roman Catholic Church or Latin Church; from Alexandria the Coptic Church; and from Antioch the Syrian or Jacobite Church, all of which soon came to possess complete independence in organization.

Further division, reflecting other regional diversity, rapidly occurred. By the fourth century, an independent Armenian (Gregorian) Church had come into existence and the teachings of Nestorius, denounced as heresy in the West, were adopted by many in Iraq, Iran and countries further east. At a somewhat later period, the sect of Maronites, followers of a Syriac monk named Maroun, came into existence in north-west Syria and established itself in northern Lebanon during the seventh to eleventh centuries AD.

Later political developments in Europe and western Asia had great influence on the various religious communities of the Middle Eastern region. The supremacy of Constantinople passed away and effective leadership amongst Christian peoples was taken over by the Church of Rome. The impact of the Muslim conquest fell chiefly on the Christians of the east, so that groups like the Copts, Jacobites, Maronites and Nestorians (or Chaldeans) dwindled to tiny communities, whilst the Churches of Rome and Constantinople continued to flourish. At a later date, further expansion of Muslim power reduced the importance of the Greek Orthodox community, but later still this decline was partly offset by the rise of Russia, which adopted the Greek faith in preference to that of Rome.

During periods of Muslim persecution, the autonomous Christian sects of the east obtained support from the Church of Rome, but often at the price of obedience to Rome. Agreements were made whereby, in return for recognition of the Pope as head of the community, local usages in doctrine and ritual were permitted to continue. Hence a number of eastern Christians broke away from sects such as the Jacobites or Nestorians and formed what are known as the Uniate Churches, communities with practices that differ widely from those of the main Roman Church, but which nevertheless accept the supremacy of the Pope. There have thus come into existence the Armenian Catholic, the Greek Catholic, the Syrian Catholic, the Coptic Catholic and the Chaldean (Nestorian) Catholic Churches. The entire Maronite Church entered into communion with Rome in the twelfth century. Formation of the Uniate Churches did not extinguish the older sects, some of which preferred a precarious independence. Hence, at the present time, representatives of both groups, Uniate and non-Uniate, are to be found in the Middle East.

Islam

Shortly after AD 600 Muhammad, a fairly poor member of the important family of Qureish, began to preach a doctrine that had come to him by divine revelation. His message was first received with indifference, later with hostility, so that in AD 622 he decided to leave his native city of Mecca and established himself at Medina. This flight or *Hegira*, from Mecca to Medina is taken as the beginning of the Islamic era, though it was not until some time afterwards that Muhammad's teachings were accepted by the Meccans. Much of Muhammad's teaching was his own, although part was derived from Christian and Jewish beliefs, with which he had a slight acquaintance. Born into a merchant community in the trading city of Mecca, he was a citizen, not a nomad and Islam as he conceived it was, by virtue of its social constraints and its spiritual demands, very much a city religion. The rhythm of Islamic religious practices is made for city dwellers.

Muhammad preached submission (*Islam*) to the will of one god, Allah, of whom Muhammad was the chosen prophet. Besides the profession of faith, the chief ritual duties enjoined on Muslims are: (a) prayer five times a day; (b) fasting during the month of Ramadan; (c) the giving of alms; and (d) the undertaking, if possible, of the pilgrimage, the Hajj, to Mecca. In addition, mention should also be made of the Muslim prohibition of alcohol and pork. Although there are other elements of belief and practice that set Islam apart from other religions, the five 'Pillars of the Faith' remain the most distinctive. Islam has no consecrated priesthood and the *imam* (leader) who conducts the faithful in prayer and the *sheikh* (elder) who delivers the sermon are laymen, not necessarily occupying positions of prominence, but generally respected for their moral and intellectual qualities. In fact, the act of prayer can be carried out anywhere, although Muslims are supposed to meet together at the chief mosque for the prayer at midday on Fridays. The mass of the people, however, found that observance of the basic duties of Islam was not enough and the worship of saints spread rapidly, often integrating local religious traditions. This innovation, which quickly became an important part of popular religious life, was eventually accepted as orthodox practice.

The five 'Pillars of Faith' are central supports of the *Sharia*, the Islamic law. Divinely inspired, the Sharia has been revealed to man through the Koran and through the recorded actions and statements of Muhammad, known collectively as the *hadiths* or traditions. The task of interpreting the legacy of the Prophet and building the Sharia was achieved during the first three centuries of Islam, mainly through the achievements of the *ulema*, Muslim theologians and legal experts. Nevertheless, although agreed on many essentials, there emerged among the *ulema* many different schools of interpretation or varying 'pathways to the truth', as they are known. They ranged from mildly liberal in their outlook to strictly fundamentalist.

Four legal schools still exist today, named after their founders: (i) the Maliki, the oldest, found in North Africa and Sudan; (ii) the Hanafi, the most flexible and the official school in most countries of the Arab east (except for the Arabian Peninsula); (iii) the Shafi, which prevails in northern Egypt; and (iv) the Hanbali, found in Saudi Arabia. The first two schools are somewhat modernistic and progressive; the last two conservative in outlook. However, in spite of these differences, each school has long accepted the others as orthodox, and in many countries a person may choose to be judged according to a school different from that officially recognized by the state.

One important feature of Islam has been the close connection between religion and civil government. To Muslims, Islam provides a full way of life, with complete interpenetration of civil, religious and even economic activities and codes of law derived from religious bases. Islam and the State, for the devout, are one, recognized by the fact that formerly the political head of state held religious functions, under the title Defender or Leader of the Faith (Caliph). This makes for great cohesiveness in society, but also poses difficulties for modernists.

Within a very few years, from being an obscure doctrine held by a small group of townsfolk and nomads in the arid interior of Arabia, Islam had spread into Palestine, Syria, Egypt, Iraq and Iran. The surprising political success of the new religion in establishing itself alongside the Byzantine and Iranian Empires, which at that time divided the Middle East, is startling when one considers how few were the followers of Muhammad and how poor they were in material resources. In order to explain how a group of shepherds, merchants and camel drivers came to wrest the greater part of the Middle East, within a few decades, from the successors of Imperial Rome and Sassanid Persia, a number of factors must be considered.

Politically and militarily, both the Byzantine Empire to the west and the Sassanid Empire of Iran were exhausted. Long-continued warfare between the two powers, with Armenia, Syria, Iraq and Palestine as a battleground, had achieved no definite results, since both belligerents were too large and too remote to be entirely overrun by the other. Border raids and skirmishes had, however, devastated much of the central part of the Middle East and the high taxation and misery that resulted had made local populations very ready to acquiesce in a change of power. Rigid social divisions and indifference to the lot of the peasant did not serve to popularize either Byzantine or Sassanid rule.

Spiritually, too, the common people of the Middle East had become weary. Unending theological debates in the Christian world with intricate discussions of forms of belief and doctrine were frequently beyond under-standing by the average Anatolian or Syrian peasant, but the charges of schism and heresy that resulted led to much general persecution and oppression, first by one sect, then by another. In Iran, the finer ideas of Zoroaster had become overlain by a number of gross practices borrowed from earlier religions, or else were too abstract in conception for the mass of the people.

Thus, the appeal of a new religion, direct and simple to understand, with a clear relationship to matters of everyday life, and which offered a broth-erhood of man in place of the formalities of Zoroastrianism or the strict class distinctions by Byzantine society, had an immediate appeal. Buttressing all this were the material gains to be made from conquest. Sectarian differences among Christians were in fact so acute that on some occasions persecuted minorities opened the gates of Byzantine cities to Muslim attackers. The rapidity with which Islam developed and spread, with enormous impact on life and culture in an area extending from China to France, is one of the most remarkable features of history.

Sectarian divisions within Islam

Over the centuries, although there have been numerous sectarian divisions in the House of Islam, these sects account for only some 10 per cent of the Muslim World and the orthodox or Sunni branch (*Sunni* means majority or orthodox) has prevailed as a powerful majority. All four legal schools, for example, are Sunni. The appearance of divisions within the Muslim commu-nity was a direct consequence of the nature of Islam itself. An important feature, already noted, was the close connection between religion and the state. In the early centuries, because Islam and the state were one, a revolt against the central power was often accompanied by a religious crisis. Indeed, as politics and religion were inextricably interwoven, religion provided the only possible expression of sustained opposition. Sectarian divisions were therefore nearly always political in origin, but later differences of doctrine and outlook developed.

Muhammad left no male heirs nor did he nominate a successor so that his death in AD 632 precipitated a political crisis. In the course of the seventh century, a struggle developed for the succession to the Caliphate between Ali, Muhammad's cousin and also son-in-law, and the younger Omeyyad branch of the Prophet's family. During these events, two main heretical groups developed, expressing in religious terms the opposition of certain parties to the existing social and political order. The earliest of the dissident sects, the *Kharijites* or 'seceders' believed that the Caliphate should be open to all Muslims and not merely men of Muhammad's own family. Their egal-itarian fervour played an important role in the Islamization of the Berbers of North Africa, but today only small scattered communities survive in the Middle East, in the Jebel Nefusa in Libya and in Oman.

The second and by far the more important opposition group was the party of Ali, commonly known as the *Shi'a* or partisans. After Ali was eventually defeated by the Omeyyads, his loyal followers refused to accept his abdication. When he was assassinated in AD 661 by the Kharijites, his supporters started a movement to restore the Caliphate to the family of Ali, and this political group rapidly developed into a remarkable and distinctive religious sect. The Shi'a were convinced that the Omeyyad and Abbasid caliphs set out systematically to eliminate the descendants of Ali, and they transformed these victims of political manoeuvring into semi-divine martyrs, known as *imams*. A decisive element of separateness became the importance attributed by the Shi'a to the imams and the virtual exclusion of Muhammad. Many other factors separated Shi'a from Sunni Muslims, but particularly important are their intense emotionalism, expressed especially during the focal point of their religious year, the first 10 days of the month of Muharram, their ingrained suspiciousness over religious matters and their intolerance born of periodic repression and persecution over the centuries. The Shi'a also have their own holy places, notably the tombs of their imams at Karbala and Nejf (Najaf) in Iraq, which are more important to the Shi'a as places of pilgrimage than Mecca or Medina.

The majority of the Shi'a recognize the existence of twelve imams beginning with Ali, and are known as *Twelvers*. The twelfth imam is believed to have disappeared and will one day re-emerge as the Mahdi or new Messiah. This group is dominant in Iran, the only state where Shi'ism is the state religion, while they represent well over half the population of Iraq. They form a compact community in southern and eastern Lebanon under the local name of Metwali and considerable numbers are also found in Bahrain.

Not all Shi'a are Twelvers. The *Zaidis* only recognize the first four imams and there is little in their beliefs and practices to distinguish them from orthodox Muslims. At the end of the ninth century the Zaidis established their control over North Yemen, and although the majority of the population remained Sunni, Zaidi imams continued to rule there until the 1962 revolution, which led to the creation of a republic.

Another group, the *Ismailis* or *Seveners*, originated in the belief by some Shi'a that Ismail, the son of the sixth imam, was unjustly accused of unworthy conduct by the Twelvers. They set out to vindicate him and during the ninth century launched a vigorous campaign to overthrow the orthodox Abbasid Caliphate. They did not succeed, but a group of Ismailis seized power in North Africa in the tenth century and established the Fatimid Caliphate which ruled in Cairo until the twelfth century. With the decline of the Fatimids, the Ismailis did not disappear. In the eleventh century, a Persian leader of the sect founded an extremist political group, the Assassins. Their executions of key political figures were acts of ritual murder. From their strongholds in the Elburz and Syrian mountains they terrorized much of the Middle East until they were suppressed in the mid-thirteenth century. A remnant of the sect still survives in north-west Syria with other communities scattered through

Iran, Oman, Zanzibar and India, and in the nineteenth and twentieth centuries they emerged from obscurity under a leader known as the Aga Khan. They now depend on business success to advance their cause.

Ismaili missionary activity also resulted in the creation of some sects incorporating so many non-Islamic practices that they are sometimes regarded as separate faiths. The most notable are the Alawi and the Druzes. The *Alawi*, worshippers of Ali, carry to the extreme the Shi'a deification of Ali. They retain certain features of pagan cults and follow a ritual adopted largely from Christianity, including the celebration of Christmas and Easter. Today, their main centre is in the Jebel Ansarieh in north-west Syria.

The *Druzes* trace their origins to an eleventh century Ismaili missionary, Darazi, whose supporters recognized the Fatimid Caliph Hakim as the hidden Imam. Druze doctrine and practice differ widely from Muslim orthodoxy. Belief in the transmigration of souls is widespread and they do not observe Ramadan or make the pilgrimage to Mecca. Polygamy is forbidden. To escape Sunni persecution the Druzes found refuge in the mountains of southern Lebanon and in the Jebel Druze in southern Syria.

A modern heretical movement in which Ismaili doctrine played a prominent role is the *Bahai* faith. The movement began in the mid-nineteenth century in Iran where a young religious teacher declared himself to be the hidden Imam. As a reaction against materialism, corruption and self-seeking, the movement won many converts in Persia, but was repressed with great severity by the government. The survivors became known as Bahis, after Bahaullah, a disciple of the founder, and are noted for their tolerance and their strong commitment to social improvement and international peace. Although subjected to intermittent persecution in Iran, the Bahai faith has spread beyond the Middle East and there are now scattered communities all over the world.

Sufism

Sufism is the spirituality of the religion of Islam. Its main tenet is belief in a mystical union of the human soul and the Deity, with the omnipresence of God's purpose and guidance. In many ways man acts as inspired directly by God and is not a wholly free agent in that there could be an element of predestination. Meditation, spiritual possession, prayer and asceticism are components of the Sufi way of life, which aims at human behaviour determined by God's will. The term *Sufi* is derived from the Arabic word for wool (*suf*) and refers to the coarse woollen garments worn by the early Sufi ascetics. Nearly two hundred major Sufi *tariqas* (orders) are recorded during the history of Islam. Each tariqa consisted of a leader (sheikh) and his disciples (dervishes) who were supported by lay members. Two distinct types of brotherhood emerged: the urban orders such as the Qadiriya and Naqshbandiya, characterized by moderation and with close links with Sunni orthodoxy; and the rural orders, the Rifaiya and Bektashi, with wider popular support. The rural orders often followed practices such as walking through fire, eating glass, self-wounding without pain, borrowed from paganism, Shi'ism or Christianity, and denounced by orthodox Muslims.

A revival of the orders occurred in the nineteenth century with the appearance of a number of new movements including the Sanussi. The Sanussi movement affected Algeria, Egypt and Arabia, but took deepest root in Libya where it played a leading part in the Arab struggle against Italian penetration after the withdrawal of Ottoman Turkish rulers in the period 1912–18. Following the Second World War, the then exiled Sanussi leader, the Emir Idris, was recognized as head of the Libyan people and, with the attainment of Libyan independence in 1951, ruled as king of Libya until the Revolution of 1969.

At the beginning of the twentieth century, the hold the Sufi orders exercised over the people was still strong, but as the century progressed their popularity declined with government action against them and the spread of secularist ideas. This process of change has greatly undermined the orders, and many have declined or virtually disappeared, though the Green Dervishes of Konya (Turkey) still have some political influence and there is a Dervish mosque in Damascus.

Wahhabism

Wahhabism began during the eighteenth century in the Arabian peninsula as a reaction to the popular practice of Sufism and its preoccupation with saints and sheikhs. The founder, Muhammad ibn Abd al-Wahab, and his followers called for the purification of Islam and a return to the primitive nature of the faith as preached by Muhammad. It was felt that there had been much backsliding. Many of the precepts of Islam, especially those relating to self-indulgence, were ignored and much superstition and unnecessary elaboration had crept into religious observance. The Wahhabi preached a return to austerity and simplicity.

At an early date, the Wahhabi formed a political alliance with the Al-Saud family who originated from Riyadh in the centre of the Arabian peninsula. The Wahhabis came into great prominence at the beginning of the twentieth century when the Saudis, under the inspired leadership of Abdul Aziz ibn Saud succeeded in conquering the Holy cities of Medina and Mecca. This compelled the rest of the Muslim world, if not accepting them, to give attention to Wahhabi doctrines. The present state of Saudi Arabia, the creation of King Abdul Aziz (Ibn Saud), remains a stronghold of Wahhabism and of Muslim conservatism.

Paganism

Older beliefs and practices have not entirely died out in the Middle East and other influences deriving from more primitive forms of religion such as fertility rites and cults, tend to occur in some rural areas alongside more orthodox religious observances. Two distinct communities practise religious observances in which both Christian and Zoroastrian borrowings can be discerned.

The Yazidi of the Jebel Sinjar region in north-west Iraq are sometimes spoken of as worshippers of Satan. However, far from being devil worshippers, their prime tenet is antidualist, denying the existence of evil, sin and the devil, and it is for this reason and not out of reverence, that they abhor the name of Satan. The Yazidi owe their survival as a separate religious community to centuries of persecution by both Kurds and Arabs, Sunni and Shi'a and to the difficult nature and isolation of the terrain they inhabit. However, what persecution failed to achieve is being rapidly realized by the modernization of Iraq. At the beginning of the twentieth century there were 150,000 Yazidi and today there are probably no more than 30,000.

The Mandeans or Sabians live in the marshes of lower Mesopotamia. About one-fifth of the community lives in the Iranian districts of Ahwaz and Khurramshahr, the rest in Iraq. Their beliefs are derived from the gnostic sects which sprang from the dying struggles of paganism with advancing Christianity, blended with some Zoroastrian elements. Their distinction seen from the West results from their discipleship of St John the Baptist and for their strict practice of baptism in flowing water. Unlike the Yazidi, the Mandeans were not persecuted by their Muslim neighbours, and like the Christians and Jews, they are mentioned in the Koran as 'people of the book'. They are famous as boat-builders in their native marshes and, in the suqs of Amara, Baghdad and Basra, as Iraq's finest silversmiths. Today, the Mandeans number no more than about 20,000.

Religion and modern society

The origin and character of certain religious communities has been described at some length, because many features of social and political life in the modern Middle East derive from religious matters. Following conquest of much of the region by the Osmanli Turks during the fifteenth and sixteenth centuries, the *millet* system was adopted as the basis of civil administration in the Turkish Empire. This system merely institutionalized and refined an arrangement which had existed since the rise of Islam.

A millet was a separate religious community with a leader who was recognized by the Ottoman Sultan as having important religious and civil functions. This arose from the fact that, owing to the view that Islam was both a religious creed and a form of civil government, the status of non-Muslims was dubious and had to be regulated in a special manner. Each head of a millet was a member of the local provincial administrative body, with right of direct access to the Sultan, and was, in addition, permitted to maintain a kind of law court of his own, which supervised such matters as marriage and dowries and even the inheritance of property among his co-religionists. A most important feature was the fact that non-Muslim communities were in some cases allowed to operate their own code of law, even when this differed from the official law of the Ottoman Empire.

A feature deriving from the millet system has been the emphasis given to religion as a basis of political grouping. Although millets ended with the fall of the Ottoman Empire in 1918, the habit of associating politics with religion still persists in the Middle East. Insistence on sectarian differences as a basis of political grouping has tended to produce an atmosphere of strife and conflict, and the restlessness, faction and extreme individualism that characterize many present-day political affairs hinder co-operation, and hence retard the development of a stable form of government.

The influence of religion on political organization still persists. The Lebanon is an extreme example. Until 1975, the structure of government itself was organized on a religious basis. There was an understanding that the president should be a Maronite and the prime minister usually a Sunni Muslim, with the other government portfolios and parliamentary seats allocated on a *pro rata* basis per religious group, so that the religion mattered more than the individual cabinet minister himself.

Other instances occur, especially in Israel where the relationship between religion and the state is particularly complex. An important factor is the existence of religious political parties of which the largest, the National Religious Party, has been a member of most governing coalitions since the state was founded in 1948. The question of just how much Judaism there should be in a modern Jewish state remains a critical and as yet unresolved issue.

Despite these examples to the contrary, the present age is of sharp religious transition and reappraisal, and one in which the influence of Western imperialism and rising nationalism has been profound. Different views regarding the role of religion and the state have emerged. In some states, notably Turkey and Egypt, nationalism is tending to replace religion as the principal socially cohesive force. Westernizing reforms have been imposed, often ruthlessly, including the nationalizing of *waqf* revenues and the introduction of modern Western-style legal and educational systems. Increasingly, identity is defined and loyalty claimed on national rather than communal lines, while criticisms and aspirations are expressed not in religious, but in secular terms. In contrast, in certain countries, religious feeling is still the mainspring of rule. In Saudi Arabia the political authority of the ruler traditionally derives from his alliance with the Wahhabi movement.

The drive for reform and for change in the Middle East is widespread. Some believe that this can only be achieved through a rejection of the West and a return to the fundamentals of Islam, while for others a secularization of society represents the only way forward. Yet, whatever the future, Islam remains a force in the Middle East. It can still be a powerful rallying cry and an attack on God and religion is likely to arouse the anger and resistance of the Muslim masses. Even those Muslims who have rejected Islam often retain Islamic habits and attitudes and new values and ideologies are sometimes merely an artificial and superficial veneer. Moreover, with the manifest limitations of present-day secular forms of government, there would seem to be a very marked and large-scale turning towards and revival of religion, especially Islam, as a political force. An appreciation of Islam is still very necessary in order to understand the contemporary Middle East and its peoples.

Demography

In the Middle East, apart from the variations among peoples, the population situation itself is of critical importance, partly because of the very unequal distribution between densely occupied areas where pressure is acute and areas still undeveloped through scanty population, and partly because of the high rates of population growth, amounting in some regions to over 3 per cent per annum. Moreover, because of differing ethnic or cultural groups, the location of population numbers can have the most important political implications, as in Israel, Iraq and Sudan. In the UAE and certain of the other Gulf States, expatriates outnumber the indigenous populations.

A further problem in the Middle East is that as a result of census limitations, details of populations are not always available. No census exists for Lebanon and Sudan and in many countries the latest census dates from the middle 1980s, for example in Egypt (1986), Israel (1985), Kuwait (1985) and Libya (1984). In the case of the UAE, the latest figures officially promulgated are for 1980. The most recent census figures available are for 1994 and are published by Jordan, Syria and Yemen. Thus, population details must, in almost every case, be estimates. Given many of the upheavals in the Middle East over the last twenty years, the reliability of such estimates must be queried. Statistics for Kuwait, Iraq, Iran and Sudan are all questionable.

Population distribution

There are considerable areas, for example the central *Kavirs* of Iran, the Rub' al Khali of Arabia, the Saharan deserts of Egypt, Libya and Sudan, and the south-centre of Sudan, where populations are non-existent or at most tiny in numbers. On the other hand, in certain of the larger and well-watered alluvial river valleys, population densities are high, with the Nile Valley of Egypt by far the most heavily peopled. Moreover, even in the desert areas there are oasis settlements that are densely populated, sometimes by an agricultural population, sometimes by mixed groups of cultivators and oil workers. Regional imbalances are thus the principal distinguishing feature, with, for example, over 99 per cent of Egyptians living in the 4.5 per cent area comprising the Nile Valley and delta, over half of the population on about 14–16 per cent of the area in Jordan and Iraq, and almost 60 per cent of the people of Israel on 10 per cent of the area. Adequate rainfall is a principal though not total influencing factor. This is well seen in Asia Minor and along the Caspian coast of Iran, but an instance to the contrary is demonstrated by the relative emptiness of the well-watered southern zone of Sudan. It is obvious also that irrigation potential as distinct from natural rainfall plays a considerable part in conditioning population distribution.

Population factors

There are considerable differences both in densities and in the absolute size of populations. By the standards of the Middle East, the populations in millions of Iran (67.5), Egypt (64.8) and Turkey (63.5) are large. By any standards, the populations of Bahrain (0.6), Qatar (0.7), Oman (2.2), Kuwait (2.0) and the UAE (2.2) are extremely small.

In the 1970s, birth rates ranged in general between 39 and 50 per 1,000 of the population. This was high, paralleling similar occurrences elsewhere in the developing world. By 1997, birth rates ranged from approximately 20 to 45 per 1,000. The highest were recorded in Yemen (44.8), Libya (43.9), Iraq (42.5) and Sudan (40.5). Of the Gulf oil states, Saudi Arabia and Oman, both with birth rates of 37.9, were easily the highest. The figures for Qatar (17.3) and the UAE (18.5) were the lowest in the Middle East. Thus, the key to the population issue, the birth rate, is showing quite clear signs of decline. Egypt, Turkey and Jordan have all introduced policies aimed at reducing the size of families. At the other extreme, Iraq, Libya and Saudi Arabia have been encouraging large families.

Other issues affecting the pattern of reproduction include the fact that there is an almost universal emphasis upon marriage and the rapid production of a family. Although Islam does not prohibit family planning, there is a general unwillingness to accept contraception on any wide scale and Western-style family planning programmes are in their infancy. In certain parts of the region, such as Yemen, the fact that a large number of men work overseas delays the age of marriage and this may reduce family size. However, in the poorer parts of the region in particular, family welfare depends in no small measure upon child labour and the number of children produced is very important. Furthermore, there is great value in having numerous adult offspring to care for the older members of the family.

Until relatively recently, death rates were also high, but as elsewhere in the world, have been greatly susceptible to improvements in public health and in social conditions. These include enhanced nutrition through better living standards, more widely available medical care, in some countries highly subsidized by the state and, perhaps above all, improved public health measures. Such developments as the chlorination of drinking water, the eradication of disease vectors by modern insecticides and the use of antibiotics together with parallel improvements in personal medical care, hygiene and education have caused reductions in death rates.

In the 1970s, the death rate per thousand ranged from 20 to somewhere below 15 throughout the region. In 1997, the highest death rate was estimated to be in Sudan (11.2), followed by Yemen (9.2) and Egypt (8.6).

At the other extreme were Kuwait (2.2), UAE (3.0), Bahrain (3.3) and Qatar (3.5). Thus, for many countries in the region the demographic pattern follows closely that which has led to population stability in Western Europe. However, the total fertility rate, the number of children per woman, remains rather higher than the Western average. The highest in the region is Yemen (7.2), followed by Saudi Arabia (6.4), Iraq (6.3), Libya (6.2) and Oman (6.0).

The effects of these demographic factors are seen in the figures for annual growth rates which in the 1970s averaged about 2.9 per cent per annum. This implied that the populations doubled in less than 30 years. While there has been, for most countries in the Middle East, a significant lowering in growth rates, there remain some very high figures. Turkey (1.6) and Egypt (1.9) are two countries which have successfully slowed their growth rates, while Lebanon (1.6) and the UAE (1.8) also have low figures. In contrast, the annual population growth of Kuwait is 5.7 per cent per annum and high figures were also recorded in the West Bank (4.3 per cent), Qatar (4.0 per cent), Yemen (3.6 per cent) and Saudi Arabia (3.4 per cent).

Two other significant factors related to population are infant mortality and life expectancy. Infant mortality, the number of deaths per 1,000 births, has dropped throughout the region, but there are still some stubbornly high figures, most notably for Sudan (74.3), Egypt (71), Yemen (68.1), Libya (57.7) and Iraq (57.5). These may be contrasted with the infant mortality in Israel which is 8.3 per 1,000 births and low figures for Kuwait (10.6), UAE (15.5) and Bahrain (16.4).

As life expectancy increases, this affects not only the total population, but also the population structure. Life expectancy, expressed in years from birth, has risen everywhere, but still remains 55.5 in Sudan. The next lowest in the region are Yemen (60.3) and Egypt (61.8). In contrast, many countries have Western-style life expectancy statistics. These include Israel (78.2), Kuwait (76.2), UAE (74.6), Bahrain (74.6) and Qatar (73.6).

Overall, therefore, demographic statistics in the Middle East, with the principal exception of Sudan and to a lesser extent, Turkey and Yemen, are fast approaching Western rates. If demographic transition has not occurred, this is frequently because of deliberate policy. However, some countries, most particularly the UAE, Qatar and Kuwait, owe a substantial part of their population and indeed its increase to immigration. Thus, a distinction must be made between natural growth and that due to in-migration.

Another aspect of Middle Eastern demographic statistics with both social and economic importance is the percentage of the population which is 14 years old or under. Of the eighteen states (together with Palestine) which comprise the Middle East, ten have 40 per cent or more of their population under 14. The only countries with significantly lower percentages are Israel (28), Qatar (28), Lebanon (30) and Bahrain (31). Thus, there is what might be termed 'youth pressure' and there must be, as a result, an emphasis upon education. However, as education develops there must also be employment at suitable levels. Labour requirements in the agriculture sector are being increasing reduced and industrialization has produced its own problems. Modern production techniques have militated against craft industries and, in many of the countries, the service sector is still being developed. One answer for many of the countries may be tourism, but while helping to solve economic problems, tourism may exacerbate social ills. With regard to higher education, in some countries the effect of over-production of college and university graduates has unduly raised expectations in a society that cannot absorb the expertise which it has trained. These are all issues which are likely to increase in severity in the future.

In contrast, throughout the Middle East the percentage of the population which is 65 years old and over is low. The problems of geriatric pressure increasingly afflicting Western Europe are some generations away from occurring in the Middle East. The population pyramid which most approaches the Western European mean is that of Israel, with 28 per cent of its population under 14 and 10 per cent over 65. Elsewhere throughout the region, the percentage of people over 65 ranges from 2 to 4 per cent, with the exceptions of Lebanon and Turkey which both have a figure of 6 per cent.

Migration

Besides internal movement which is chiefly from rural to urban areas, from poorer areas to oil-rich territories and from war-torn regions, there is a fairly considerable outward migration. Yemen has traditionally supplied not only nearby regions of the Middle East such as Saudi Arabia, the Gulf states, Iraq, Jordan and Syria, together with Eritrea and Somalia, but also further afield including the East Indies, East Africa, Pakistan and India. The emigrants mostly work as labourers, traders and seamen, with a minority as officials. Lebanon and Syria were also traditional sources of emigrants. However, the Gulf War resulted in the forced return of some one million Yemenis from Saudi Arabia and the recurrent crises within the region have severely affected the migration patterns.

Another feature has been the movement of well-educated Arabs to other Arab countries, where they fill professional and technical posts. Many Egyptian teachers staff schools in Libya, Saudi Arabia and the Persian–Arabian Gulf states. Oil companies recruit throughout the Middle East and businessmen from Syria and Egypt are widely scattered. Palestinians find employment in professional occupations and administration, and as skilled workmen in many Arab countries. It should be remembered that two major displacements of Palestinian Arabs occurred in 1948 and 1967.

The establishment of the Israeli state also led to considerable in-migration of Jews from other parts of the Middle East, from Iraq and Yemen especially, but overall involving some 90 per cent of Middle Eastern Jews. Total annual immigration into Israel has varied considerably, the largest single number in one year being 239,000 (1949).

While several areas of the Middle East have specialized in supplying migrants, the Gulf states and Saudi Arabia, with a desperate need for labour, have attracted migrants. A small percentage of these are professional expatriates, principally from North America, Western Europe and Japan, but the bulk provide the middle rank clerical staff and most of the manual workers. These were originally drawn from Palestine, Philippines and the Indian subcontinent. Following the Gulf War, recruitment has been focused upon the Muslim populations of the Indian subcontinent. The result is that all the Gulf states have a relatively high proportion of expatriates in their population, while in the cases of the UAE and Qatar, this proportion exceeds the total of the indigenous population. Although the expatriates have been responsible for constructing the infrastructure of these states, their presence on such a scale can lead easily to social and economic problems.

Current inward migration figures per thousand of population are dominated by Kuwait (42.3), Qatar (26.6) and the West Bank (10). In the case of the UAE, the figure is 2.5 and for Saudi Arabia, 1.5. The major losses have been sustained by Iran (−4.9) and Jordan (−6.1). Interestingly, the inward migration figure for Israel is +6.1.

Summary

The considerable anomalies in distribution, density and living standards that now clearly exist between various human groups in the Middle East invite the question of how far more positive action to spread population and amenity would be desirable and feasible. There are certain obvious areas where population pressure is great, as in the Nile Valley, and in parts of Asia Minor, but equally, there are areas, such as the Mesopotamian lowlands of Iraq, and less certainly, parts of Iran and the coastal uplands of Libya, that are undercultivated. There are the phenomena of already highly urbanized areas drawing even larger numbers most of all into the primate cities, and then into rather smaller urban centres, with all the strains on infrastructure and provision of social benefits that this rapid urban growth entails. Furthermore, there are over 2.5 million refugees, which in a region where several states with populations of less than one million exist, represents a problem that would be extremely acute for any state, but is desperate in the context of the political pattern of the Middle East.

There is also wealth, mostly from oil, but even its redistribution, were this to be achieved, seems unlikely to be able to solve the population problems of the Middle East. Changes are occurring with urbanization and industrialization, but as problems of population growth rates are lessened, they are replaced by other issues such as youth pressure.

8   Society

For centuries human society in the Middle East has been divided into three different types of community: nomads or *Bedouin*, settled cultivators or *Hadhar* and townspeople. Through history, writers and travellers have been impressed by the contrasts in ways of life and social organizations shown by these three groups and the inter-relationships between them. All three communities have of course evolved over time, but it is only since the dawn of the oil era and particularly from the 1960s onwards that the rate of change has been exponential. Nor do these changes apply only to the oil-rich states; the effects of oil wealth have been felt throughout the region.

For the remaining Bedouin (Figure 8.1), apart from Sudan a maximum of 5 per cent of the population in any state, oil has had profound effects, most obviously in the reduction of people leading a nomadic life. For the settled cultivators change in many parts of the region has been more modest, but in many of the oil-rich states and particularly those of the Gulf, the rural area is effectively obsolete. Food can be imported on a large scale and the dependence of town upon village has ceased. Worries about food security have witnessed the growth of agri-business rather than the rejuvenation of villages. For the townspeople, life has in many ways changed immeasurably with the provision of modern services and infrastructures. Everywhere, but most notably in the Gulf states, there has been vastly increased urbanization and with it have come the economic and social strains commonly encountered in similar circumstances elsewhere.

Rapid as these changes have been, they have lasted only a few decades and have not in any sense obliterated the effects of the threefold pattern of life set over time. Indeed, they have not, as yet, obliterated the pattern itself. In that human society in the Middle East has been founded upon the three different types of community, the contribution of each needs to be examined. Thus, with a few additions to illustrate modernity, the three lifestyles in their traditional forms are described. An added advantage of this approach is that what are some of the more interesting sections in *The Middle East* (Fisher 1978) can be retained.

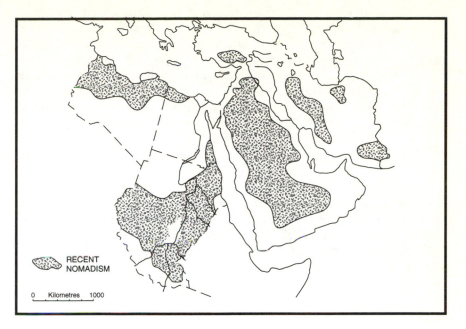

Figure 8.1 Middle East: areas of recent nomadism.

The Bedouin

The main feature of Bedouin life is regular movement in search of pasture for animals. In the main, this movement is from one district to another and is therefore true nomadism, but in mountain regions transhumance occurs and different height levels in the same district are occupied successively. Nomadism is, in effect, horizontal movement, while transhumance involves a change in altitude. The Bedouin of Arabia may be cited as true nomads, whilst certain of the Kurds of Anatolia and the western Zagros tend rather to practise transhumance.

With rainfall and therefore pasture both scanty in amount and liable to much variation from one year to another, the nomads have found that experience provides the best means of survival. Thus, to the oldest members of the group is given the task of making decisions which guide the activities of the entire community. In the face of sudden crisis, such as the failure of wells or an attack from outsiders, rapid decision is called for and one man alone rather than a group is looked upon as supreme leader. This type of social organization, termed patriarchal, places the life of a community under the control of a chosen man and is highly characteristic of nomadic pastoralists. The chief, or sheikh of the community, has considerable personal power, which is tempered only by precedent, the collective experience of the group as a whole, and the opinion of older members of the tribe. There is little place for individuality or for innovation and patriarchal communities have kept their way of life largely unaltered through many centuries.

A clear limit to the size of social groupings amongst nomads is set by natural resources, which in a given district can support only a relatively small number of people. The unit amongst Bedouin is the tribe, a group large enough to profit by advantage of numbers, yet small enough to exist under desert conditions. Family relationship is strong and perpetuated by emphasis on intermarriage within the group. Yet although tribal solidarity and discipline are conspicuous features, it has been the weakness of Bedouin society throughout the ages that political combinations larger than the tribe tend to have been achieved only for a short time.

An outstanding leader has brought together a number of tribes into one larger unit, but union has proved temporary and more often than not, on the death of the leader his organization has dissolved again into smaller groups. Large-scale political organization is not, therefore, a feature of Bedouin society and resulting intertribal feuds and political instability have had an effect not merely on life in the desert and steppe, but also on that of the settled lands of the arid borders.

To each tribe belong certain rights of pasture and occupation. The limits of each tribal territory are carefully defined and generally comprise a summer and a winter camping ground. The exact size of territory usually has a certain relation to the physical force and prestige commanded by each tribe and limits, therefore, vary from one period to another. Weaker tribes seek the protection of stronger tribes.

Environmental conditions within desert areas vary considerably, with a resulting variation in influence on the way of life of the inhabitants. For example, the northern part of the Arabian desert is a vast open tableland, dissected into occasional shallow valleys or closed basins, whilst further south, topography ranges from a succession of jagged uplands, lava-fields and highly eroded basins and valleys, to a more subdued relief eastwards. Climate, especially rainfall, also shows significant variation and in western Arabia there can be sufficient for sporadic cultivation without irrigation.

Human response to these conditions is extremely varied in character. Near the Euphrates, relatively short-distance movement with sheep- and goat-rearing is the rule, with an approach to transhumance rather than full nomadism in the extreme north on the edge of the mountain rim of Turkey–Kurdistan. Further south, in the open zones between Damascus, Baghdad and Jordan, more extensive migration based on the use of the camel is characteristic. In Arabia proper, there are all stages from extensive movement, involving camels as the basis of the economy, to partial nomadism only, with an approach to cultivation and semi-settlement. A similar state of affairs exists in the deserts of Libya, while nomads of Iran may sometimes cover 2,000 km in an annual movement between the interior deserts and the Elburz uplands. On the Jebel Akhdar of Cyrenaica, nomads have a complicated annual routine involving a double seasonal movement to and from the uplands with an approach to cultivation. Further south, movement is around wells within the more arid zones, whilst the presence of oases allows the existence of completely settled agricultural communities.

Since camels can go for several days without watering, sheep four and cattle only one or two, it is possible to draw a summary distinction between camel-nomads, who usually cover great distances; shepherds whose territory may be more confined; and semi-nomadic or transhuman groups, often with cattle, who practise some degree of cultivation. This pattern is in constant fluctuation, partly as the result of climatic vagaries and also with changes in numbers and influence. Where tribal territories cross international frontiers, special problems arise and whilst agreements give to nomads temporary nationality of the country in which they find themselves, this arrangement is increasingly less agreeable to the governments involved. However, the recent boundary agreement (1990) between Saudi Arabia and Oman specified a porous boundary with a grazing zone stretching 20 km either side for tribal grazing land.

Because of the necessity for constant movement, the material culture of the Bedouin is limited. The chief possession of the tribesman, after his animals, is a tent, usually black or brown and woven of camel or goat hair. When in good repair, these tents are quite waterproof and the size of the tent, shown by the number of tent poles used, is an indication of the affluence and social standing of the owner.

The clothing of the Bedouin is simple. A long robe of thick material is the principal garment, with in winter, a waterproof cloak of woven camel hair. A voluminous headcloth, held in place by a rope or band, is wound round the head so as to give protection to the face and neck, and rawhide, heel-less sandals are worn on the feet.

Water is usually too precious for much washing and indeed, the Koran permits the ceremonial use of sand in the daily ablutions which are enjoined on all Muslims. Bedouin food is monotonous and scanty. Most important are milk products: curds, buttermilk, various kinds of cheese, of which *labne*, a kind of cream cheese, is the most widespread, and *samne*, or butter. In addition, wheat and barley and occasionally a little rice are obtained from agriculturalists, or else are grown very sporadically by the nomads themselves. Small amounts of dried fruit, usually dates, are also eaten. Meat is provided only as a great luxury, since animals themselves are in effect a kind of fixed capital. Owners must live from the yield, not upon the animal itself. Apart from occasions of high festival, when special slaughtering takes place, only those animals that die naturally are eaten. In general, the standard of nutrition is low and, as a result, the Bedouin are small and lightly built, though their physical powers and endurance are great. Life is very hard and by 40, particularly in the case of women, old age has begun. Fifty years is a long life for a nomad.

it is a bitter, desiccated land which knows nothing of gentleness or ease. Yet men have lived there since earliest times . . . men lived there because it is the world into which they were born; the life they lead is the life their forefathers led before them; they accept hardships and privations; they know no other way.

(Sir Wilfred Thesiger 1959, Prologue to *Arabian Sands*)

One means of supplementing the deficiencies in arid areas is by raiding, although this is now very rare. Amongst the Bedouin the *ghazu* is a recognized activity, amounting almost to sport, but it also fulfils a definite economic function. The effects of a bad season, as when rains have failed in the desert, are mitigated. The Bedouin have a choice of taking what they require by force from other people, or of starving. The word 'sport' has been used because although fundamentally a serious matter, Bedouin raids are conducted under certain conventions, almost rules. Rapid, unlooked-for coups, using cunning and guile are most favoured and bloodshed is as far as possible avoided, although goods and formerly women and children may have become the property of the victors. Successful leadership in tribal raids is the means by which personal reputation and power are built up. Bedouin raids may be directed against other tribes, but frequently toll is levied on agricultural settlements on the fringe of the desert. Less able to resist, and with more at stake than the nomad, settled cultivators sometimes remain under permanent tribute to Bedouin tribes and supply a quantity of needed foodstuffs to their overlords. The name *khaoua*, meaning tribute of friendship, is applied to such transactions. Raiding has now greatly declined, because of increasing control by local governments, but in addition to the political problem presented in controlling outbreaks, there is the economic factor of providing an alternative occupation to ensure a livelihood.

Strict discipline, necessary both in everyday routine and because of liability to attack, has left its mark on Bedouin ways of thought. In religion, there is little room for compromise or doubt and a strong, vivid and intolerant faith is held. Similarly, there is a strict code of behaviour. Hospitality, in an environment without fixed routes or settlements, is a highly regarded virtue, and ordinarily social intercourse has been developed into an elaborate code of manners and conduct. In view of the vagaries of human existence amongst the Bedouin, it is hardly surprising that superstition and fatalism are also strongly marked. Men, animals and even motor vehicles carry charms. Possibly as a result of reduced physical strength, and their preoccupation with household tasks, the status of women is relatively low, although they are not veiled, like many of the women of the settled areas.

Nomadism is a special response to environment, by which the frontier of human occupation is pushed further within a region of increasing difficulty. In the past, nomadism has often been regarded as an unsatisfactory alternative to agriculture. It was believed that primitive man passed from a life of hunting and collecting first into a stage of pastoralism and later, in places where conditions were favourable, to the fullest development of a life based on cultivation. In contrast, there is the view that the three different types of community are mutually dependent, each being necessary for the maintenance of the others. However, the benefits of any interdependence were unequally divided and urban dominance is central to the concept of the ecological trilogy. Furthermore, recent changes have disturbed the equilibrium even further and the forces of modernization are producing new patterns and new relationships.

In certain instances, it is clear that pastoralism may represent a development from cultivation, a dynamic response to environmental and social conditions, and not necessarily a regression or halfway stage between hunting and gathering and settled cultivation.

At a particular period of history or in a specific locality, it is possible to observe the shift towards nomadic pastoralism in response to various pressures. At the present time, however, the tendencies are strongly in the opposite direction, towards the sedentarization of nomads. This is apparent in almost all areas, due to the 'pull' of increased opportunities of employment in towns and higher urban standards, and the 'push' by reduction of natural grazing grounds as these are taken over for other purposes, and the pressure by governments which often find nomads unsatisfactory and unreliable citizens. In the recent past, forcible measures have been taken, for example in Iran and Turkey, to settle nomadic tribesmen, but it was often found that poverty and disease increased to such an extent as to bring about the virtual extinction of some communities.

Despite these considerable changes, it can be argued that pastoral nomadism has a continuing part to play, though of greatly reduced significance. This is because, in certain localities, pastoralism is still the only possible means of land evaluation. Indeed, the Bedouin lifestyle may represent the most effective and efficient form of development in which a stable environment can be retained. The introduction of cultivation, in such instances, may result in soil erosion, land degradation and eventually land abandonment. In such cases where cultivation may not be possible, or other resources are non-existent, semi-arid lands can produce at least a proportion of the wool, milk and meat now increasingly required in towns. To achieve this, an enlightened policy is necessary, including: the provision of rural credit, grading and marketing schemes for animals, the development of co-operatives, the provision of health education and veterinary services, and above all an acceptance by the central government that nomads are not to be regarded as second class, or even potentially dangerous citizens.

The traditional Bedouin lifestyle is clearly in decline and it is likely that the provision of education and other services will hasten the process. Certainly, it will change the basic migratory patterns. However, many of the Middle Eastern governments are aware of the need to conserve the Bedouin, because they alone can operate an economy which is in dynamic equilibrium with the harsher landscapes. Furthermore, underlying government support is the awareness of the psychological significance of the Bedouin. In many parts of the Middle East, particularly in the Arabian Peninsula, the Bedouin are viewed as the only true embodiment of the national spirit.

Bedouin now number perhaps four million, predominantly in Turkey, Iran and the Arabian Peninsula, with possibly as many again in Sudan. Virtually all have adapted in some way to the modern world if only through the use of plastic domestic utensils, nylon rope and other pieces of household equipment. In some ways the most obvious change is through the introduction of the motor vehicle. The Toyota pick-up is now more influential than the camel and indeed camels can be seen loaded into pick-ups for the journey to the grazing areas. A Bedouin tent may now contain an array of electronic equipment including a television and a portable telephone. Thus, although the Bedouin may continue to live in a harsh environment, their lifestyle has been changed to accommodate many of the facilities of modern living.

Villages and cultivators

The decline in nomadism, especially since the late nineteenth century, has led to an increase in the settled rural population in the Middle East, and today, despite recent rural to urban migration, well over half of the region's population live in village communities, the vast majority as cultivators.

From time immemorial the village has been the typical form of rural settlement in the Middle East, and there are few isolated farms and farmhouses. Dispersed settlement is impractical in many parts of the region because of the scarcity of water, the primary determinant in the distribution of population and the location of villages. Lack of security and protection from attack and the common practice of collective ownership of land, where lands were held in the name of the whole village, were also factors favouring nucleation.

Other influences have contributed to the survival of this pattern of settlement. The big landowners and their agents could supervise and control the tenants and sharecroppers more easily and effectively in nucleated villages than in scattered farmsteads. A communal pattern of agriculture and co-operative forms of production long prevailed in some parts of the region, especially in Iran, with open fields for grazing and frequent redistribution of holdings by the landlords. Government taxes were levied on the village as a unit and not on individuals. In some areas, notably in Egypt, where cultivable land is in short supply, construction beyond the limits of the village was actually forbidden. Over the centuries, social customs developed in such an environment have become obstacles to any change to a more dispersed pattern.

Most Middle Eastern villages are small, the average size being about 400–500 people. However, where population densities are unusually high, the villages are larger and more densely distributed. Villages in Egypt, for example, are generally larger than those in other Middle Eastern countries and village population size ranges from 300 to 20,000 people; in contrast, the majority of Iranian villages have an average population of only 160 inhabitants.

In the plains and along the major river valleys, settlements are compact clusters of low dwellings normally built of mud-brick and separated from one another by narrow, winding alleys and unpaved streets. They are situated on the least productive plots of ground, on rocky outcroppings, or along the edge of the desert and the sown land, in order to avoid wasting valuable agricultural land. The traditional type of dwelling usually consists of a number of dark rooms built around an open courtyard which the family sometimes shares with its animals. Most homes are meagrely furnished one-storey structures. Many have no furniture except mats made of straw or blankets which, together with a few cooking and eating utensils, comprise most of the possessions. Although the kerosene stove has appeared widely since the late 1960s, water may have to be carried considerable distances and is thus too precious for much washing. Dried animal dung may still be used as fuel for cooking and the family may keep warm in winter from a pan of hot charcoal. In the mountains, villages are usually less compact and often comprise a number of loose clusters of dwellings with stone replacing mud-brick as the main building material. Courtyards are a less common feature, because of the shortage of land suitable for building, and houses sometimes have a second storey.

Few amenities are found in the villages beyond the mosque, a bathhouse and a few shops, usually located around a central square which functions as a meeting place for all the inhabitants and is sometimes the site of the weekly market. Some villages also have one or more guesthouses which act as a convenient meeting place for the men. The potentially monotonous appearance of the village is only broken by the variety of building materials and external decoration employed, and particularly by the different roof types found. Where a single building material is used, however, one village looks very much like another.

Until the last thirty years, the tone and pattern of village life in the Middle East had changed little for centuries. Poor health, high infant mortality rates, illiteracy, oppressive forms of tenancy, crippling debts through high interest rates, poverty and low standards of living were widespread. The majority of villagers lived in ignorance and isolation, hunger and servility, beneath the sway of the big landowners.

Rents and taxes often absorbed as much as five-sixths of the total produce of a holding so that the tenants were forced to turn to the money-lenders who might demand rates of between 50 and 200 per cent, if not more. Traditionally, the right to profit from agriculture rested on provision of five elements: land, water, seed, implements (including animals) and labour, each rating sometimes 20 per cent. Thus, a landless tenant cultivator might be regarded as entitled to as little as 20 per cent if he supplied only his own labour. Because of poor communications all the produce of one district tended to find its way to a single market where in times of plenty the price fell because of a glut, and high prices occurred only when crops failed. Under such conditions, monopoly control by middle men could easily take place.

Today, although many villagers still live near the margins of subsistence, rapid, irreversible and fundamental changes are occurring in the rural areas, changes which are usually for the better, but not always. Major road-building programmes have improved communications between many villages, the main towns, and smaller administrative and market centres. Journeys which once took many hours, even a whole day, along poor tracks, often impassable for part of the year, can now be undertaken in a fraction of the time by car, taxi or bus along metalled roads. A growing number of the villages now have piped drinking water and many also have electricity. Medical centres of various kinds have been established in rural areas and the health of the villagers is improving, if only slowly. Since the Second World War, many new state elementary schools have been opened in rural areas. Much progress has been achieved and the attitude of the villagers to public education is changing with growing awareness of the importance of formal education in occupational success.

Due to the influence of the mass media, especially the transistor radio, battery-operated and most recently clockwork, and lately the television, the horizons of the villagers are beginning to expand beyond their own small community. The radio can be seen everywhere in the villages and it is particularly effective because it overcomes the barrier of illiteracy and it can be considered a fundamental in the process of change. It brings the villagers information about markets, about life in the towns and about events in other countries, and it has been used by the governments of the region not only for political and ideological communication, but also for promoting various social and welfare programmes such as family planning. The radio has brought new ideas and an increased political awareness among the villagers. Television is becoming equally prominent and in the future its influence on attitudes and behaviour will be profound.

New crops, improved agricultural technology and new techniques of cultivation are modifying the agricultural bases of village life. In most countries, land reclamation and irrigation schemes both large and small are increasing the cultivated area and permitting more intensive cropping. In many areas, cultivation is no longer merely for subsistence. Foodstuffs, particularly wheat and barley, still remain the most important crops for the majority of villages, but the cultivation of cash crops, for example, cotton, vegetables and fruits, for export has greatly expanded. With the introduction of fertilizers, insecticides and improved equipment, the villagers are now able to achieve much higher yields for their crops. Following major land reform programmes and the breakup of multi-village estates, some cultivators have become landowners for the first time, although their holdings are often very small. Co-operatives have been set up to provide credit and technical advice, to arrange for the marketing of agricultural produce and in some cases to supervise farm production.

In some cases, agrarian reform has merely aggravated existing social divisions, resulting from the introduction of money and market principles and produced new tensions in rural society. A modest improvement in the income of the small landowners, especially the beneficiaries of land reform, has occurred and tenant farmers also appear to have made some gains from the reforms, whether they operate under cash rents or as sharecroppers. The most unfortunate group remains the landless labourers who have often suffered a loss of income and employment as a result of the reforms. Among the landless, the position of casual labourers, hired for short periods often under extremely oppressive contracting arrangements, is particularly acute.

Amid these changes, kinship ties and religious observances remain important among the cultivators. The family: husband and wife, their married and unmarried sons and unmarried daughters, functioning within the clan structure, is still the basic social and economic unit. Obedience and respect to father and elder brother as well as to clan elders led by the clan head and loyalty to close relations and clansmen are basic rules of social life. In villages of Muslims and most Christian sects, marriage within the clan is generally preferred, and in many places there is a traditional hostility between different clans of a single village. Disputes over land, water or the appointment of the village headman can occur with sometimes violent consequences. Indeed, it is generally recognized that loyalty to clan and religious sect has served to undermine the unity of the Middle Eastern village community.

However, changes are certainly taking place. New ideas and resources from the outside are altering existing social relationships within the village. Common residence among kinsmen, although still widespread, is no longer the dominant pattern everywhere, and the number of nuclear family residences is increasing. Fathers now have decreasing control over their sons, the young are generally less deferential and less obedient to the older members of the village and women are less submissive to their menfolk. The appearance of political parties has also begun to undermine the system of kinship solidarity and rivalry. In contrast, in a number of important areas, marriage, childbearing and religion, there is considerable evidence that change is resisted.

In spite of a number of very real material improvements in the lives of the cultivators, the gulf between rural areas and the town is still very wide. At the same time, growing contact with the world outside the village has given rise to aspirations and expectations among the cultivators that the limited opportunities available within the village cannot fulfil. Indeed, in most villages the falling death rate is causing an increase in the number of mouths to be fed, in people to share land resources and in hands available for work. One result has been the acceleration of labour migration to nearby towns, to neighbouring countries and even to the West, a process which in itself has become one of the most important factors in change in the Middle Eastern village. The remittances which the migrants send back to their villages are increasingly vital in the support of those who remain at home. Some migrants eventually return to the village, but many others settle permanently in the towns, where they are joined by their wives and children. Links with their native village are nevertheless maintained and regular visits are customary, reinforcing and increasing the villagers' knowledge and experience of urban life and in this way contributing to further permanent migration.

For many of the countries, however, it has been the draw of the oilfields and oil-related industries which have most profoundly affected village life. The more ambitious and fitter members of village society migrate and, while the village benefits from the remittances, it loses in other ways. In many Gulf state villages, the cultivation is carried out at least in part by expatriates from the Indian subcontinent. As this process continues, there is a danger that village society becomes moribund. The population is increasingly dominated by the old and the infirm and even they are excluded from the economy as traditional crafts wither in the face of the onslaught of Western-style materials, purchased with remittances. For example, when modern household equipment can be purchased at the urban supermarket and brought to the village, there is little future for many of the traditional crafts which formerly provided such equipment.

In a variety of ways the Middle Eastern village is becoming increasingly integrated into a national system of government and into a national society. State intervention has increased steadily in recent years and the growing importance of the central government and its changing relations with the rural areas is one of the major developments. If representatives do not live locally, there are regular visits from the police, party officials, agricultural officers, bank and co-operative officials and health workers. Local branches of national political and economic organizations have been set up in the villages. The teachers in the state elementary schools in the villages follow a national curriculum and bring national values to the children. In many states of the region all men must serve in the armed forces. Thus, the major forces that have set in motion the process of change have come from outside the village and consequently more progress is made for the villagers than is made with their participation. Government intervention, moreover, although aimed at reducing inequalities in the rural areas and narrowing the gap between town and countryside has, in fact, produced new tensions within the rural community.

Towns and townspeople

Urban civilization began in the Middle East and no other region in the world has had such a long and venerable tradition of urban life. At the height of the Islamic Empire such centres as Cairo, Baghdad and Damascus outshone any of their European rivals in the development and sophistication of their intellectual and artistic life.

Through its need for permanent settlement and security in which the exchange of goods can take place, commerce is generally regarded as a main contributing factor to the rise of urbanism. Situated at the junction of three continents and fringed by areas of sea or desert, the Middle East region has developed extensive trading relations, not only within itself, but more important, with China, India, Europe and, to a lesser extent, Africa. From earliest times commerce has played a conspicuous part in Middle Eastern affairs and the existence of a merchant community with special outlook and interests has had considerable repercussions on cultural development. Because of the unusual extent of commercial relations in many parts of the Middle East over many centuries, urban life showed exceptional development and the region contains some of the earliest and largest continuously inhabited town sites in the world.

An active and evolved urban life related to continent-wide trading has often, also, contrasted with conditions in the countryside where standards of living have for long been relatively low. Moreover, trading involves acquaintance with many different communities and ways of life. Hence, open-mindedness, adaptability and receptiveness to new ideas tend to be characteristic of commercial peoples, particularly in towns, where wealth accumulated and material progress generated considerable cultural development.

Besides commerce, other factors have contributed to the extraordinary development of urban life in the Middle East. First, because of deficiency or salinity of water, people have been forced to congregate in a few favoured spots where water supplies were available. Second, the course of Middle Eastern history reveals that time and time again, a vigorous yet small community has asserted political dominion over a large area, but because of numerical inferiority has been unable to undertake extensive colonization. Instead, a hold has been maintained on towns, from which an alien countryside has been ruled. The ancient Persians, Greeks and Romans, the early Arabs and the Turks all followed this plan, so that a tradition of rule and dominance has come to be characteristic of the towns.

Third, there is the religious aspect. The great religions of the Middle East spread from small beginnings and this spread has been easiest and most marked among urban populations, who by their receptiveness to outside ideas provided a favourable milieu for the propagation of new creeds. Among a conservative and isolated peasantry, propagation was slower. Even now Islam and Christianity as practised in remote rural areas are often very different from the same religions in towns. To a great extent, therefore, religious life has come to be associated chiefly with towns and cities, key examples being Jerusalem and Mecca. The growth of religious traditions has, in turn, further stimulated commercial activity, especially through pilgrimage.

It is interesting to observe that, with a basis of trading activity, administrative control and religious association, the cities of the region have been able to maintain an uninterrupted tradition over seven thousand years. Of course, the fortunes of individual cities have fluctuated over the centuries. Baghdad, with an estimated 1.5 million inhabitants in the ninth century, was devastated when it fell into the hands of the Mongols in the mid-thirteenth century. Under the Ottoman Turks, Istanbul experienced a remarkable revival after centuries of decline so that by the early sixteenth century it was the largest city in the Middle East and Europe, containing some 400,000 people. Throughout these vicissitudes, the importance of towns and the deeply rooted urban pattern continued. Political groupings have come to an end, but, in the Middle East, cities have outlasted empires.

During the last two or three decades, all parts of the Middle East have experienced a considerable increase in the absolute and relative numbers of town dwellers and rapid urbanization is one of the most striking developments in recent years. At the end of the last century, less than 10 per cent of the population lived in towns. Today, with the exception of areas such as Yemen, most countries in the region are at least one-third urban, while some countries are more than two-thirds urban. Furthermore, the great majority of urban dwellers are found in cities with more than 100,000 inhabitants. At the present time, Greater Cairo has a population of 15 million and Tehran 10 million. The population of Istanbul is 7 million and those of Ankara and Baghdad are both well in excess of 3 million. All of these are major conurbations.

The growth of many of these cities owes much to migration from the villages in the surrounding regions, as in the case of Ankara, Baghdad and Cairo; or to in-migration from other parts of the Middle East, as in the main towns of the Gulf states. Nevertheless, natural increase is also an important, and in may cases the predominant, component of urban growth. In the past, cities were centres of disease and high mortality. Today, mortality, especially infant mortality, is much lower than in rural areas as a result of better working and housing conditions, enormous improvements in sanitation and hygiene, and a marked concentration of medical facilities in urban centres. At the same time, fertility remains high, in some cases higher than in rural areas.

The gulf which has always separated rural from urban has been greatly accentuated in recent years, because it is the towns and cities which have been the main beneficiaries of modern social and economic development. The town dweller has by far the larger share of available amenities in buildings, communications, public health, education and entertainment, and relatively few professional men practise outside the towns. The town represents power and wealth, while rural communities remain effectively disinherited and this distinction has been enhanced even more rapidly since the oil era began. However, many Middle Eastern towns, especially those without major international trading links, continue to function as the focus and outlet of their region. Indeed, in some senses and in some countries, there is more obviously a rural–urban continuum than a sharp separation between urban and non-urban life.

The morphology of the Middle Eastern towns has undergone striking changes in recent decades and in many cases there is a clear juxtaposition of two distinct urban styles: the traditional form with the architecture of the pre-industrial city or 'medina' and the Western-style layout and buildings of the modern suburbs or extensions. In some cases, such as that of Cairo, the old is effectively continuous with the new whereas in others, such as Jerusalem, there is a clear spatial distinction between the two. In the modern cities of the oil-rich states such as Riyadh, fragments of the original architecture remain as museum pieces. Elsewhere, as in Muscat, the old city has undergone an expensive modern refurbishment.

The physical form of the traditional city is best understood by reference to the relationship between social organization and spatial patterns. In most cities there were districts which were primarily commercial and essentially public, and others that were primarily residential and private. The commercial areas, the bazaars or *suqs*, consisted of a complex network of narrow streets and alleyways, sometimes open, sometimes covered, lined with the workshops of the artisans, storerooms and shops, stalls and alcoves where goods were displayed for sale. Within the bazaars, shopkeepers and artisans were grouped by trades with whole streets where the same commodity such as grain, copperware, cloth and jewellery was produced or sold. The religious and political life of the community was also concentrated here. The suqs had mosques and *madrasas* (religious schools) and in some cities the main markets were situated close to the citadel or fortress.

Public and private space was clearly differentiated, and in the residential quarters the emphasis was on privacy and security. Here was a maze of narrow alleys, twisting lanes and cul-de-sacs surrounded by the high compound walls of the houses. Life centred on the inner courtyard or garden completely shut off from the street. Houses tended to present few openings to the outside, with solid bare walls devoid of ornament, heavy doors and window shutters. Within these quarters, the citizens were grouped by religious sect or community, rather than by wealth, resulting in the close juxtaposition of large and humble dwellings.

It was common for Jews, Armenians, Greeks and Europeans and the various Muslim sects to occupy a distinct quarter of the city, in some cases walled off from the others. In Antioch, perhaps the classic example, forty-five such quarters have been identified. This pattern derives in part from the 'millet' system, or from forcible colonization by an autocratic ruler, and partly from the general need for solidarity and protection. Most streets were too narrow for wheeled vehicles and offered access only to pedestrians. Until recently, drinking water could be scarce and water-borne sanitation non-existent. Civic feeling often had difficulty in finding adequate expression and it is only within the last few decades that there have arisen municipal councils with elected representatives and general powers over urban development. Nevertheless, the pre-modern Middle Eastern city was far from being a formless jumble of houses and streets. There was a definite and logical organization of space.

Beginning in the late nineteenth century and gathering momentum since the Second World War, the traditional urban setting in many, particularly the large, cities has been reshaped and often radically transformed by physical expansion and the introduction of Western-style town planning, Westernized architecture and building techniques, and new urban attitudes. Wide avenues have been laid out for motor vehicles and the number of cars, taxis and trucks has increased enormously. Modern buildings have sprung up along these new thoroughfares, including multi-storey office blocks for government and private business, high-rise apartment blocks, department stores, hotels, sophisticated restaurants and cafes, cinemas and theatres. Street lighting has been introduced, together with piped water supplies and public sewage systems. In some areas the scarcity or absence of indigenous urban centres resulted in the establishment of entirely new towns such as Dhahran in Saudi Arabia and Abadan in Iran, both built by foreign oil companies.

Under the impact of Westernization, the centre of gravity of urban life has moved from the market areas of the medina to the central business district of the modern city, which has absorbed most of the evolving modern functions. There has been a movement of the urban élite out of the traditional quarters into the new suburbs where houses and villas are less crowded together and street patterns more regular. In turn, the older quarters have been occupied by rural migrants, often too poor to maintain the houses, which rapidly fall into disrepair. Increasingly proletarianized and overcrowded, and congested with traffic that it was never designed to accommodate, in varying degrees the medina has become peripheral or marginal to modern economic activity.

As a result of rapid urban growth, the supply of housing has not kept pace with demand. Average occupancy densities have increased steadily and shanty towns have proliferated on the outskirts of many Middle Eastern cities, often without the permission of the authorities. These squatter settlements, built of traditional materials, may often be areas of second settlement for migrants who have moved out of overcrowded slums of the inner city. Generally regarded as undesirable by urban planners, they represent, nevertheless, an attempt by the new urban dwellers to provide minimal shelter and solve their housing problems by their own efforts. Unfortunately, the location of these settlements is rarely in line with the overall pattern of planned urban expansion, whilst high densities and the absence of basic services present severe health risks and serious social problems.

Amidst rapid urban growth and the changing functions and technology of modern urbanism, the long tradition of the quarter as a corporate entity, based on ethnicity, religion or occupation, and of co-operation within the separate quarters of the city, appears to have survived to a remarkable degree in the Middle East. In new urbanizing districts there is evidence that the inhabitants of squatter settlements try to organize themselves according to social affinities: kinship, religious sect, ethnic identity or village of origin. However, there is increasing alarm about the degree of imbalance in the urban hierarchies of the Middle East, the growing concentration of industry, services and government in a few major centres or primate cities, at the expense of the small and medium-sized towns, and the striking intra-urban contrasts emerging within the region.

One view is that there is a need for decentralization policy to encourage the growth of small towns and for vigorous regional development programmes in order to reverse this trend. A diametrically opposed view is that the tendency for urban populations and functions to gravitate to the largest metropolitan centres, leading to the development of primate cities, is a normal and healthy development, reflecting the increased scale of contemporary society and appropriate to the new technological situation. Whichever is correct, there are already potential metropolitan regions, for instance those linking Cairo with Alexandria and Baghdad with Basra.

In recent decades, urbanization in the Middle East has outstripped both economic growth and industrial development so that a significant part of the urban population is unemployed, under-employed or forced into that expanding group of activities known as the black economy.

That unprecedented urbanization has occurred and that the major cities at least have changed beyond measure is an incontrovertible fact. The most obvious generator of what is virtually a second urban revolution was the oil economy, but increasingly the situation is further exacerbated by the development of quaternary industry and by globalization in general. Whether individual countries, or even the Middle East as a whole, will benefit is a question of vital importance, since by the end of this century, well over half the region's population will live in towns and cities.

9   Economy

Since the Second World War, the Middle East in general has enjoyed a period of unprecedented economic growth. While in most of the countries social development has paralleled growth, the discrepancy between the rich and the poor has generally increased. Apart from the phenomenal growth of the oil industry, economic advances can be attributed to the large amounts of foreign aid received in the region. The result has been the rapid development of infrastructures and the enhancement of human resources. While the economic transformation has been felt throughout the region, the rate and pattern of progress varies considerably from one country to another. Furthermore, the political structure varies throughout the Middle East and this has greatly influenced the direction of economic development pursued.

Within the region, the two major resources are petroleum and agricultural land. Twelve of the countries export oil and oil products. In ten of these it is the dominant export and in seven it is virtually the only export. Among the major oil economies, only Iran and Bahrain have a significant range of non-oil exports although petroleum products nonetheless account for 85 per cent of exports from Iran. While diversification is being pursued in all the Gulf oil states, only Bahrain has so far produced any significant change in its list of exports. This observation excludes Iraq which is totally constrained in its exports by the UN sanctions.

Real growth rates in the region vary from –1 per cent to –2 per cent for the West Bank and Gaza Strip (Palestine) and 0 per cent for Iraq to 6 per cent for Saudi Arabia, 6.5 per cent for Oman and 7 per cent for Turkey. Only six countries together with the Turkish Republic of Cyprus have growth rates of under 3 per cent and it must be remembered that these figures have been generated in a period of low to very low oil prices.

As would be expected, the structural balance varies considerably across the Middle East. In terms of contribution to GDP, Bahrain, Israel, Jordan, Kuwait, Libya, Oman, Qatar, Saudi Arabia and the UAE all exhibit very much a Western pattern. Agriculture provides under 10 per cent and in all cases except Libya, Oman and Saudi Arabia, services predominate. If the labour force is analysed, a rather different pattern emerges. For example, in Bahrain only 8 per cent of the labour force is in services and government which generates 61 per cent of the GDP. Other major mismatches include the case of Oman where agriculture provides 3 per cent of the GDP but occupies 37 per cent of the labour force. Agricultural work still dominates the labour force in Egypt (40 per cent), Sudan (80 per cent), Turkey (47 per cent) and Yemen (over 50 per cent). A particularly high proportion of the labour force is found in industry and commerce, in Bahrain (85 per cent), Israel (over 50 per cent), the UAE (56 per cent) and Palestine (West Bank and Gaza 55 per cent). However, comparable figures are not available for Kuwait, Oman and Qatar. This survey indicates that a Western pattern of structural change is occurring throughout the Middle East with the notable exception of Sudan which has 80 per cent of its labour force in agriculture, 10 per cent in industry and 6 per cent in services. The other notable feature about the region is the high proportion of expatriate labourers, particularly in the oil states. Within the 15–64 age group the percentage of non-national workers is 44.5 in Bahrain, 72 in Kuwait, 83.5 in Qatar, 40 in Saudi Arabia and 76 in the UAE. In addition, there are estimated to be a million non-nationals working in Lebanon and, before 1990, there were 1.6 million in Iraq.

Economic transformation has been greatly influenced by government policy. Throughout much of the Middle East and particularly in the GCC countries, there has been a significant move away from public sector dominance towards privatization. In Saudi Arabia, approximately 40 per cent of the GDP comes from the private sector and further privatization is underway. Since the Gulf War (1991), Kuwait has been encouraged by the World Bank to move ahead with privatization. There is also a marked trend, particularly among the oil-rich economies, towards diversification to lessen dependence upon the petroleum market. In Bahrain, the most diversified of the GCC economies, advantage has been taken of a highly developed communication system and the country is now a base for numerous multinational corporations. Apart from services and banking, other elements of diversification have included an increase in the downstream processing of petrochemicals and the development of tourism. The petrochemicals industry has been developed in all the GCC states although there have been major difficulties in entering the global fertilizer market. Issues of food security, raised at the time of the two major price rises in the 1970s, have resulted in policies which aim for self-sufficiency. Again this has been most obvious in the oil-rich countries where there has been a significant increase in agricultural output. The longer-term limitation for such increases is the water supply and there are indications that the emphasis will be increasingly upon high-value crops.

Among the non-oil-rich economies, developments in Egypt illustrate the problems and some of the possible measures. In the face of virtual social and economic crisis, Egypt undertook macro-economic stabilization and structural reform measures in 1991. This has been supported by three International Monetary Fund (IMF) arrangements together with massive debt relief following the Gulf War (1991). By 1986, the foreign debt had fallen considerably, budget deficits had been slashed while foreign reserves were at their highest ever. The country is moving towards a decentralized market-orientated economy but advances are always likely to be nullified by the rapid rate of population growth.

Growth throughout the region is highly dependent upon political and economic stability. Oil prices affect not only the major producers but, as a result of trade and remittances, can influence the entire region. Over the past decade several conflicts, notably those in Lebanon, between Iran and Iraq and in Yemen, have ceased but the impact of the Gulf War (1991) on the Middle East has been considerable. For the social and economic development of Iraq it has been catastrophic but it has also exercised a dampening effect upon the economy of Jordan. The loss of Iraq as a major oil producer has been offset by increasing production in other parts of the world and prices have, by 1999, reached their lowest levels for more than 10 years. Furthermore, the departure of foreign workers from Kuwait not only exacerbated the problems of Jordan but profoundly influenced the global Palestinian community. The long-running peace process between Israel, Palestine and the neighbouring Arab states has also affected development regionally. The obvious inequities in Western approaches to Iraq and Israel together with continuing US and UK militancy have brought added stress to the area. Socially, the semi-permanent basing of US troops in Saudi Arabia has been disastrous, while each renewed US/UK strike against Iraq further strains the economy of Kuwait.

Other factors influencing growth have been the impact of US sanctions on Iran and UN sanctions on Libya (lifted in 1999). Both have locations more favourable to sanctions resistance and a sufficiently strong oil base to their economies so that the levels of privation visited upon Iraq were never a possibility. Nevertheless, sanctions have markedly hindered economic and social growth and produced regional economic distortions. Of particular concern for Iran is the routing of pipelines from the burgeoning Caspian Sea oilfields which, should they traverse the country, would bring considerable benefits.

A further factor influencing economic diversification has been the actions of freedom fighters or terrorists, characterized generally by the West as Islamic violence. Already this has severely handicapped the tourist industry of Egypt and the effects have also been felt in eastern Turkey and Iran. The tourist industry to Iraq has been virtually obliterated.

In the face of these instabilities, there has been relatively little successful large-scale co-operation. Apart from a number of UN organizations, such as the Economic and Social Commission for Western Asia (ESCWA), the membership of which comprises twelve states of the region together with the PLO, the GCC is the only functioning body. However, since its foundation in May 1981 to promote regional co-operation in economic, social, political and military affairs, progress has been slow. Regular meetings at various levels have produced some economic and social co-operation but, given the obvious potential, results have been disappointing. Disagreements within the region, well illustrated by all the potential geopolitical problems, militate against co-operation whether between contiguous countries, within drainage basins or according to shared interests. Even OPEC, a predominantly Middle Eastern body, has declined sharply in significance while OAPEC has never achieved great prominence. Meanwhile, conflict on an almost daily basis continues in Israel and the surrounding area, in east Turkey and in Sudan.

In summary, it can be seen that the Middle East has undergone massive economic change since the Second World War but development has been uneven with a contrast between the oil-rich and the non-oil-rich states. Development has been hampered and continues to be hampered by a variety of tensions and, in some cases, open conflict. In attempting to discern some economic pattern among the complexities of the Middle East, Richards and Waterbury (1996) have indicated a five-fold classification of states based upon a rather more sophisticated appreciation of the oil dichotomy:

1   The abundant oil, low population states of Libya, Kuwait, Oman, the UAE, Bahrain and Qatar. These states are almost entirely dependent on oil together with earnings from overseas investments.
2   The oil industrializers: Iraq, Iran and Saudi Arabia. These countries have sufficient oil exports and large enough populations together with other resources to make industrialization an option.
3 . The limited resource–high technology states: Israel, Jordan and Syria. These countries have limited natural resources and have focused upon human capital and skill-intensive manufacturing.
4   The newly industrializing countries: Turkey and Egypt. Turkey has no oil and Egypt has insufficient for long-term growth but both have large populations, relatively good agricultural land or potential and long-term experience with industrial production.
5   The agro-poor: Sudan and Yemen. These are the least developed countries in the region and, despite the oil reserves of Yemen, are likely to look to agricultural development-led growth.

Of the other Middle Eastern countries, Cyprus is effectively two economies in one and progress is hampered by the lack of overall co-operation. Palestine is as yet dominated by Israel and Lebanon is undergoing a period of resurgence. The rebuilding of Lebanon, despite the presence of a variety of warring factions, after decades of almost continuous conflict, provides a beacon of hope for the region in general.

Agriculture

In *The Middle East* (Fisher 1978) the section on agriculture begins: 'Agriculture is by far the most important economic activity in the Middle East ...'. This statement was based upon the significance of agriculture throughout the region, other than in the main Gulf oil-producing states and Israel, for both employment and export earnings. Since 1970 (FAO 1975), the statistics for which produced the basis for the statement, there has been change, often massive, throughout the region. The one country still virtually totally dependent upon agriculture is Sudan. In 1970, 82 per cent of the economically active population was in agriculture and this figure had declined to only 80 per cent by 1997 (CIA 1998). While there has been a significant decline in the figures for Egypt, Iran, Iraq, Libya, Syria and Turkey, the major change has occurred in Jordan in which there has been a major change from 33.7 per cent in 1970 to 7.4 per cent in 1997. The general fall in agricultural employment over the 27 years in shown in Table 9.1.

This change parallels that which has already occurred throughout the developed world. With regard to export earnings Fisher (1978) states:

> With the exception of the main oil producing states and Israel, earnings from agriculture constitute the major source of foreign exchange: in Egypt, Sudan, Syria and Turkey about three-quarters of export earnings come from agriculture.

Today, food and related products as a percentage of exports have been reduced in importance in almost all of the Middle Eastern countries by petroleum and petroleum products and manufactured items. Only in Cyprus, Sudan and Palestine does the entire export list comprise agricultural products. In Turkey, food accounts for only 20 per cent of the value of exports while in Syria food and live animals total 16 per cent. In Egypt, cotton products are still important but they are second to crude oil and petroleum products. In Yemen, crude oil is the highest valued export commodity but agricultural

Table 9.1 Economically active population in agriculture

Countries	Economically active population in agriculture (%)	
	1970	1997
Cyprus	38.5	G 13; T 23
Egypt	54.4	40
Iran	46	33
Iraq	46.1	33
Israel	9.7	3.5
Jordan	33.7	7.4
Lebanon	19.6	12
Libya	32.1	18
Sudan	82	80
Syria	51.1	40
Turkey	67.7	47
Yemen	N 64.5; S 79.2	50+

G = Greek Cyprus, T = Turkish Cyprus, N = North Yemen, S = South Yemen.

products including cotton, coffee, hides, vegetables and fish predominate. Other than in Sudan and the West Bank, the contribution of agriculture to GDP is under 30 per cent. In both Sudan and the West Bank the figure is 33 per cent and apart from these two, the highest are Syria with 28 per cent and Iran with 21 per cent. In ten of the seventeen countries, the contribution of agriculture is under 10 per cent. Nowhere is agriculture the largest contributor to GDP and, indeed, only in Sudan, Syria and the West Bank does agriculture exceed industry in its contribution.

Land reform

The land tenure system in many parts of the Middle East, and particularly the areas which are now Turkey, Iraq, Syria, Lebanon, Jordan and Israel, as well as Cyprus, was based, until the introduction of land reforms in the 1950s, on principles which were influenced by Ottoman land law. It is useful to describe the main features of this system, although it is important to realize that the actual situation varied considerably, according to custom and to type of land, both within and between Ottoman provinces. It is also important to stress that the Ottoman Land Code of 1858, which summarizes the various forms of tenure and defines methods of registration of title, had little relevance to local conditions in many areas, although later generations of administrators were to treat the provisions of the Code as if they did reflect the actual situation.

There were four main categories of landholding, *mulk*, *matrūka*, *waqf* and *mīrī*, of which *mīrī* was by far the most important. Briefly *mulk* is equivalent to freehold, and was largely restricted to buildings and gardens; *matrūka* (lit., set aside) is land reserved for public purposes, such as roads and government buildings; *waqf* is land left for religious or charitable purposes, generally administered through a department of state or ministry; and *mīrī* covers the rest. The basic premise of Ottoman land law was that the State was sole landlord, and that it alone possessed *raqāba*, possessory rights. All grants to individuals were thus theoretically simply of *tasarruf*, usufruct, rather than of actual property rights, although the practical implications of ultimate state control were rarely felt. The state would grant out this usufruct to lessees, *multāzimin* or *muhāssilīin*, who could acquire the leases by making bids at periodic auctions. This was known as true *mīrī* or *mīrī sirf*, and such arrangements were, apart from customs and excise, the Ottoman government's main source of revenue, amounting for instance, to 42 per cent of all income from the three provinces which are now Iraq in 1911. The arrangement was a cash commutation of the former *tīmār* system under which leases had been given to individuals in return for the provision of soldiers for the Ottoman army.

In most of the Ottoman Empire, however, the code did not fit around the kind of corporate communal ownership which prevailed. Both the *mushā'* system in Syria and the *lazma* system in Iraq were forms of joint customary ownership which were alien both to Islamic and to Ottoman law, neither of which recognized the existence of legal entities, or corporations. This meant that, in the future, leases could be given to individuals, and not to a particular village or tribe. Furthermore, it must also be remembered that in the second half of the nineteenth century much of the Middle East was gradually moving from an economy based on subsistence farming or subsistence stock-rearing to cultivation for a cash market, either at home or abroad. Realization of the fact that there were profits to be derived from agriculture contributed to encouraging the sedentarization of nomads. Questions of boundaries and leases, and conditions of tenure began to assume a greater importance than in the past. This was especially true of areas which came under cultivation for the first time, the potential of which was vastly increased by the extension of mechanized irrigation after the First World War.

These changes, which were in response to external economic pressures, were to cause, in the late nineteenth century and in the first half of the twentieth, immense changes in the structure of landholding throughout the Middle East. Over the period, an original society of generally free tribesmen was transformed into groups of serfs, bound to the soil by debt and inertia, where traditional leaders and 'new' landowners gained unprecedented legal and economic powers over their peasantry. Communal land ownership disappeared, and tribal sheikhs and absentee landlords in the towns became the unchallenged owners of vast estates.

Since the early 1950s the Middle East has experienced a number of major land reform programmes. Their primary aim was political rather than economic, but the social and economic consequences have been far from negligible. In Egypt, Syria and Iraq, land reforms were introduced following political revolutions with the immediate objective of breaking the power of the old ruling oligarchies dominated by large landed interests. Their long-term aim was to bring about a restructuring of social and economic relationships in the rural areas. In South Yemen, the first land reform measures were implemented immediately after independence in order to destroy the power and influence of those landowners who had worked for the British. In contrast, land reform in Iran (the 'White Revolution') was carried out to strengthen the existing regime and give it a broader popular appeal by securing the support of the peasantry.

The land reform laws imposed limits on the size of holdings but the official ceiling varied from country to country and between irrigated and rain-fed land. The impact of land reform has been far reaching, but its consequences must be viewed in perspective. The proportion of the cultivated area affected by the reforms varies considerably and not all the large estates have disappeared. The effect of land redistribution on employment opportunities overall has thus been negative. In the future and in spite of continued rural–urban migration, increasing population pressure in the countryside will probably accentuate the problems of disguised unemployment and open unemployment among the rural population. These problems will only be avoided by the creation of additional employment opportunities in agriculture.

Land use

The Middle East is the global region with the least amount of arable land although only Kuwait, Oman and the UAE claim to have no arable land at all. Land use is set out in Table 9.2 (CIA 1998) and summarized in Figure 9.1 and it can be seen that only Turkey, Palestine, Syria and Lebanon have significant areas which can be classified as arable. Permanent crops are defined as those which are not replanted after each harvest and therefore include dates, citrus and olives. Clearly in the figures in Table 9.2, there are variations in the interpretation of this definition. It would appear that dates and olives are not included. On the other hand, citrus crops are produced in Libya while limes and bananas are grown in Oman, both of which countries claim no area of permanent crops. By far the highest figure is the 39 per cent for the Gaza Strip which is a noted producer of citrus fruits. Otherwise, only Lebanon, Cyprus, Israel, Syria and Turkey are of any significance for the production of permanent crops.

Similarly, there are problems over the definition of permanent pasture. This is defined as land permanently used for herbaceous forage crops. In the more temperate and wetter areas of Iran, there is permanent pasture and there are areas of permanent grassland in Sudan which might be characterized in the

Table 9.2 Land use

Country	Percentage of land use			
	Arable	Permanent crops	Permanent pasture	Forest and woodland
Bahrain	1	1	6	0
Cyprus	12	5	0	13
Egypt	2	0	0	0
Iran	10	1	27	7
Iraq	12	0	9	0
Israel	17	4	7	6
Jordan	4	1	9	1
Kuwait	0	0	8	0
Lebanon	21	9	1	8
Libya	1	0	8	0
Oman	0	0	5	N/A
Qatar	1	N/A	5	N/A
Saudi Arabia	2	0	56	1
Sudan	5	0	46	19
Syria	28	4	43	3
Turkey	32	4	16	26
UAE	0	0	2	0
Yemen	3	0	30	4
Palestine				
West Bank	27	0	32	1
Gaza Strip	24	39	0	11

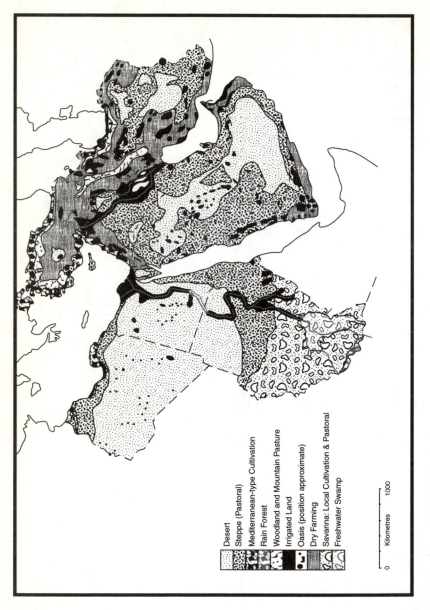

Desert
Steppe (Pastoral)
Mediterranean-type Cultivation
Rain Forest
Woodland and Mountain Pasture
Irrigated Land
Oasis (position approximate)
Dry Farming
Savanna: Local Cultivation & Pastoral
Freshwater Swamp

0   Kilometres   1000

Figure 9.1 Middle East: generalized land use.

same way. However, the large percentages claimed by Saudi Arabia, Syria, the West Bank and Yemen must refer to what in temperate climates would be called rough grazing. As a comparison, the percentage of permanent pasture in the UK is 46. Furthermore, in countries such as Bahrain, Kuwait, Libya, Oman and Qatar, the figure probably refers to the growing of forage crops such as alfalfa. With the exception of the few extreme figures for permanent pasture, it can be seen that the countries best developed for agriculture are Cyprus, Iran, Iraq, Israel and Turkey. The key determining factor is climate and, in particular, rainfall except in the case of Israel in which there has been large-scale investment in irrigation.

Forest and woodland coverage throughout the region is very low and is in most cases decreasing. Population pressure has resulted in the felling of much of the natural forest within the region as the requirement for agricultural land grows. Furthermore, in the poorer areas in particular forests are denuded for firewood. In the predominantly desert countries, for obvious climatic reasons, the area of forest is extremely limited. The mountains of Cyprus and Turkey are still relatively well forested as is the southern wetter part of Sudan. The historic groves of cedars of Lebanon have now been largely decimated but most of the countries have afforestation programmes. For example, Oman, predominantly a desert state, has focused upon the replenishment of its *Prosopis cineraria* groves.

Irrigated land is defined as that artificially supplied with water. Therefore, it takes no account of the type of irrigation which may vary from drip or bubble to inundation. Furthermore, areas which are naturally irrigated by river inundation are excluded. Given the climatic constraints of much of the region, irrigation is vitally important for the development of agriculture. By far the largest area under irrigation in the Middle East is that in Iran which is estimated to be $94,000 \text{ km}^2$. Egypt has over $32,000 \text{ km}^2$ and Turkey almost $37,000 \text{ km}^2$. The other major areas are found in Iraq ($25,500 \text{ km}^2$) and Sudan ($19,500 \text{ km}^2$).

As a percentage of the land area of the country, by far the highest percentage, 33 per cent, is estimated for the Gaza Strip. By percentages, the following states have significant areas of irrigation: Israel (9), Lebanon (8.5), Iran and Iraq (6), Syria and Turkey (5), Cyprus (4) and Egypt (3). Irrigation is insignificant in the UAE, and accounts for 0.1 per cent of the area in Kuwait, 0.2 per cent in Saudi Arabia and 0.3 per cent in Libya and Oman. Lebanon, Iran, Iraq, Syria and Turkey all include the basins of the major permanent rivers of the region. In addition, Lebanon, Turkey and Iran also receive, in many areas, comparatively high rainfall.

Israel alone among the states with major areas of irrigation receives generally very low rainfall and is not situated in a major catchment. The flow of the Jordan is at best 2 per cent of the flow of the Nile or the Tigris–Euphrates and irrigation has been facilitated by the construction of the National Water Carrier and through the use of groundwater. In Israel, agriculture accounts for approximately 70 per cent of water consumption.

The major irrigation scheme in the region and perhaps the largest in the world is the South-East Anatolian or GAP project. This combines eighteen major irrigation and hydro-electric power programmes utilizing water from both the Tigris and the Euphrates. Apart from the Ataturk Dam, a key element in the GAP project, the other particularly controversial hydraulic structure in the Middle East is the Aswan High Dam. In many ways, the effects of this have been to change the agricultural face of Egypt. Instead of natural inundation with the deposition of clay nutrients, water released from the dam is largely sediment-free. Therefore, the requirement for fertilizers has risen considerably with the result that many areas once used for subsistence can now only grow cash crops. This has affected the dietary condition of many of the rural areas particularly in Upper Egypt. The current major project in Egypt is the linking of a series of oases which stretch westwards from Lake Nasser into the Western Desert by a canal system. In this way, it is hoped to expand considerably the irrigated agricultural area.

The major scheme in the Middle East dependent upon groundwater, in fact groundwater from deep aquifers, is the Great Man-Made River of Libya. This is being used to irrigate the coastlands of the Gulf of Sirte.

Agricultural products

From the lists of the main agricultural products (CIA 1997), it is possible to discern a number of well-marked agricultural groupings. Mediterranean polyculture is fully developed only in Cyprus and in favoured areas of the coastal lowlands of southern Turkey, the Levantine coastlands and Libya. The key product by which this regime is defined is the olive which features prominently in the agricultural produce inventories of Cyprus, Lebanon, Libya, Turkey and Palestine (both the West Bank and the Gaza Strip). The olive is, in fact, tolerant of aridity and can grow on the borders of the desert provided there is some irrigation. The vine is found in similar areas but its cultivation is restricted not only by climate but also by the Islamic prohibition on the consumption of alcohol. Nonetheless, wine is made and grapes are consumed for dessert and as sultanas. The main areas of production are Cyprus, Lebanon, Israel, Jordan, Turkey, Iran and coastal Egypt. Cultivation is best developed in the more elevated areas of Cyprus, western Anatolia and the Levant. Other associated fruit trees include fig, apricot, peach and pomegranate. In the drier interiors, nuts including almonds, pistachios and walnuts are characteristic and are a particularly significant product of Iran. On the higher hill slopes, apples are of increasing importance.

A second grouping, which overlaps the first, is characterized by citrus production using irrigation. The countries chiefly involved are Israel, Jordan, Lebanon, Libya, Oman, Saudi Arabia, Turkey and both parts of Palestine.

The grouping of countries with more extreme aridity is characterized by commercial date production. Dates are significant agricultural products of Iraq, Libya, Oman, Saudi Arabia and the UAE. Sudan, with its production of sorghum, groundnuts and millet, has more in common with African agriculture and is significantly different in this respect from the other Middle Eastern countries.

With the exception of Sudan, wheat and barley are the commonly grown cereals throughout the region. In the case of Saudi Arabia, wheat has been grown as a form of agribusiness. Wheat and barley production is important in Cyprus, Iran, Iraq, Jordan, Libya, Saudi Arabia, Syria and Yemen.

The other category of crop of increasing note is vegetables. Almost all the countries of the region list them as an agricultural product. Most significant are onions, beans, aubergines, peppers, cucumbers, pumpkins and more recently, tomatoes and potatoes. Certain countries, particularly Lebanon, Israel and Egypt have developed an export trade. Moreover, vegetable-growing is now relatively important in countries which have only restricted areas of agriculture such as Bahrain, Qatar and the UAE.

Industrial crop cultivation is relatively limited but locally very important. The most important product is the long staple cotton from lower Egypt but cotton is also grown as an important export crop in Sudan, Turkey, Iran, Iraq, Israel and Syria. Tobacco is grown widely throughout the Middle East for local consumption but there is an important export trade of quality tobacco from Turkey. Sugar beet is grown in the temperate areas of Turkey and Iran while sugar cane has effectively displaced much of the original subsistence agriculture in upper Egypt.

Other crops, the importance of which is limited to the specific countries, are opium (Turkey and Iran), hashish (Lebanon), *qat* (Yemen), gum arabic (Sudan) and coffee (Yemen). The first three are all drugs. Opium is the exudate from the unripe seedpod of the opium poppy and is used to produce morphine for medical purposes and most importantly heroin, a narcotic. Hashish is the resinous exudate of the cannabis or hemp plant and *qat* is a stimulant from the buds or leaves of *Catha edulis*.

In Turkey, the production of opium is strictly controlled and is focused on medical products. In contrast, the western end of the 'Golden Crescent' extends into Iran from which illicit drugs are trafficked. The use of *qat*, either chewed or drunk as tea, is virtually ubiquitous in Yemen and additional supplies are flown in from the Horn of Africa. The demand is so great that *qat* has increasingly displaced other crops with the result that Yemen has become a major food importer. Furthermore, land once used for export crops such as cotton, fruit and vegetables has been turned over to *qat* production. However, coffee, the world-famous Mocha coffee, remains a significant export.

Livestock production

In *The Middle East* (1978) Fisher wrote:

> Two distinct systems of livestock production may be identified in the Middle East – pastoralism and livestock farming – and the traditional

separation of these systems from each other has accentuated the inherent differences between them.

This distinction is still fundamental although both forms of production have changed. Pastoralism has, throughout the region, continued in inexorable decline so that today in Saudi Arabia the nomadic part of the population dependent upon livestock is no more than 5 per cent. The major change in livestock farming has come with the development of agribusiness. Largely as a result of factory farming most of the Middle Eastern countries are self-sufficient in poultry products while stall-feeding of cattle has resulted in a flourishing dairy industry, even in the arid Gulf states.

Nomadic pastoralism is still found throughout the states of the Middle East and represents the most efficient use of the vast rangeland areas. For many countries, the spiritual and psychological significance of this lifestyle far outweighs the importance of any contribution it might make to the economy. For people who are basically tribal, the nomads represent a repository of their basic social values. In the Arabian Peninsula, the oil-rich states make special financial provision to sustain, as far as possible, the Bedouin. However, not only is the Bedouin culture among the most rigorous of all lifestyles but it is now subject more than ever to competition from elsewhere. Menial employment in the petroleum industry yields far greater financial rewards and the products of pastoralism have to compete with high-quality imports and increasingly the output from agribusiness. Indeed, the trend towards intensive urban agriculture has rendered much of the rural landscape of the Arabian Peninsula all but obsolete.

Livestock farming is sedentary and is independent of precipitation and rangeland plant growth. The animals are mostly stall-fed or at least confined to relatively limited areas and subsist on fodder crops such as alfalfa grown under irrigation. In contrast to pastoralism, therefore, livestock farming has added significantly to water demand. Mixed farming as in Western Europe, using permanent pasture, is found only in the more temperate areas of the region, in parts of Israel, Egypt and Sudan.

The changing pattern of livestock production has obviously affected animal numbers and, while sheep and goats have retained their predominance throughout the region, numbers of cattle have increased but the herds of camels and of buffaloes have declined. Only five countries: Egypt, Iran, Iraq, Turkey and Yemen have what might be considered balanced livestock production with more than a million each of cattle, sheep and goats. Turkey, with almost 12 million, has the largest cattle herd in the Middle East. It is followed by Sudan, Iran and Egypt. Iran has the largest flock of sheep, estimated at over 50 million but every state in the region has significant numbers. After Iran, the largest numbers of sheep are found in Turkey, Sudan and Syria. The number of goats has declined over the recent past although Iran maintains a herd of almost 26 million. Significant herds of goats are found also in Turkey, Sudan, Saudi Arabia, Yemen, Egypt, Syria and Iraq. However, Turkey now has more cattle than goats.

The trend towards modern livestock production is illustrated by Israel which is the only state with more cattle than sheep and which has a relatively modest herd of goats. As the ability to travel considerable distances between pastures, a prerequisite of grazing in semi-arid conditions, declines in importance, cattle numbers are likely to increase everywhere while numbers of sheep and particularly goats will diminish. However, while all three animals produce milk and meat, sheep and goats provide respectively wool and hair for which there is both a specialized and a local market. Furthermore, mutton and lamb are still much preferred to beef by most Middle Eastern consumers.

Poultry are ubiquitous throughout the region and most countries have achieved a high degree of self-sufficiency in poultry meat. Owing to Islamic and Jewish prohibition, pigs are relatively unimportant although they are found in small numbers in Israel, Iran, Lebanon and Turkey. Numbers of buffaloes have declined everywhere and Egypt, with over 3 million, has the only herd of any size. Iran now has less than half a million and the numbers in Turkey and Iraq are considerably lower. The buffalo remains important in the marshes of Lower Egypt and Iraq, the Caspian Provinces of Iran and the alluvial coastlands of Turkey.

Large numbers of donkeys remain in the Middle East and for the developing areas of the region, they are the principal draught animals. However, with modernization, particularly that based upon petroleum wealth, donkeys are being replaced everywhere by motorized transport, particularly the Toyota Pickup. In the GCC countries, there are now herds of feral donkeys in the rangelands. In contrast, donkeys remain a common sight in the settlements of Yemen.

Camels have also declined in importance and only in Saudi Arabia and Sudan are numbers significant. In Israel and Lebanon there are no camels while in Turkey there are now under 2,000. However, like the nomadic lifestyle, camels are redolent of past glories and it is likely that they will for a long time to come remain features of the Middle Eastern landscape. Not only is camel meat highly prized on the major feast days but camel racing is growing in importance as a spectator sport. In Iran, there is the Bactrian or two-humped camel but throughout the remainder of the region the Arabian one-humped camel predominates. Like the donkey, the camel has been largely replaced for transport purposes. Similarly, the horse is rarely used as a beast of burden but is kept for recreational riding and racing. Indeed, using irrigated pasture, Dubai has become a global centre for the horse-racing industry. It was, of course, Arabian horses which gave rise to the Western blood stock industry.

Agricultural summary

Agriculture and agricultural progress can be summarized by countries, bearing in mind the major distinction between those states in which the petroleum industry dominates the economy and those which are more diversified. In the GCC countries, although the agricultural areas themselves are small, some 70 per cent of the agricultural land is irrigated, mainly through groundwater. Agriculture counts for a high proportion of water consumption, for example 68 per cent in Qatar and 80 per cent in the UAE. Although as a proportion of GDP, agricultural production is everywhere small, there is a strong desire to increase the level of self-sufficiency. Nonetheless, most agricultural commodities have to be imported despite self-sufficiency in some countries in specific vegetables and fruits.

With strong government support, Saudi Arabia has been characterized by high rates of agricultural growth over the past decade. Between 1984 and 1992, wheat production increased from 1.4 million tons to 4 million with a surplus for export. However, there is a need to ration water consumption in agriculture and to initiate allocation between agriculture and the other sectors of the economy. At present, between 85 and 90 per cent of water consumption is for agriculture and there is a heavy reliance upon non-renewable aquifers.

Bahrain has diversified the agricultural sector so that more than 50 per cent of demand for poultry products is now met. Meat- and date-processing plants have been opened and there is a national dairy pasteurization plant. Many sectors of agriculture remain subsidized and one notable development is the use of recycled water for irrigation. Qatar has extremely limited water sources and little land of good quality for agriculture. It has become largely self-sufficient in a range of vegetables but still has a high reliance upon food imports, particularly livestock. Agriculture in Kuwait and the UAE follows a similar pattern although production in the former remains small and in the latter has shown significant recent development.

Oman has a rather more diversified agricultural economy with considerable cultivation in the major wadi valleys and livestock production in the mountains of both the north and the south. Libya has the stated aim of agricultural self-sufficiency but is still reliant on imports for over 80 per cent of its food. The major irrigation schemes around the Gulf of Sirte will lead to a considerable change in production over the next few years.

Among the more diversified economies, Egypt has made significant progress with a growth rate for agriculture of some 3–4 per cent annually. This has resulted from advances in practice, improved infrastructure, land reclamation, land reform and improvements in the educational level of the labour force. However, there are problems such as high rates of population growth, fragmented land holdings and inefficient marketing which remain. There is a surplus of fresh vegetables and fruit while cotton remains crucial as a source of foreign exchange. Increases in the irrigated area together with regulations constraining building construction on good agricultural land give rise for some optimism.

In Iraq, the effects of sanctions have been disastrous upon agriculture. Imports of machinery, fertilizers and spare parts have been severely constrained and agricultural production has shown a sharp decline in both quantity and quality. Damage to both irrigation and sewage systems has hindered production and malnutrition is widespread. As a result, there has been a significant move from cash crop to subsistence production and exports have been greatly reduced.

Surprisingly, given the constraints of water and soil, Jordan has considerably improved the production of its agricultural sector. The application of modern technological farming systems and increases in capital investment have led to significant improvements in both quality and quantity. The key area is the Jordan Valley in which fruit and vegetables are produced on capital intensive, high-yield irrigated farms. Cereal production in the non-irrigated upland areas has changed little. The main obstacle to further development is water shortage.

In contrast, Lebanon is one of the few Middle Eastern countries in which water for agricultural development is not the major concern. Of the total area, 31 per cent is cultivated and 8.5 per cent, by far the highest in the Middle East other than the figure reported for the Gaza Strip, is irrigated. Agriculture contributes almost 20 per cent to the value of exports, a statistic which has increased even during the period of intense conflict. Lebanon remains the traditional source of fruit and vegetables for the GCC countries.

Increases in irrigation and reforms in agricultural policy have resulted in significant progress in Syria. Wheat imports ceased in 1996 and agricultural products account for some 20 per cent of total exports. In the West Bank and Gaza, agriculture was considerably hampered by Israeli occupation and production has fluctuated. Furthermore, there has been instability in both the internal and the external markets. In general, productivity levels are below those of Israel but Palestine competes successfully with neighbouring Arab states. The foundation of agriculture is the olive which generates approximately a quarter of agricultural income. In sharp contrast to Palestine, the agriculture of Israel is the most highly dependent upon research and development in the Middle East. However, as other sectors of the economy have increased in importance so the agricultural share has declined. Furthermore, agriculture continues to depend upon government support, particularly for water and in providing a protected market.

Since the end of the war with Iraq, agricultural output in Iran has risen but management remains haphazard and the infrastructure requires development. The aim is to increase food exports and develop further processing industries. In Turkey, with some 40 per cent of the work-force in agriculture, the sector remains highly significant. Given the variety of climates and soils, the range of products is wide varying from cereals to citrus fruit, grapes, nuts and a variety of raw materials for industry including cotton, sugar beet and tobacco. As the South-East Anatolian Project comes on stream, so the face of Turkish agriculture will be changed and it will become an increasingly important source of food for the Middle Eastern countries.

Sudan is completely different from any other country in the region in that 80 per cent of the work-force is in the agricultural sector. Already there is a large area of irrigation but Sudan still does not utilize to the full its share of the Nile water. The range of agriculture is far greater than that in any other Middle Eastern country varying from the typical African savannah agriculture of the south through high-technology irrigation to nomadic pastoralism in the north. The principal cash crop remains cotton although this has been affected by climatic variations and pest infestations. Recently, sesame has been a major source of export earnings. In the livestock sector, there have been marked improvements as a result of better veterinary practices and extended credit facilities. Further modernization is required but, with Turkey and Iran, Sudan remains a major agricultural producer.

Fishing

There are no landlocked countries in the Middle East and there is access to the Black Sea, the Mediterranean Sea, the Red Sea, the Persian–Arabian Gulf and the Indian Ocean. However, fishing is relatively little developed except where it is part of local tradition. Turkey alone has a catch of over half a million tons and, with the exceptions of Egypt and Iran, no other country has significant fisheries. In Sudan, Egypt, Iraq, Israel and Syria the catch is dominated by fish from inland water rather than the sea. With a few exceptions, the sector is dominated by artisanal fishing using perhaps modern additions such as ice boxes. For example, much of the fish caught on the north-eastern coast of Oman is exported to the UAE in large ice boxes lashed to the back of Toyota Pickups.

The petroleum industry

It is a reasonable assumption that for most people oil and the Middle East are synonymous. Global interest has focused upon the region for a variety of reasons but most obviously because of oil. Many states of the Middle East came into being purely as a result of oil and, without it, the region could not have become the key global geopolitical flashpoint.

> The eight OPEC countries in North Africa and the Middle East represent more than ⅔ of the world's proven crude oil reserves and about ⅓ of the world's proven natural gas reserves. On the other hand, these countries currently represent less than a ⅓ of the world's total crude oil production and less than 10 per cent of the world's marketed natural gas. . . . This suggests that the OPEC member countries of North Africa and the Middle East have significant opportunities ahead of them and the potential to increase their market share in the oil and gas sectors in the medium and long-term.

Thus Dr Rilwanu Lukman (1998), Secretary General of OPEC, summarizes the fundamentals of the petroleum industry. His statement covers all the major Middle Eastern oil producers although there is substantial non-OPEC oil and gas output from Oman, Syria, Yemen and Egypt. Furthermore, Algeria is included as a Northern African OPEC member. However, Algeria is responsible for only 1.8 per cent of global oil production and 3 per cent of global natural gas production.

Despite its current global dominance, the vast structure of the petroleum industry is of very recent origin. Oil was discovered in the Middle East at the beginning of the twentieth century and by the early 1940s the area was responsible for only 5 per cent of the world's oil. This figure had become 15 per cent by 1950, 25 per cent by 1960 and 39 per cent, the highest percentage yet achieved, by 1979. The industry first achieved prominence and the region first became geopolitically sensitive in 1911 when the bunkering of Royal Naval ships was changed from coal to oil. From this time until the early 1970s, the entire industry was controlled by initially seven and later eight major international corporations. Known as the 'seven sisters', the original seven were Standard Oil of New Jersey (later Esso/Exxon); Royal Dutch Shell (Shell); British Petroleum (BP); Gulf Oil (Gulf); Texas Oil (Texaco); Standard Oil of California (Socal/Chevron); and Mobil Oil (Mobil). Later, Compagnie Française des Pétroles (CFP) became the eighth member of the 'Majors'. During the 1950s and 1960s, the all-embracing control of the Majors was somewhat loosened by the entry of the so-called 'Independents', approximately thirty relatively small corporations, some private and some state-owned. Prominent among these were Getty and Ente Nazionale Idrocarbui (ENI). By offering better terms to the producer countries, the Independents were influential in breaking the total dominance of the Majors. However, although in modified form, oil imperialism remained (Anderson and Rashidian 1991).

The development of the oil industry in the Middle East was governed almost entirely by the international oil corporations, all of which were fully supported by their own governments: British, American, Dutch and French. The producers may have been politically independent but they were denied any semblance of economic independence. All facets of production refining and marketing were controlled by the corporations and the producers were discouraged from acquiring skills in the field or from showing any initiative.

Indeed, questionable methods were used to enforce the ascendancy, as for example in Iraq:

> Iraq for instance was threatened with dismemberment of the Mosul Province from its territory unless the government agreed to grant a concession to the then Turkish Petroleum Company, later renamed the Iraq Petroleum Company (IPC).
>
> (al-Otaiba 1975)

Intrigue was not only restricted to local events but was also international as the British government made great efforts to keep the US interests out of the Middle East. However, in 1930 the USA gained concessions in Bahrain and later the Arabian–American Oil Company (ARAMCO) Group was awarded a concession which covered the entire territory of Saudi Arabia.

Oil imperialism was never more obviously exercised than in the concession arrangements. The exact terms varied but concessions were generally very extensive, covering half or more of an entire country and were all-embracing. Terms were either a fixed royalty per ton of oil, usually 20–25 cents, or a percentage royalty on net sales. The duration specified was usually up to 75 years and the privileges included exclusive rights to control all aspects of the oil and related products industries from the initial prospecting to the export of refined products. Furthermore, within the producer countries, the corporations enjoyed a range of rights and privileges which rendered them free from local government control (UN 1951). The chief points of criticism therefore were:

1   the financial terms were too low;
2   the national state had no control over rate of prospecting or of exploitation and some known oil deposits were left for future use;
3   the concession areas were too large and on too long a lease; and
4   the operating countries employed a majority of expatriate personnel who operated the concessions as 'states within a state', exporting through company terminals and importing goods sometimes paying no customs duties.

However, from the beginning of the Second World War until the foundation of OPEC in 1960 the situation gradually changed. As a result of negotiations or nationalization, the power of the multinationals and their home governments was reduced. The principle of nationalization was established by Iran with the nationalization of the Anglo-Iranian Oil Company's (AIOC) concession in 1951. At the time, the UK was in the throes of nationalizing its own industries but the British government together with the oil Majors declared what was effectively an embargo on Iranian oil. An accommodation was reached in 1954 but the principle remained and an agreed 50/50 per cent share-out between the new consortium consisting of British and American interests, and the Iranian government was reached. Conditions then ameliorated with the entry of the Independents and both the geographical scale and the duration of concessions were reduced.

The country which was most obviously instrumental in developing the new system was Libya where oil was first discovered in 1957. Concessions comprising fifty-one separate areas were awarded to seventeen oil-producing companies, many of them comparatively small. The concessions were granted for relatively short periods and, if undeveloped, were returned to the government. Thus, the corporations were changed from concessionaires to contractors and this whole movement towards a fairer return on resources for the producers was given impetus by the creation of OPEC in 1960.

With the emergence of OPEC, politically independent countries also became effectively economically independent and were able to control their own destinies in the petroleum industry. Following oil price reductions by the Majors in 1959 and 1960, which took the posted prices below those which had obtained in 1953, the five key exporting countries: Saudi Arabia, Iraq, Iran, Kuwait and Venezuela formed OPEC. With the increasing dominance of Arab states in the industry, the OAPEC was founded in 1968.

Meanwhile, the Arab–Israeli war of 1967, although illustrating the limited power of OPEC, greatly influenced the oil industry through the closure of the Suez Canal. The result was very much to the benefit of Western Hemisphere producers together with Nigeria and Libya, the oil from none of which had to transit the canal. Following civil war in Nigeria, Libya was left in the most favourable position.

In 1969, Libya increased prices and threatened reductions in production and this set the trend for the next few years. From the end of the Second World War until about 1970, the price of oil in real terms had remained generally stable and had even declined. During the Arab–Israeli war of 1973, OPEC imposed a price rise of 70 per cent and this was followed later in the year by a further rise of 128 per cent. Even then, the price stood at $11.65 per barrel, a figure which was hardly excessive in the light of the vast profits being made by the oil corporations.

The effect of the new prices in the Middle East was dramatic and resulted in a massive transfer of wealth. With their low absorptive capacities, the countries could not directly accommodate the influx of petro-dollars and very large surpluses were invested in Western banks. Thus, the oil-rich countries of the Middle East became influential in the world banking system and this fact has had a long-term effect on the use of the oil weapon. If oil sales are manipulated to damage Western economies, the key Middle Eastern producers are likely to suffer.

The effect upon the Western world was equally dramatic in that oil prices affected directly or indirectly everybody. The concept of resource geopolitics was born. The response was to initiate conservation measures and to redouble efforts to locate other sources of oil. The significance of both these moves became apparent in 1979 when, following the Revolution, Iranian oil production was suspended and prices were again raised to reach in early 1980 $43 per barrel. The combination of enacted conservation procedures and supplies from new sources meant that the global effect did not compare with the events of 1973–74 and since then, the behaviour of OPEC has been tempered.

However, another factor was activated signalling the rise of the spot market in importance. Increasingly, the price of oil was being set in the Rotterdam Market. This spot market allowed short-falls and surpluses to be accommodated with the result that the price of oil became defined as the price at which marginal barrels changed hands. With the sudden disappearance of Iranian production in 1979, surplus oil was swiftly purchased and as a consequence the spot price rocketed. The long-term result has been that the corporations have effectively surrendered control of oil prices at the margin and have become more reliant upon the spot market. This has led to the emergence of the 'Wall Street refiner', the risk manager, a new category of oil man. As more sophisticated measures have been introduced including options and futures, new volatility has come to oil prices. Oil prices are now dominated by a trading system in a manner incomprehensible to most people. Prices depend not upon oil availability but upon psychology. Meanwhile, to develop economically and politically, the Middle East requires stability and this is in no small measure related to the stability of oil prices. Yet again power would appear to reside far from the oilfields.

Reserves, production and consumption

Whatever the effects of globalization in moving capital around the world and in its selective use of labour, resources are located within states and are under the jurisdiction of those states. Thus, despite current volatility in oil prices, the oil-rich producers of the Middle East remain, in the long term, in control of the industry. This is illustrated by current figures from *BP Statistical Review of World Energy* (1998). The complete pre-eminence of the Persian–Arabian Gulf producers is illustrated in Table 9.3.

The R/P (reserves to production) ratio is the accepted measure of longevity of the reserves. It is calculated as the number of years which the reserves will last at present rates of extraction. Figures are likely to change because reserves may be reassessed and extraction methods are likely to be improved. However, the present situation is that, in the context of requirements, the reserves of the USA and the FSU are very limited. Other than the countries of the Persian–Arabian Gulf, major potential suppliers are Mexico, Venezuela, Nigeria and Libya. Mexico and Venezuela have comparatively large reserves but both are developing countries and will increasingly need to use their own oil. Nigeria is also developing and, over the recent past, has been politically unstable. Libya has for some while been considered, as a result of its leadership, a maverick state and has only recently been released from UN sanctions. At present, the only other reserves of significance are in China

Table 9.3 Proved oil reserves 1997

Country	Percentage of world reserves	R/P ratio
USA	2.9	9.8
FSU	6.4	24.7
Persian–Arabian Gulf	65.2	87.7
Mexico	3.8	33.6
Venezuela	6.9	59.5
Nigeria	1.6	20.2
Libya	2.8	55.6

R/P ratio = reserves to production ratio.

which is desperately searching for other sources of supply. The next major reserves in order of size are those of Norway and Algeria. Thus, as oil conditions become tighter, options outside the Persian–Arabian Gulf are very restricted.

Within the Middle East there are six countries with major proved reserves (Table 9.4).

Saudi Arabia has more than a quarter of the world's proved reserves of oil while Iraq, Iran, Kuwait and the UAE all have about 9–11 per cent. Both Iran and Iraq have recently reappraised their reserves resulting in a relative downgrading of Kuwait which, until 1987, was clearly the second most important country in the world. In the remainder of the Middle East, Oman has 0.5 per cent of proved world reserves while Qatar, Yemen, Egypt and Syria have smaller amounts.

Given their reserve base, the countries of the Middle East are all producing well below capacity (Table 9.5).

The figures for Iraq are, of course, constrained greatly by UN sanctions. In 1989, Iraq produced 4.5 per cent of the world's oil, a figure which could be compared with 4.6 per cent by Iran in the same year. Table 9.5 needs to be examined in the light of the fact that the USA produced 10.9 per cent of world oil production and the countries of the FSU 10.5 per cent. Thus both are well out of phase with their reserve base. However, the major contrast can be seen in the pattern of consumption (Table 9.6) below.

Table 9.4 Proved reserves

Country	Percentage of world reserves	R/P ratio
Iran	9.0	69
Iraq	10.8	100+
Kuwait	9.3	100+
Saudi Arabia	25.2	79.5
UAE	9.4	100+
Libya	2.8	55.6

R/P ratio = reserves to production ratio.

Table 9.5 Production

Country	Percentage of world production	Percentage change over 1996
Iran	5.3	0.2
Iraq	1.7	94.3
Kuwait	3.0	0.3
Saudi Arabia	12.9	3.5
UAE	3.5	2.9
Libya	2.0	1.6

Table 9.6 Consumption

Country	Percentage of world consumption	Percentage change over 1996
Iran	1.8	2.3
Kuwait	0.2	10.4
Saudi Arabia	1.6	3.8
UAE	0.5	—

The USA consumes 24.9 per cent of world production and the states of the FSU 5.9 per cent. In marked contrast, the states of the Middle East are all very minor consumers. The difference between production and consumption indicates the imports required and therefore, in geopolitical terms, vulnerability. If vulnerability is assessed in the light of R/P ratio, it will be seen that both the USA and the countries of the FSU must become increasingly dependent upon oil from elsewhere, eventually the Persian–Arabian Gulf. The situation is even more dire for the countries of Europe which possess 1.9 per cent of the global reserve base, while producing 9.4 per cent and consuming 22.1 per cent. This analysis reveals clearly the current importance and future enhanced importance of the Middle East in the oil industry. Furthermore, oil from the GCC countries is the cheapest in the world to produce.

Refining capacities

Several countries in the region are currently pursuing a policy of expanding refinery capacities to reorientate their exports towards refined oil products. Since 1980, the refining capacity has risen by more than 70 per cent, but by world standards the total remains low. At present, Kuwait can refine some 40 per cent of its daily oil production, Saudi Arabia over 20 per cent but the UAE under 10 per cent. As listed in the *BP Statistical Review of World Energy* (1998), only five of the main Middle Eastern oil-producing countries have significant capacity (Table 9.7). To these can be added Turkey which has 0.8 per cent of world capacity (675,000 barrels per day).

Table 9.7 Refining capacity

Country	Thousands of barrels per day	Percentage of world capacity
Bahrain	250	0.3
Iran	1,305	1.6
Iraq	530	0.7
Kuwait	800	1.0
Saudi Arabia	1,720	2.2

The total Middle Eastern capacity is therefore just over 8 per cent and this figure can be compared with 3.7 per cent for the whole of Africa, 21.1 per cent for Europe and 23.9 per cent for North America.

With approximately 70 per cent of global proved oil reserves, the level of refining capacity is very low. In addition to attempting to close the gap between production and refining capacity, several countries, including Saudi Arabia, Kuwait, Oman and UAE are moving downstream through the purchase of joint ventures. It is expected that both upstream and downstream capacities will be significantly increased in the region as oil prices recover.

Oil trade

Every country is affected by the oil industry, which is now so vast that its network of inter-relationships influences the entire world. It is in effect the only truly global industry. As a result, oil also dominates world trade. This is apparent from the foregone analysis which shows clearly that major areas of oil production and consumption are separate. In general, the developed world is the dominant consumer while producing relatively modest amounts. The focus of inter-area trade is the Middle East. Table 9.8 shows, for each major area of consumption, the percentage of its oil imports from the Middle East and from North Africa. Each of these percentages is also tabulated as a percentage of the total exports of either the Middle East or North Africa.

The Middle East is by far the dominant supplier of oil to Japan and the other developing industrialized countries of Asia-Pacific apart from China. For Western Europe, Middle Eastern oil predominates while for the USA it is second to production from South and Central America. North African oil, most importantly that from Libya, is mostly sent to Western Europe. Thus, the global trading pattern merely underlines the present and long-term importance of the Middle Eastern region as a whole for the global oil industry.

Table 9.8 Inter-area trade

Countries	Percentage from Middle East	Percentage of Middle East oil exports	Percentage from North Africa	Percentage of North African oil exports
USA	18	10	3	12
Western Europe	40	21	21	72
China	28	2	—	—
Japan	77	24	0.4	1
Other E. Asian and S. Asia	79	33	1	3

Source: BP Statistical Review of World Energy (1998) in which Middle East and North Africa are separate, the former being the countries of the Persian–Arabian Gulf.

Oil by country

The economy of Saudi Arabia is, like most of those of the GCC, completely dominated by the petroleum sector which produced 75 per cent of government revenue and almost 90 per cent of export earnings. Saudi ARAMCO is the largest oil company in the world and the country has over 25 per cent of world proved reserves and over one-third of the reserves of OPEC. Reserves and production are concentrated in al-Hasa Province in eastern Saudi Arabia, the location of the four largest fields. Ghawar is the largest oilfield in the world and Safaniya is the largest global offshore oilfield. Together with Abqaiq and Berri these fields total approximately 45 per cent of reserves and 85 per cent of production capacity.

Nonetheless, despite its dominance of the industry, exploration has continued in Saudi Arabia and a new group of fields south of Riyadh, producing high-quality light oil, came into production in late 1994. The area of the Empty Quarter has also received attention and in particular it is expected that production will begin for the end of the century in the Shaybah field. Development in the southern part of the Empty Quarter can be related to the emphasis being placed upon the delimitation of the boundary with Yemen and also upon the possibility of the construction of a pipeline directly to the Indian Ocean coast. Such a pipeline would, of course, allow access to the global market avoiding the choke points of Hormuz, Bab el Mandeb and the Suez Canal. At present, the major pipeline is that linking Abqaiq and Yanbu across the peninsula which has the capacity of 5 million barrels per day.

The economy of Kuwait is even more heavily dependent upon oil earnings than that of Saudi Arabia. Kuwait has concentrated on developing an extensive network of offshore refineries and product outlets which now market some 40 per cent of exports. During the 1990s, the industry has of course been constrained by the damage to refineries sustained during the Iraqi invasion.

Production in the UAE is dominated by Abu Dhabi which has about 90 per cent of the resources. New fields have been found particularly offshore, and proved reserve levels are continually rising. There has also been a significant expansion in refining capacity, although this is still low, and in the petrochemical industry.

In Libya, oil accounts for about 95 per cent of export earnings and 30 per cent of GDP. The industry originally developed in the east of the country, inland from the Gulf of Sirte, but recent discoveries have concentrated efforts offshore to the west in the neighbourhood of the boundary with Tunisia. Although not so seriously affected, like Iraq the oil sector has suffered from UN sanctions. In a similar manner to Kuwait, Libya has expanded its offshore refining and distribution network, particularly in Western Europe.

Before the Gulf War, oil counted for some 95 per cent of Iraqi foreign earnings. But since then output has suffered from sanctions, war damage and

the increasing obsolescence of equipment and infrastructure. Foreign investment is being sought and contracts for oilfield development have already
been signed with Russia and China while France is waiting in the wings.

Following the war with Iraq, Iran has begun strenuously refurbishing its
oil industry. There are at present eight refinery complexes and further development is planned. Trade links with Russia have been fostered but the country
is still embargoed by the USA. Thus, major improvements are likely to be
limited and any share in the profits from the Caspian Basin development will
be severely restricted.

Of the other producers in the region, Qatar and Oman are relatively small
producers with limited reserves. Development in Qatar is directed offshore
but is hampered by its continuing dispute with Bahrain over the Hawar
Islands. In Oman, several smaller fields to the north of Marmul, the main
producing area, have been developed but the most commercially viable oil
is found in the south-west. One oil refinery is on the north coast at Mina
al-Fahal but plans are being studied for a second refinery in the south near
Salalah.

Of the other Middle Eastern producers, the most interesting is Yemen,
the one country in the Arabian Peninsula not to have benefited significantly
from the oil boom. Both North and South Yemen, then separate countries,
became producers in 1987 but efforts are now concentrated predominantly
in the south and east at Masila and around Shabwa. At present, production
is concentrated in the Marib Basin from which there is a pipeline to the Red
Sea. However, further development is hindered by the lack of an internationally recognized boundary with Saudi Arabia.

Natural gas

At present, the Middle East plays only a minor part as a gas producer but
this masks immense potential (Abi-Aad 1998). The main discernible trend
in the sector at present is the vast increase in home consumption of
natural gas throughout the region. The domestic use of natural gas frees
increasing amounts of crude oil for export. However, during a period of low
oil prices the economic effect has been disappointing. Furthermore, foreign
consumers have, where possible, used natural gas to reduce dependence upon
oil imports.

The Middle East, including Egypt and Libya, has 35.1 per cent of natural
gas proved reserves. There are nine super-giant fields, defined as those with
over 1,000 bcm of gas, and reserves have doubled since 1987. The predominance of the region is shown by the fact that there are only twenty super-giant
fields world-wide. The proved natural gas reserves of the major producers
are shown in Table 9.9.

The table illustrates clearly the regional dominance of Iran and, in contrast
to its oil production, the importance of Qatar. These six major producers
account for approximately one-third of world natural gas reserves. However,

Table 9.9 Proved natural gas reserves

Country	Percentage of world reserves	R/P ratio
Iran	15.8	100+
Iraq	2.2	100+
Kuwait	1.0	100+
Qatar	5.9	100+
Saudi Arabia	3.7	100+
UAE	4.0	100+

R/P ratio = reserves to production ratio.

rates of production are limited, a point clearly indicated by the R/P ratios. In North Africa, Egypt has 0.5 per cent and Libya 0.9 per cent of global reserves. Since, globally, oil and gas reserves are roughly equal, it would seem that natural gas reserves in the Middle East have been severely underestimated. At present, the ratio of oil reserves to gas reserves in the region is more than 2:1. Therefore, the potential for new discoveries of associated gas must be large. Furthermore, world-wide the ratio of non-associated gas reserves to associated gas reserves is 5.7:1. In the Middle East, the ratio of known resources is approximately 2:1 and therefore it would be expected that significant volumes of non-associated gas remain to be identified.

This increasing focus upon natural gas represents a major change in the region in which exploration has concentrated on the search for oil. As late as 1976, nearly 80 per cent of the region's total production of natural gas was flared. It is now particularly important as a feedstock for the petrochemical industry.

For the global economy, natural gas is of increasing importance. For the region, a vital factor is that gas diplomacy is largely free of the Cold War, East–West, conflict which so bedevilled the development of oil. The development of reserves, and the construction of pipelines, liquefaction units and power plants have been aided by Western transnational corporations.

Natural gas by country

After the Russian Federation which alone has proved reserves equal to those currently attributed to the whole of the Middle East, Iran has the world's largest natural gas reserves. Supplies are both associated and non-associated and are found in fields onshore and offshore. Agreements have already been reached for supplies to southern Russia and there is huge potential in the market of the Indian subcontinent. The future will depend, among other factors, upon improved relationship with the USA and upon developments in the Caspian Basin. Turkmenistan, the key gas producer to the north, has already agreed to links with the Iranian pipeline network.

Qatar has the third largest reserves but development is so far modest. Exploitation has been hindered by the Gulf War but expansion is underway with the construction of a liquefaction plant. Elsewhere, reserves are by present estimates more modest but in many cases little effort has been made to estimate production potential. In general, demand has increased following expansion in the petrochemical sector.

Oil and development

For the countries of the Persian–Arabian Gulf, oil has been critical in development. This is, however, something of a mixed blessing in that the volatility of the oil market has at times resulted in major economic problems.

Broadly speaking, the 1970s was a decade of surplus, that of the 1980s was one of deficit and during the 1990s the signals have been mixed. During the 1970s in particular, the rate of change was astonishing. In one generation many inhabitants of the GCC countries passed from the camel to a Mercedes and such changes proved difficult to absorb. The result, depending upon the particular stratum of society, was reflected in excesses of materialism, religion and nationalism (Anderson and Rashidian 1991).

However, despite the well-publicized abuses, economic development on a grand scale did take place. In particular, urban living burgeoned as older mud-brick structures were rapidly replaced by lavish modern development. Transport infrastructures, from roads to airports, were vastly improved and at the same time large amounts were spent on education and social services. Agriculture and industry were greatly assisted by subsidies. Thus, while living conditions undoubtedly improved significantly, any underlying economic problems never needed to be addressed.

With the 1980s came cut-backs all round following reduced levels of revenue available from the petroleum industry. There was a positive feedback effect described by Hunter (1986): 'These developments have induced an exodus of immigrant workers which, by reducing economic demand, rents and real estate values, has further depressed business'.

At the same time, the Iran–Iraq War, in which the GCC states supported Iraq, further drained resources. As economic expectations declined, so social divisions became exacerbated. In general, the vast revenues received had masked mismanagement and led to a disregard for the norms of economic discipline. However, many of the lessons have been learned and during the 1980s and 1990s government restructuring has allowed the imposition of economic discipline resulting in more long-lasting development.

Nonetheless, it has to be included that while oil has provided massive development opportunities, as yet the full potential has not been realized. Oil and natural gas production offer enormous benefits to the producers but dependence upon them seems likely to enhance the range of cleavages already apparent in their societies.

Mining

The mining industry is concerned with the extraction of minerals from the ground and it therefore subsumes the petroleum industry. However, oil and natural gas are so obviously important in the Middle East compared with any other mineral commodities that they have been considered separately. This section is concerned entirely with non-fuel minerals.

In the regional reports of the US Bureau of Mines, minerals are subdivided into metals and industrial minerals (Bureau of Mines 1993a, 1993b). However, in the Middle East there is the more obvious distinction which can be characterized as 'geological', between hard rock minerals and evaporites. If these two categories are used, a further broad geographical division follows. Only two countries in the region, Turkey and Iran, have extensive upland areas comprising rocks from across the geological time-scale. These two are the major producers of hard rock minerals. Saudi Arabia also has relatively extensive uplands as does southern Sudan but, for different reasons, neither country has as yet developed its non-fuel mineral industries. Indeed, since the mineraliferous complex which has given rise to the Rand Mining Industry of South Africa is replicated in southern Sudan, there must be a significant potential for development. The remaining countries of the region, while in many cases having some upland areas, are broadly plains or low plateaux covered in recent sediments. It is in the semi-arid and arid low areas of these countries that evaporites have been deposited.

On a global scale, the only non-fuel mineral industry of importance within the Middle East is that of Turkey. A wide variety of minerals is present but the most notable developed so far have been bauxite, iron ore, copper, chromium, zinc, lead and mercury. Of these, since accessible world supplies are relatively limited, the most significant is chromium. Indeed, chromium is considered a strategic mineral, possibly the most strategic, and Turkey is the only European supplier of any significance.

Iran has a less-developed industry but an inventory of forty-three minerals. Of these, chromium, copper, gold and iron are the most important. Saudi Arabia produces bauxite, iron, gold, silver and zinc while Oman has an age-old copper industry.

The main evaporites produced are gypsum, phosphate, potash and salt. The major producer of phosphate and potash, key fertilizer raw materials, is Israel, followed by Jordan. Syria is the third largest producer of phosphate rock. The most important sources of salt are Turkey and Israel while Iran dominates gypsum production.

As yet, the non-fuel minerals of the Middle East have been under-exploited and, for most of the area, there has been relatively little exploration. As more accessible global sources decline, more interest will be shown in Middle Eastern deposits, particularly those of Turkey, Iran and Sudan. Furthermore, as part of their diversification strategy, the states of the region are looking to increase mineral prospecting.

Manufacturing industry

Since the earliest times, craft industries have been widespread throughout the Middle East. In particular, metal working gained an international reputation and the ancient tradition continues, often in modern workshops. As tourism grows throughout the region, other craft industries are likely to be resurrected as in Oman where, until recently, the production of traditional rugs had ceased.

The most obvious link between craft industries and the modern era is provided by textiles. Cotton, flax, silk, wool, camel hair and goat hair, are all produced within the Middle East and artificial fibres have been introduced as part of the downstream development of the petroleum industry.

Among the better known projects are mohair from the Angora goat which is produced in Turkey and carpets and rugs for which parts of Iran and Turkey are world famous. The highest quality carpets are tinted by hand-made dyes and have become objects of investment as they retain their value. The textile industry in modern form, an important producer of industrial goods, is listed for every country in the region other than the members of the GCC and Jordan (CIA 1998). Indeed, textiles have been represented at each stage of industrial development from craft through early processing of agricultural products to the introduction of modern light industry. The only other industry common to the same list of countries is food processing.

Textiles and food illustrate the obvious division between the countries of the region into: (a) those with long industrial histories and (b) those reliant almost totally upon oil. The former group may include petroleum among its industrial products but tends to have a diversified list. The countries of the GCC are virtually limited to petroleum products, evaporites in most cases and cement. Jordan has only evaporites and cement.

Industrial diversification among the major oil-producing countries has resulted from the desire to obtain maximum return for oil products. This has led to downstream production and to the use of surplus energy for other industries such as aluminium smelting which is found in Bahrain.

From the mid-1970s, the GCC countries in particular began the development of their petrochemical industries to broaden their industrial bases, to enhance the profits from oil and to invest surplus revenue. Downstream development also presented an opportunity to use rather than flare off natural gas and appeared timely given the contemporary optimistic forecasts about world agriculture and the need for fertilizers. A further bonus was that many of the major oil companies were keen to co-operate in joint ventures and as a result petrochemical complexes were established in Saudi Arabia, Qatar, Kuwait, Iraq, UAE and Oman. However, by the late 1970s, there was a decline in the industry as the world economy had gone into recession and there was over-capacity. More fundamental was the fact that the development of a petrochemical industry in the Gulf was a very high-cost endeavour. All key components had to be imported and, in many cases, there was little available infrastructure, a shortage of skilled labour and surprisingly frequently a lack of feedstock. Since much of the natural gas of the region is associated with oil, cutbacks in oil production means that less gas is available. Most critically, as the industry had moved downstream, it encountered competition from the industrially developed world. The less sophisticated products required by the developing world were not espoused and there was little co-operation within the GCC with the result that there was also competition within the Gulf itself.

The current position is that there is, throughout the region, a new outward-looking industrial strategy seen in the diversification of markets, the developments of new projects, economic deregulation and the introduction of economic reforms. Given current global economic optimism, expansion can be expected in the textile industries together with those producing intermediate oil and gas products, petroleum refining, petrochemicals, fertilizers and aluminium. The transformation of the entire globe to a market economy has opened an increased range of opportunities but trade barriers remain and there is likely to be increased competition from Asia-Pacific.

Manufacturing industry by country

Among the countries with a long manufacturing history and well-established diversification, Turkey has been particularly buoyant. While textiles and food remain important, there has been a move into higher technology industries including electronics and car production. The largest sector remains textiles which is responsible for approximately 20 per cent of manufacturing output. The development of the car industry has been particularly significant because of low labour costs and access to the Western European market.

However, the state which has depended most heavily upon research and development and the use of technology is Israel. The country first developed a high-class arms industry and has subsequently specialized in niche markets. Research and development is now concentrated on electronics and computing in particular and the country is a world leader in fibre optics. As a result, Israel alone among the countries of the Middle East has reached what might be described as developed country status. In sharp contrast, the manufacturing sector in Palestine has been seriously limited by political difficulties experienced with Israel. These include border controls and a decrease in subcontracting.

Since 1995, the restoration of law followed by the rejuvenation of the infrastructure and economic activities has led to accelerating manufacturing industry growth in Lebanon. There has also been rapid growth in Syria resulting from an increase in investment which followed the Gulf War. However, modernization is desperately needed before the country can attain its full industrial potential.

The countries most requiring industrial rejuvenation and modernization are Iran, Iraq and Libya. The long war with Iraq severely restricted development in Iran, particularly through large-scale disinvestment. Sanctions in Iraq have almost totally constrained manufacturing which has undergone shrinkage and have limited development in Libya.

Yemen is at an early stage of industrial development and the main industries are connected with agricultural products, fisheries and oil. Egypt has diversified its manufacturing industry but production has slowed. Economic restructuring has boosted the private sector as has the move towards manufacturing under international franchises. The basic industries remain food processing, textiles, chemicals, engineering and the production of electrical equipment, while the vehicle assembly sector has been rejuvenated. The defence industry has increased considerably in size but this has coincided with a downturn in defence equipment procurement. In contrast, Jordan has a narrow manufacturing base with a particular focus on chemicals and fertilizer production. With so few resources, it seems unlikely that Jordan will ever develop a strong manufacturing base.

The least significant industrial sector in the Middle East, whether judged in terms of contribution to GDP or percentage of the work-force employed, is that of Sudan. There are shortages of raw materials, fuel, trained labour and investment while the infrastructure is generally deficient. In many sectors, for example textiles, the industry is running well below capacity. Joint venture companies may provide one answer but it will be difficult for Sudan to face global competition.

Among the oil-rich countries, Bahrain and Kuwait have managed the greatest diversification. In Bahrain, the focus is upon aluminium smelting, the alumina being imported from Australia, and the petrochemical industry. In Kuwait, the range of products is much wider but the smaller units in the manufacturing base were adversely affected by the Gulf War. Food processing, textiles and wood working were mainly serviced by Palestinians and other skilled foreign workers, virtually all of whom left because of the war.

In Saudi Arabia, the main expansion in manufacturing results from extensions to the petrochemical industry. This, with steel production, is almost all in the hands of the Saudi Basic Industries Corporation (SABIC). The engineering, metal processing, food and chemicals industries are mostly in the private sector and there is an ongoing process of privatization. The UAE is similar in that the industrial base is still dominated by chemicals and plastics. Oman is significantly different being in the early stage of industrial development and possessing a traditional manufacturing base including shipbuilding and weaving. The small internal market and competition, particularly from the GCC countries, has limited diversification.

The Middle East is not a major manufacturing region. Craft industry survives as does some traditional production such as *dhow* building. Throughout much of the region, light industry predominates particularly in the form of food processing and textile manufacture. The other key component is the petrochemicals industry together with associated downstream development but this faces strong global competition. Indeed, in all the countries of the Middle East there is some economic restructuring but if the region is to become competitive in an age of globalization, far more effort in this direction will be required.

Service industries

Three countries in particular have developed service industries: Bahrain, the UAE and Lebanon. Bahrain has 170 international and domestic financial institutions and, as oil resources are finally exhausted this economic base is likely to expand. Bahrain is ideally located as a distribution centre for the Gulf and, during the constant conflict in Lebanon it proved an attractive alternative to Beirut. Recently, competition from Cyprus and Dubai has increased but Bahrain's close geographical and political ties with Saudi Arabia and Kuwait are likely to be a continuing advantage. In the UAE, Dubai has grown as a centre for regional trade particularly through its free-port area at Jebal Ali. Re-exports are now the mainstay of its trading system. Lebanon is recovering from continual conflict and Beirut is being rebuilt. Before the war, it was the premier financial centre in the Middle East and the rejuvenation of this sector is a key government priority. With the reconstruction, foreign banks are now returning and the stock exchange has reopened.

Tourism

Of particular relevance within the service industry sector is tourism, although for much of the region this has been blighted by terrorism and various forms of violence. Tourism is vital for Egypt but, following the 1997 bombing in Luxor, tourism declined steeply. Like Egypt, Jordan has relatively inadequate tourist facilities, but it has benefited from the peace with Israel and could benefit further through bilateral agreements with that country. Israel itself has suffered almost continuous violence of some form or another and this has limited tourist development. Following a successful peace process, tourism would be one of the first sectors to benefit. This also applies to Palestine, the growth of tourism in which has been hampered by the political situation. Lebanon, once the tourist focus of the eastern Mediterranean, has virtually lost the industry as a result of constant warfare.

Syria has immense tourist potential but like so many of its neighbours requires peace and an upgrading of facilities and the infrastructure. In Libya, tourism has been decimated by sanctions while in the Sudan no improvement can be expected until the political situation stabilizes.

The one major tourist industry which has, with a few setbacks, developed continuously over the recent past has been that of Turkey. The country is ideally located to draw tourists from east and west and can combine a basic package of sea, sun and sand with cultural tourism.

In the countries of the Arabian Peninsula, tourism has, for a variety of reasons, been little developed. Political factors have hampered the sector in

Bahrain and Kuwait while in Saudi Arabia tourism other than for the annual Hajj is barely encouraged. The one oil-rich state in which efforts are made to encourage tourism is the UAE, particularly Dubai. The attractions are sand, sun and sea together with a relatively free society and tax-free shopping. Furthermore, there has been great expenditure upon the improvement of service and leisure facilities.

In contrast, Oman, with a great deal to offer the cultural tourist, has been slow in developing the sector. Quality tourism is now being actively promoted to help underpin economic diversification but tourism is still on a small scale. Like Oman, Yemen also has great tourist potential but facilities are undeveloped and the infrastructure is poor. The other major problem has been security but, with long-term political stability, Yemen could become a significant destination on a global scale for tourism.

10 The states of the Middle East

The states of geopolitics

In *The Middle East* (Fisher 1978), there was no section devoted specifically to states. The geography of the Middle Eastern countries was woven into regional geography, Part III of his book. The broadly based regions were effectively formal regions in that they were defined essentially by macro-landforms. This pattern accorded with the thinking within geography of which regions were taken to be characteristic. In the same way that history deals with periods of time which can be built up into a total chronology, so geography is concerned with areas of space which can be aggregated to produce a world picture.

Since the 1970s, regional geography has largely been superseded as a primary geographical approach. As the question of application has increasingly arisen, so it has been realized that the use of formal regions has a number of limitations. Most significantly, and this problem was debated endlessly throughout the earlier part of this century, no two geographers could agree on the exact alignment of the most appropriate boundaries. Indeed, it was only with the large-scale regions that there was anything approaching accord. The core of a region, encapsulating its key characteristics, did not present great problems of identification, but the position of the line separating two core areas was almost impossible to demarcate in any accurate fashion. Thus, in effect, formal regions remained a question of dispute among geographers and a mystery to everybody else.

However, even should they be delimited in any generally agreed fashion, because they rarely coincided with political boundaries it was extremely difficult to obtain reliable data for the areas covered. Partly to overcome this problem, but also as a result of the development of systematic geography which in many senses replaced regional geography, the concept of the functional region was developed. This comprised the area of influence or operation of any geographical entity from a farm to a city or a state. For functional regions, there were many objective measures of where the boundary should be placed. In the case of states, government control and territorial integrity are the most significant. At this point, it should perhaps be indicated that

Fisher (1978) identified nine regions within the Middle East, but five of those – Iran, Cyprus, Egypt, Libya and Sudan – were also states. Thus, even within a regional approach, there was the need for something of a hybrid system.

The basic belief behind regional geography was to distil the major characteristics of a region and, as a result, to be in a position to make some rational judgements and even predictions about that region. Functional regions allow the same purpose to be achieved, but more reliably. If states are the functional regions, then the key elements of the state can be identified and it is possible to make a reasoned assessment of the position and potential role of the state as an actor upon the world stage. It is realized in making any assessment that geography as such is the stage and influences the performers on it, but does not provide the script.

This role can be characterized as the field of geopolitics, which examines the potential influence of the current environment upon political decision-making. Geopolitics therefore involves the interplay of environmental, political, social, economic, historical and cultural factors, any or all of which may be influential. In Chapters 2 to 9 of this book, all these variables have been examined in the context of the Middle East as a whole. In Chapter 10, for each country of the Middle East, the relevant elements of each of these factors which comprise geopolitics will be considered. Thereby, a geopolitical portrait of each country will be produced and this will provide some basic understanding of the actions of each country when the key geopolitical issues which affect the Middle East are discussed in Chapter 11. However, it is accepted that this geopolitical approach to international relations (IR) cannot be complete in that the actions of factors external to the state, together with those of individuals within the state, may affect decisions and the outcome of events. Therefore, the portrait of each state has been produced to include these factors. It is intended thereby to provide some increased understanding of what happens in the Middle East, but there is always the possibility of something which was least expected. Therefore, conclusions drawn are more likely to be at the level of anticipations, rather than forecasts or predictions.

The geopolitical elements

Although they are by no means alone, states can be envisaged as the major actors on the world stage since others, in many cases equally powerful, such as multinational corporations, lack political legitimacy. The pattern of states provides the relatively stable system within which international interaction occurs. Despite its characterization as a turbulent area, the state framework of the Middle East has seen only three fundamental changes in the past twenty years. In 1979, following the revolution, Iran became a theocratic republic. However, this change affected style rather than pattern, since Iran and Persia before it, has a history of statehood from antiquity. Geopolitically, the most significant changes were the unification of the Yemen Arab Republic and the People's Democratic Republic of South Yemen in 1990 and the establishment of Palestine in 1994 as a separate political entity.

The dynamics of the system are provided by the interactions between the states, which range from trade and aid to intelligence and drug trafficking. These flows produce changes, sometimes rapid and sometimes gradual, which result in enhanced stability or instability. The controls are provided by the UN, the ICJ and as the one superpower, the USA.

This model, at the international level, is repeated at national and local levels. Within each state, key actors provide the relatively fixed elements and the dynamism is reflected in the exchanges between them which lead to change. However, in both the state and the local models, there is a legitimate system of control, which does not occur at the global level. The outputs from local level interactions may exert influence at the national level, the outputs from which, in turn, may affect the global stage. Although ripples from local or national level may be of international significance, in general, the focus of geopolitics is likely to be at the global level where states and groups of states are the dominant actors.

In the Middle East, the nineteen states provide the stable framework. Flows between them, such as trade, the movement of people and diplomatic signals introduce the dynamic elements which result in change. In 1990, the result was, in one part of the Arabian Peninsula, a merger between the two Yemens and at the opposite end, the invasion of Kuwait by Iraq. In the latter case, control was exercised by the USA, acting through the UN and later, through the UN Commission, which re-surveyed and demarcated the boundary between Iraq and Kuwait.

The aim is to produce a synopsis of the key facts about each state which influence its ability to act upon the Middle Eastern or world stage and thereby to provide illumination for the sections on geopolitical issues which follow in Chapter 11. The factors identified are therefore indicators of power, defined very broadly. Indeed, the whole approach must be broadly based in that the portrait which emerges must be far more significant than any of its constituent elements and that each state is eventually defined, however inexactly, as a geopolitical entity.

To characterize each state geopolitically, eight key factors, each with various modifications, have been selected. Since the intention is to identify the ability to act within the Middle Eastern context, statistics are totals, rather than *per capita* values and each is put in its regional setting. Clearly, events in the Middle East are not only controlled by the actors within that region, but their relative abilities need to be assessed. Potential external controls, particularly the more powerful such as the UN, the USA, the EU and Russia, have a greater influence upon international relations within the region, but their effect is a given, common to all states of the Middle East. That is not to say that the control exercised by the USA over Iraq and Israel is in any sense equal, but that both are subject to a degree of US control.

To select and describe the eight factors, a wide range of source material has been used. For the most current data, the following, each of which is part of a continuing series, were the most important:

1 Bureau of Mines (1993) *Mineral Industries of the Middle East.* Washington DC: Department of the Interior, Bureau of Mines.
2 Bureau of Mines (1993) *Mineral Industries of Africa.* Washington DC: Department of the Interior, Bureau of Mines.
3 Central Intelligence Agency (1998) *The World Fact Book 1997.* Washington DC: CIA.
4 Dostert, P.E. (1997) *Africa 1997.* Harper's Ferry, WV: Stryker-Post Publications.
5 Russell, M.B. (1997) *The Middle East and South Asia 1997.* Harper's Ferry, WV: Stryker-Post Publications.
6 The International Institute for Strategic Studies (1998) *The Military Balance 1997/98.* London: Oxford University Press.

Other notable sources include:

1 Brough, S. (Ed.) (1989) *The Economist Atlas.* London: Hutchinson Business Books Ltd.
2 Carpenter, C. (Ed.) (1991) *The Guinness World Data Book.* Enfield: Guinness Publishing.
3 Dempsey, M. (1983) *The Daily Telegraph Atlas of the Arab World.* London: Nomad Publishers.
4 Sluglett, P. and Farouk-Sluglett, M. (1996) *Guide to the Middle East.* London: Times Books.

At least since Kjellen (1917) there has been discussion on which elements are crucial in defining the power of states to act upon the world stage. The most basic are position with regard to other states, size, resources, including population and level of organization. However, apart from economic strength and although the two are often closely related, there is the question of military strength. At the present time, Japan is extremely powerful economically but this is not matched by its military strength. In contrast, the Russian economy is in some chaos but it remains highly significant militarily. The importance of individuals must also be taken into account. Turkey was resurrected from the ashes of the First World War by Ataturk, while the actions of Saddam Hussein have resulted in the destruction of much of the Iraqi economy. External support is also vital, the best example being Israel which is supported both economically and politically by the USA.

In selecting the key factors it was considered important that they should provide indicators of all the major characteristics of each state which could in any way be considered related to power projection. On the other hand, given the array of national characteristics and for each the depth of analysis possible, the accent was on clear, and if possible, measurable variables:

1   Location: the significance of geopolitical position relative to other states and potential problems.
2   Recent history: events of relatively long-lasting significance.
3   Area: total area in square kilometres, also expressed as a percentage of the total area of the Middle East. This variable has been modified by a short description of the main physical elements which comprise the national landscape.
4   Population: the total population also expressed as a percentage of the total Middle Eastern population. For this there were a number of possible modifiers, but percentage literacy was considered the most significant.
5   Economic strength: expressed principally as the GDP in dollars and that figure expressed as a percentage of the total Middle Eastern GDP. This was modified by a number of considerations, including particularly indigenous resources and external debt.
6   Military strength: indicated by the defence budget in dollars and that figure expressed as a percentage of the total Middle Eastern defence budget. This was qualified, where relevant, by comments on the size and quality of the military.
7   External support: potential support by other states, local, regional or global.
8   Individuals influential beyond their own states: either those of such significance earlier in the century that their influence remains, or those of recent or current vintage.

Following this framework, there is a brief assessment of all the main geopolitical issues which are of relevance to the particular state. Finally, the linkages between the states most obviously concerned with each issue are discussed.

Geopolitical setting

Geographical illumination of the political pattern of the Middle East can best be initiated by an examination of the overall spatial arrangement (Figure 10.1). The core land area of the Middle East has always been the land bridge, linking Asia, Europe and Africa, the crossroads of the world. Geopolitically, rather than rehearse the arguments for the full extent in any direction of the Middle East, it is more appropriate to identify the hub, the approximate rectangle of land, comprising the states of Iraq, Syria, Jordan, Lebanon, Israel and Palestine. In land area, Iraq and Syria are relatively extensive, while Palestine is a potential micro-state. In terms of power projection, Israel has been, since 1948, predominant. From this hub, the remainder of the Middle

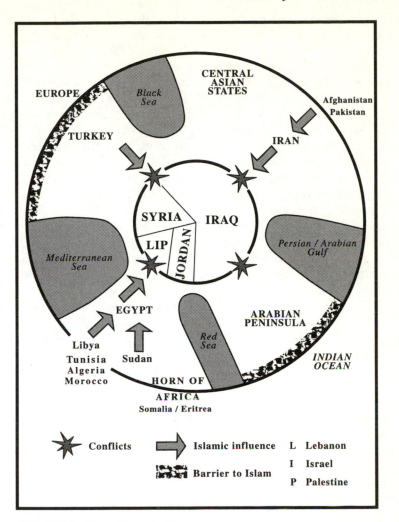

Figure 10.1 Middle East schematic diagram.

East can be envisaged as four axes, each trending in an approximately inter-cardinal direction:

1 to the north-west: Turkey;
2 to the north-east: Iran, linking with the Muslim states of the Trans-Caucasus and south-central Asia;
3 to the south-east: the Arabian Peninsula; and
4 to the south-west: Egypt, linking with Sudan and the Maghreb.

It can be seen that to the north-west and south-east, the Middle East ends abruptly, in one case against non-Muslim Europe and in the other, against the Indian Ocean. Towards the north-east and the south-west, the Middle East becomes attenuated, forming what has become known as the 'arc of crisis'. To the north-east, the influences in either direction are predominantly Muslim, to the south-west, they are Muslim and Arabic.

Thus, while the four axes appear to indicate a fundamental symmetry, there are obvious asymmetries. Middle Eastern relationships are far more extensive along the approximate south-west to north-east axis, than along any other. However, even in this, there is a major difference. The states beyond Iran, largely through Islam, tend to have close ties with the Middle East, but are not normally considered part of the Middle East. Beyond Egypt, Libya and at least northern Sudan are certainly considered Middle Eastern, while cases have been advanced for Mauritania as the western outpost of the region.

Three states adjacent to the hub: Turkey, Iran and Egypt, are, in terms of population and potential influence, major actors. To the south-east, despite the oil wealth, the total population of the Arabian Peninsula approximates to only half that of each of the other three states and there is a clear-cut power imbalance.

Furthermore, despite the preponderance of Arab rather than Muslim influence as a defining factor for the Middle East, only one of the major actors, Egypt, is an Arab state. Turkey looks towards Western Europe and towards the Middle East. It was a member of the Baghdad Pact, later the Central Treaty Organization (CENTO), and of NATO. As a secular Muslim state, it provides a model for the newly independent Muslim republics of the FSU. At the opposite pole, the Middle East terminates abruptly as the Arabian Peninsula is, in effect, a cul-de-sac. It is basically a wholly Arab environment, but it is influenced by labour migration, in the case of some states to a conspicuous degree, by non-Arab, frequently non-Muslim countries of south and south-east Asia. When the high Western profile is added to this, it can be appreciated why what in essence is the most, can often seem the least Middle Eastern region.

Immediately to the east and the west of the hub, are respectively the Persian–Arabian Gulf and the eastern Mediterranean Sea. The littoral states of both, other than along their northern shores, are Arab and in each case the dominant state to the north is non-Arab, but Muslim. There is in this similarity between the positions of Turkey and Iran, although the former shares the northern coastline of the eastern Mediterranean with non-Muslim states. Both water bodies have disputed islands, Cyprus in the case of the Mediterranean and many smaller islands in the Gulf.

To the north of the hub is the Caucasus, a region of significant Muslim influence between the Muslim-dominated Caspian Sea and the Black Sea. To the south, the over-riding feature is the Red Sea, since the independence of Eritrea, a sea with a totally Muslim littoral. To the south again is the Horn of Africa, like the Caucasus, a region of continuing conflict.

Conflict, or the potential for it, is evident along each axis, for example:

1   to the north-west: Turkey and Greece;
2   to the north-east: Afghanistan;
3   to the south-east: Yemen; and
4   to the south-west: Sudan, Libya and Chad, Algerian and Western Sahara.

However, although the Aegean Sea and Afghanistan, following the invasion of the Soviet Union in 1979, could be considered global flashpoints, the major foci of conflict have been around the periphery of the hub. Within the hub itself, the three largest states in terms of land area – Iraq, Syria and Jordan – have each had large-scale disagreements, but it is at the four corners of the rectangle that the most obvious confrontations have occurred:

1   in the north-east: originally Hatay and later the Kurdish problem;
2   in the north-west: Iran and Iraq;
3   in the south-east: Iraq and Kuwait; and
4   in the south-west: Israel with all its Arab neighbours and the Palestinians.

By simplifying the geography, the key factors in Middle Eastern geopolitics can be emphasized.

The region can be appreciated in relation to the core or hub, which comprises six states. Around the periphery of this are the major geopolitical flashpoints and from it extend four axes, each of which exercises a distinctive influence upon the Middle East.

Bahrain

Bahrain (Figure 10.2) comprises a group of islands located centrally off the western coastline of the Gulf. This is a highly strategic location with regard to the major global petroleum production and reserves. The surrounding states are Saudi Arabia, Kuwait, Iraq and Iran, all major, or in the case of Iraq potentially major, oil producers. To the south is Qatar which, with Iran, is a key source of natural gas. The main tanker routes from all of the four key oil states pass near Bahrain. Given the global significance of the region, oil is likely to be a benefit and also a disadvantage.

Bahrain has undoubtedly benefited from its proximity to Saudi Arabia, particularly since the opening of the causeway in 1986. With a population which is 40 per cent Sunni and 60 per cent Shi'a, there must remain a lingering fear of Iran, even though all official claims to the territory of Bahrain have been abandoned. There are also potential problems resulting from location in that the only exit from the Gulf is through the Strait of Hormuz and that is likely to remain a contentious choke point. Locally, Bahrain has benefited from being an offshore island, developed early as a trading centre, as this has allowed it to diversify its industry and commerce in the light of the decline in its oil resources. In the future, Bahrain might develop a Singapore- or Hong Kong-type status.

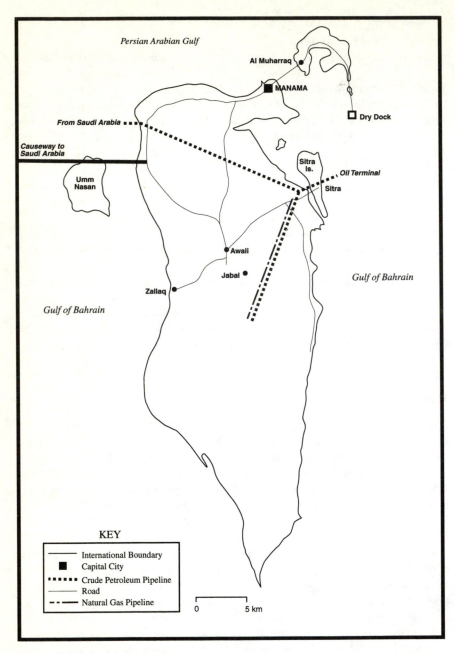

Persian Arabian Gulf

Al Muharraq

MANAMA

Dry Dock

From Saudi Arabia

Causeway to
Saudi Arabia

Umm
Nasan

Sitra
Is.

Oil Terminal

Sitra

Awali

Jabal

Gulf of Bahrain

Zallaq

Gulf of Bahrain

KEY

— International Boundary
■ Capital City
••••• Crude Petroleum Pipeline
— Road
—•—• Natural Gas Pipeline

0   5 km

Figure 10.2 Bahrain.

The main events in recent history have been the Gulf War and periods of internal unrest. Of all the Gulf sheikhdoms, Bahrain is the only one to have suffered a series of serious riots. Given the general downturn in the economy and the make-up of the population, further unrest cannot be discounted. Bahrain became independent in 1971 and has come to be one of the more economically and socially advanced states in the region.

With an area of under 100 km², approximately 0.006 per cent of the Middle East, Bahrain is a micro-state, comprising predominantly fragments of a low desert plain. The population of 600,000, 0.18 per cent of the total Middle Eastern population, confirms the micro-state status. However, 85 per cent of the population is literate, one of the higher figures in the Middle East. Nevertheless, in the working age group some 45 per cent of the population is non-national.

The economic strength of Bahrain was built upon oil resources which were exploited early and are now in sharp decline. However, diversification has been successful, supplemented by services such as offshore banking and tourism. The GDP of $5 billion represents 0.7 per cent of the total Middle East GDP and is therefore relatively high in proportion to the size of the country.

With an active armed forces membership of 11,000, Bahrain has a small army supported by a very small navy and air force. It is a micro-scale, but effective force. However, in any major emergency it would require support which is available from Saudi Arabia locally, but also from the USA and the EU. Originally, Britain provided the major military support for Bahrain, but this role has been superseded by the USA. The defence budget of $260 million represents 0.6 per cent of the total Middle Eastern defence budget and is broadly in line with other national characteristics. The ruling Al Khalifa family has long held sway, but its influence beyond Bahrain has been negligible.

Geopolitically, Bahrain is a micro-state with certain indigenous problems, but a sound economy. It is chiefly of global significance for its location which is highly strategic and which ensures external support should problems arise. The major geopolitical issues result from this location which means that Bahrain is always likely to be implicated in any Gulf conflict over oil. The Sunni–Shi'a balance in the population provides grounds for continuing internal unrest, although official Iranian claims may have ceased. There is a continuing boundary dispute with Qatar over the Hawar Islands which are claimed historically by Bahrain, but abut closely on to the coastline of Qatar. Their significance is in relation to petroleum resources.

Other potential geopolitical weaknesses include dependence on Saudi Arabia for oil, for refining and more particularly for water and dependence upon expatriate labour. The 27 km long causeway which links Bahrain and Saudi Arabia is itself geopolitically significant. In the same way that the channel tunnel has affected, politically and psychologically, the links between the UK and mainland Europe, so the causeway means that Bahrain is no longer insulated from the Arabian Peninsula. The manner in which each country chooses to use the causeway will dictate its long-term effect. Should Bahrain become, like so many offshore islands in a similar position, the centre for various forms of illicit activity, then relations with the Kingdom will sour.

Cyprus

In a strategic position, long considered key to the eastern Mediterranean, Cyprus (Figure 10.3) is the only country which could be considered Middle Eastern, but is neither Arab nor predominantly Islamic. It is located in Asia, but through its strong Greek connections is effectively European. Evidence of the significance of its location is provided by the two sovereign bases, covering 255 km² and centred on Akrotiri and Dhekeilia, which are still retained by the UK. The pull between Europe and Asia, between specifically Greece and Turkey, came to a head in 1974 from which date the island has been divided.

The key dates in its recent history are 1960 when, with safeguards for the minority population, it became independent; 1974 when the island was divided; 1975 when the Turkish area proclaimed self-rule from the republic of Cyprus; and 1983 when the Turkish Republic of Northern Cyprus was declared. Prior to independence, there was very strong pressure from the Greek community, 78 per cent of the population, for ENOSIS, or union with Greece. This was equally strongly resisted by the Turkish community, 18 per cent of the population, which was in favour of partition. During the run up to independence, British armed forces were called upon to contain the National Organization for Cypriot Struggle (EOKA), a Greek Cypriot underground organization.

Cyprus has an area of just over 9,000 km² or 0.08 per cent of the area of the Middle East. Other than Bahrain and the embryonic state of Palestine, it is therefore the smallest state of the Middle East with a marginally smaller area than Lebanon and Qatar. The landscape comprises mountains to the north and south, but substantial areas of plain which have a dry Mediterranean climate. The population of just under 800,000 represents 0.24 per cent of the population of the Middle East, but literacy at 94 per cent is the highest in the Middle East, save only for that of Israel.

The economy is small but prosperous with a GDP $7.9 billion, 1.1 per cent of the total Middle Eastern GDP. There is a sizeable agricultural sector, but the economy depends on services, particularly tourism, which is highly vulnerable to swings in response to geopolitical events. The relatively high GDP masks a sharp difference between the two sectors of the island, the Greek GDP *per capita* being three and a half times that for the Turks. However, there is an external debt factor which, for the Greek sector is $1.8 billion, small by Middle Eastern standards in the context of the overall GDP.

The armed forces of Cyprus comprise a national guard of 10,000, almost 90 per cent of whom are conscripts. The defence budget totals $0.36 billion

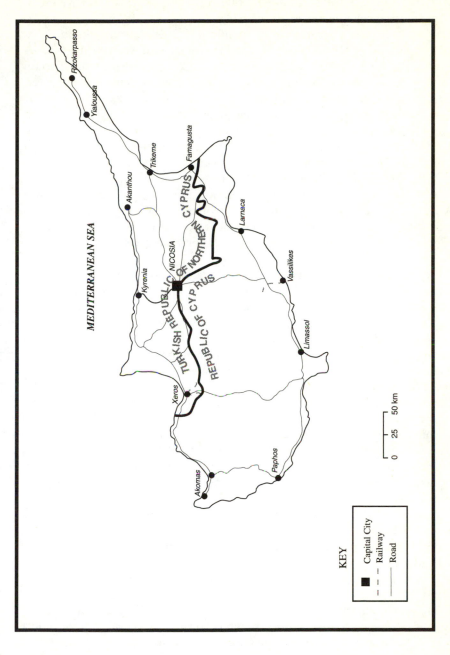

KEY

■ Capital City
- - - Railway
——— Road

MEDITERRANEAN SEA

Rizokarpasso
Yialoussa
Trikeme
Akanthou
Famagusta
Kyrenia
Larnaca
NICOSIA
CYPRUS
TURKISH REPUBLIC OF NORTHERN
REPUBLIC OF CYPRUS
Vassilikes
Xeros
Limassol
Akomas
Paphos

0   25   50 km

Figure 10.3 Cyprus.

or 0.7 per cent of the Middle Eastern defence budget. The forces of the Turkish republic of Northern Cyprus, listed separately in *The Military Balance 1997/98* (The IISS 1998) comprise 4,000 conscripts. Both forces have small-scale land capability and a maritime wing.

While Cyprus is a micro-state of significance chiefly through its location, it is symbolical of the long-lasting conflict between Greece and Turkey. Its position is of course very much nearer to Turkey from which the Turkish sector receives massive support, including regular water deliveries. However, Greece is equally vociferous in its support for the Greek sector and the Republic of Cyprus has been encouraged by the EU to apply for membership. This is likely to exacerbate the internal problems as Turkey itself has been denied EU membership. In a limited manner through the sovereign bases, the UK also provides external support.

Since the division of the island, a number of individuals have come to relative prominence, but the only significant figure on the world stage was Archbishop Makarios, initially the leader of the Greek Cypriots, but from independence until the partition, President. Given events since 1974, the effectiveness of Makarios in smoothing inter-community relations has become very apparent.

Cyprus is a small state with a small population, but a geopolitical importance which far outweighs this status. Given the increasingly close links between Turkey and the Turkish sector and the delivery of missiles to the Greek sector, Cyprus must remain a key global flashpoint.

Geopolitical issues are dominated by the division of the island into two *de facto* autonomous areas, a Greek area comprising 59 per cent and a Turkish area covering 37 per cent of the island's area. The division between the two communities, known as the Atilla Line, is within a UN buffer zone which comprises 4 per cent of the island's surface and is known as the Green Line. Since 1964, the UN Force in Cyprus, UNFICYP, has attempted to reduce violence, arrange ceasefire agreements and maintain patrols between the two sides. The occurrence of two UK sovereign bases is of considerably less significance, but, in the context of global policing can be compared with the US Guantanamo base in Cuba as a possible flashpoint of the future.

Another issue is the use of the island for the distribution of illicit drugs including heroin, hashish and probably cocaine via air routes and container traffic from Lebanon and Turkey to Europe. With its location psychologically if not physically between the Middle East and Europe, its long coastline and its limited military force, Cyprus is ideally placed for the movement of illicit materials.

The other obvious potential geopolitical factor is the problem of water resources. There have long been water problems, particularly in the north of the island and the use of Operation Medusa, used in this case for the first time, has strengthened the dependence of the Turkish sector upon Turkey.

Egypt

As befits a place central in the region and settled since antiquity, Egypt (Figure 10.4) occupies the nodal position in the Middle East. It provides the link between the Mashreq to the east and the Maghreb to the west. It is the bridge between the Middle East and Africa with strong connections through the Mediterranean to Europe. Furthermore, Egypt occupies the valley of the only river of any size in Africa north of the Sahara. The Nile Valley provides an entry to Africa and questions of water security have long linked Egypt with the full range of Nile Valley riparian states, particularly Ethiopia. Egypt also controls the isthmus of Suez, including the Canal, the most important in the world, and the SUMED oil pipeline. Also under Egyptian control is the Sinai Peninsula, itself of fundamental geopolitical significance, a fact demonstrated under the 'British protectorate' and subsequently during the Arab–Israeli wars.

Egypt became independent in 1922, earlier than any other modern state in the Middle East. However, independence was first clearly demonstrated in 1956 with the nationalization of the Suez Canal by President Nasser who had assumed power in 1954. While not the originator, he was the chief protagonist since the Second World War of Pan-Arabism and in 1958 the United Arab Republic was formed between Egypt and Syria. Although losses were sustained in the wars of 1949 and 1956, the 1967 war was catastrophic for Egypt.

After the eventual loss in the 1973 war, President Sadat signed the Camp David Accords in 1979. This was a crucial event, since after four conflicts with Israel, Egypt effectively withdrew to the sidelines. Given Egypt's central role in the Arab world, this move was not well received and Egypt was expelled from the Arab League, not returning until 1989. During the 1980s and 1990s, there has been an increasing problem of internal terrorism, generated by extreme Islamic groups. At the same time, as a result of elections, the Muslim Brotherhood has become integrated into the political system. In 1991, Egypt placed itself on the side of the Alliance in the Gulf War and was in fact the third largest force present. The effects of this were that in the post-war period Egypt was greatly favoured by the USA and the Arab oil states but its ties with Iraq, developed during the Iran–Iraq war, were severely damaged. As a result, the traditional competition within the Arab world between Cairo and Baghdad has been renewed.

Egypt has an area of 998,000 km², 8.6 per cent of the Middle Eastern land area. This comprises a desert plateau of which only 3.5 per cent, comprising the Nile Valley, the Delta, the areas newly irrigated from Lake Nasser and a few oases, is cultivable. Population is almost entirely located in and around these cultivable areas and thus, despite the size of the state, there is in much of the country extreme overcrowding. The population is 64.8 million, which represents 19.7 per cent of the Middle Eastern population and 26 per cent of the Arab world. The potential significance of the large population is partly nullified by the literacy rate of 51.4 per cent.

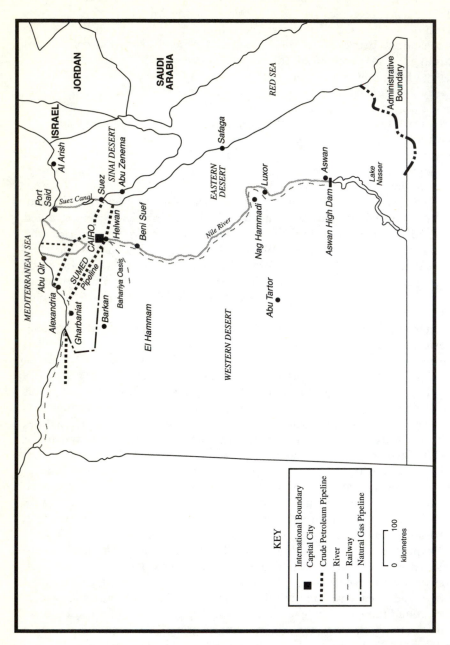

Figure 10.4 Egypt.

With a GDP of $56 billion, Egypt has the fifth largest economy in the Middle East, accounting for 8 per cent of the total. Economically, it ranks just below Iran, but well below Turkey, Saudi Arabia and Israel. Despite its low literacy rate, one of the lowest in the Middle East and its large agricultural sector, Egypt nonetheless supplies large numbers of professional people, particularly to the oil-rich Arab states. The economy depends predominantly upon tourism, Suez Canal dues, petroleum and refined products and remittances. All have been affected by recent events. The Suez Canal was blocked as a result of the wars of 1956 and 1967, oilfields were temporarily lost during the wars, remittances were greatly reduced as a result of the expulsion of Egyptians from Kuwait in the Gulf War, and tourist returns have been severely hit by internal terrorist activity. Egypt has an external debt of $31 billion, or rather more than half the annual GDP. Thus, there are elements of actual and potential strength but Egypt has been severely handicapped by geopolitical events throughout the region.

With an active membership of 440,000, the armed forces of Egypt are among the most substantial in the Middle East, being marginally smaller than those of Iran and dwarfed only by those of Turkey. There is a very large and well-equipped army, a small navy with some modern vessels and a relatively large and well-armed air force. The defence budget totals $2.4 billion, 5.1 per cent of the Middle Eastern defence budget. This is by regional standards relatively modest, particularly when compared with the defence spending of Saudi Arabia and Israel, but Egypt must be considered militarily powerful.

Since the Camp David Accords, Egypt has received large-scale US aid, second only to that provided for Israel. Initially, this allowed the country to weather the loss of support from the oil-rich Arab states, following the signing of the Accords. From the time of the Gulf War in 1991, the status of President Mubarak has risen dramatically and this has enabled Egypt's foreign debt to fall as a result of debt relief under the Paris Club arrangements. Economic support has also been forthcoming for a number of mega-projects. Apart from President Mubarak, chiefly responsible for restoring Egypt's fortunes in the Arab world after the Camp David Accords, other influential individuals were the two previous Presidents. Sadat signed the Accords and thereby changed forever the dimensions of the Arab–Israeli conflict. Nasser provided leadership for the Arab world as a result of his nationalization of the Suez Canal and is chiefly significant as the instigator of modern-day Pan-Arabism.

As a result of its central location in the Arab world and the Middle East in general, its control of the Suez Canal, its large population and its strong and relatively well-equipped military, Egypt is without doubt the most powerful Arab state and one of the three major actors in the Middle East. This position has been enhanced by the demise of Iraq and by the restoration of the status of the country following the Gulf War.

The major geopolitical issue facing Egypt concerns water. With a population which increases by approximately a million every eight months, Egypt is implicated in a wide range of actual and potential geopolitical issues, but pride of place is taken by water. The rapid rate of population growth generates a continuing need for agricultural land and irrigation, but already Egypt is using its agreed share of Nile water, together with an unused portion of Sudan's share. Its territory provides no input to the Nile and Egypt does not geographically control the headwaters. Given the requirements of the upstream riparians, particularly Ethiopia, for increased irrigation, problems over the distribution of Nile water must be resurrected. These may be addressed in a number of ways, but underlying any negotiations must be the fact that Egypt is the most militarily powerful state in the basin. Allied to the water issue is the question of food security in the face of urbanization, salinization and desertification in the agricultural areas.

As a front-line state with Israel, there must remain the potential for tension, if not actual conflict, particularly given Egyptian sympathy for Palestine. Recently, conflict has been apparent only within the state, with increasing terrorist activity directed at the government through the medium of attacks on tourists. Egypt has land boundaries with only four countries: Sudan, Libya, Israel and Palestine. There remains an international boundary dispute with Sudan over the Hala'ib Triangle and maritime boundaries have yet to be agreed. As an exporter of labour, Egypt is likely to be involved in any disputes over the transboundary movement of people and indeed the returnees from Kuwait have produced internal stress. Since it is the controller of the Suez Canal, one of the key global choke points, Egypt has a continuing interest in what must for the foreseeable future remain a global flashpoint.

If Pan-Arabism can be seen as of potential geopolitical significance, although as a concept it carries little weight at present, Egypt was and probably remains the source. Allied to terrorism is the fear of extreme Islamic activity known as Fundamentalism, although it must be stressed that the Muslim Brotherhood is essentially non-violent. The other aspect of the macro-political agenda which involves Egypt is the transfer of illicit drugs. Egypt is a transit point for south-west Asian and south-east Asian heroin and opium destined for Europe and North America. It is said to be a popular transit point for Nigerian couriers and there is, of course, a large domestic consumption of hashish.

Iran

The eastern boundary of Iran (Figure 10.5) marks the eastern edge of the Middle East, while its western boundary with Iraq defines the edge of the Arab world. Although it is in a sense peripheral, it is highly influential in that it provides the Middle Eastern links with the Trans-Caucasus, Central Asia and the Indian subcontinent. It also shares a boundary with the other major Middle Eastern non-Arab state, Turkey. The boundary with Iraq, certainly at its southern extent along the Shatt al Arab, has been of major geopolitical significance. It is Iraq's longest boundary and that fact has been

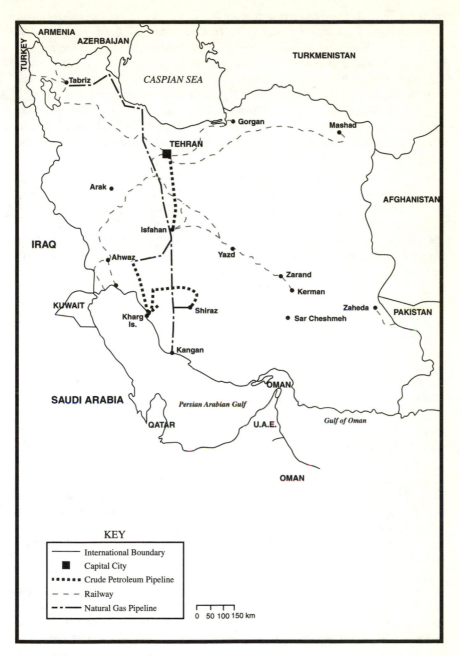

Figure 10.5 Iran.

of importance during Iraq's recent troubled past. Iran shares boundaries with seven states and, across the Gulf, six further states, and no other Middle Eastern state has as many neighbours. This provides of course opportunities for conflict and diplomacy. A further crucial factor is Iran's location with regard to the Gulf. It stretches the entire length of the Gulf, has easily the longest Gulf coastline of any state and is in a position to exercise a controlling influence over the Strait of Hormuz. Iran also has a share of the Caspian Sea coastline, bringing it further into contact with Central Asia and the Russian Federation. In the north-east, Iranian Kurdistan shares, in a measure, the Kurdish problem.

The watershed in recent Iranian history occurred in 1979 with the Revolution, followed by the proclamation of the Islamic republic of Iran, a theocratic republic. Officially, the independence date for Iran, the successor to great empires of the past, is 1979. In 1980, war with Iraq began and continued in virtual stalemate for almost eight years. In the Gulf War (1991), Iran officially sided with the Allied forces and supported the UN resolutions, but was important for its humanitarian treatment of refugees from Iraq, rather than for any military action. The resulting improved standing with the UN and the West in general was rapidly dissipated in the case of the latter, by rumours concerning nuclear weapons production. In 1995, the USA imposed an embargo on US trade with Iran, but this acted in some ways as a stimulus. Iran continues as a bastion against the encroachment of Western influence, although a thaw in relations has been evidenced by President Khatami's visit to Italy in March 1999.

With an area of 1.648 million km², Iran occupies 14.2 per cent of the Middle East, a figure exceeded only by Sudan, Saudi Arabia and Libya. The landscape comprises a mountainous rim surrounding high central deserts with small plains along both coasts. The most favoured area with regard to relief and climate is the north-west. The population of Iran is 67.5 million, the largest of any country in the Middle East. It accounts for 20.5 per cent of the total Middle Eastern population and has a middle rank figure for literacy of 72.1 per cent. In terms of both area and population Iran is a potential Middle Eastern superpower.

However, like the Arab potential superpower Egypt, Iran's GDP of $62.5 billion is not in proportion to its size and population and amounts to only 9 per cent of the total Middle Eastern GDP. Its external debt of $30 billion is approximately half the GDP. There is a strong agricultural sector, but industry is more significant and the entire economy is dominated by oil, earnings from which account for approximately 85 per cent of total export revenues. The territory includes five of the twelve mega-oilfields located in the Middle East. Increasing levels of technology are indicated by the development of the armaments industry.

The size of the armed forces, numbering some 513,000 active personnel, is exceeded in the region only by Turkey. The defence budget of $2.5 billion, or 5.3 per cent of the total Middle Eastern budget, is virtually the same as that of Egypt and marginally less than that of Iraq. It is dwarfed by those of Saudi Arabia and Israel. The relative sizes of the army, navy and air force are very similar to those of Egypt, but the quality is not as high, following catastrophic losses during the Iran–Iraq war. In addition, there is a revolutionary guard corps (the Pasdaran) of 120,000 members which comprises ground forces, naval forces and marines. Thus, within the Middle East, the status of the Iranian armed forces has been reduced considerably over the past twenty years, but the current forces remain formidable.

However, the ability of Iran as an actor on the world stage is gravely handicapped by a lack of any consistent, powerful external support. Since it is feared by the Arab oil states, Iran has at various times had to obtain assistance, whether in the purchase of armaments, the sale of oil or merely political backing, from such unlikely sources as Israel, Russia and China. Having through actions and circumstances alienated the USA, Iran lacks the superpower support which underpins Egypt. The influential individual, since the latter part of the 1970s, has been Ayatollah Khomeini whose spiritual legacy still dominates the state.

Despite the depredations on its economy and military standing, in the wake of the Revolution, Iran remains one of the three major powers of the Middle East. It occupies a crucial location, abutting on to Central Asia, south Asia, Turkey, Iraq and, across the Gulf, all the states of the Arabian Peninsula, except Yemen. Although appearing marginal geographically, Iran must remain interested in all the major events of the Middle East.

With such a large range and number of transboundary contacts, it is hardly surprising that Iran is involved in a wide variety of geopolitical issues. The most significant have been the conflicts, particularly those concerning Iraq. With scarcely a break between 1980 and 1991, the Gulf has been the scene of continuous tension and conflict involving Iran. Within this, the key has been the relationship with Iraq and the Arab world in general. To the northeast there has been sporadic, but continuing conflict with the Kurds, while to the west Afghanistan has been in a state of almost perpetual turbulence since 1979. Allied to the various conflicts has been Iranian involvement in and support for terrorism, well beyond the confines of the Middle East. However, it is possible that Iranian complicity has been exaggerated as the USA seeks to identify pariah states.

The Shatt al Arab has constituted a long-term boundary dispute between Iran and Iraq. In return for a qualified degree of support, Iran was able to persuade Iraq to recognize a *thalweg* boundary, the key issue over which the whole of the Iran–Iraq war had been fought. There remain boundary issues to be negotiated in the Caspian Sea and the issues of the ownership of three Gulf islands: Greater Tunb, Lesser Tunb and Abu Musa. The islands are strategically placed in the approaches to the Strait of Hormuz, which itself remains a potential global flashpoint.

Iran is also implicated in many transboundary issues, most of which form part of the macro-political agenda. Since it has remained for so long shut off from legitimate sources, arms transfers have been particularly significant, especially from China and the states of the FSU but also from the USA as the Iran–Contra incident indicated. Iran is also heavily implicated in illegal drugs trafficking and is a producer of opium. Equally important, it is a key trans-shipment node for the movement of heroin from south-west Asia to Europe. As with many Middle Eastern countries, refugees have also provided a significant part of transboundary movement. In particular, refugees from Iraq entered Iran during the Gulf War, while during the Soviet–Afghan campaign, refugee villages were established along the Iranian side of the boundary.

Although Iran is not involved as obviously in hydropolitics as Egypt, Israel or Iraq, it is the source of major tributaries to the Tigris and therefore must be part of any Tigris–Euphrates basin management scheme. While Iran is and will remain one of the central actors in global oil geopolitics, the more specific transboundary issue concerns the routing of pipelines from the Central Asian oilfields. Given the distances involved and terrain, the potential problem of congestion in the Turkish straits, the desire of many of the producers to avoid transiting the Russian Federation and the US embargo, this is a complex problem. Furthermore, should the pipelines take what is to many the most obvious route, through Iran, the terminals will be in the Gulf, thereby enhancing the status and vulnerability of that waterway and, in particular, the Strait of Hormuz.

Fundamentalism as a geopolitical factor is associated first and foremost with Iran because of its mode of operation as a theocratic republic. Fundamentalism is not, of course, restricted to the Shi'a sect, nor indeed to Islam, but could be applied to any religion. However, in returning to what are taken to be the basic fundamentals of Islam, the government has spawned a good deal of violence and has projected its concepts well beyond the boundaries of Iran. There has been an Islamic resurgence elsewhere but most of it can in no way be related to events in Iran.

Iraq

Iraq (Figure 10.6) is located in the core of the Middle East, but with boundaries with the two major non-Arab Middle Eastern powers, Turkey and Iran, represents the edge of the Arab world. With Jordan, Iraq is the only Middle Eastern country which is all but landlocked. The coastal length of Jordan is 26 km and of Iraq 58 km. However, this does not take into account the discrepancy in both area and population. Jordan has an area of 3,431 km² and a population of 166,332 per km of coastline. The respective figures for Iraq are 7,535 km² and 383,091 population per km of coastline. Iraq abuts on to six states and, during the Gulf War, all but Jordan were hostile. Thus, the prosecution of the Gulf War and the subsequent imposition of sanctions were greatly facilitated by geopolitics.

At either end of its short coastline Iraq has an exit point with ports and, depending upon relations with Iran and Kuwait, has developed either the

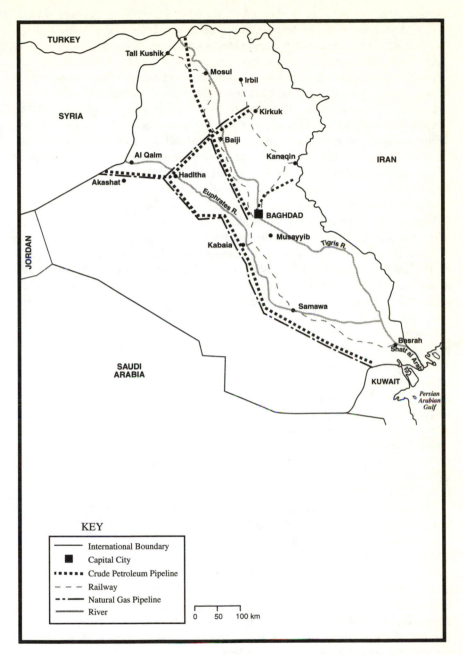

KEY

────── International Boundary

■ Capital City

▪▪▪▪▪▪ Crude Petroleum Pipeline

– – – Railway

–▪–▪– Natural Gas Pipeline

〜〜〜 River

0   50   100 km

Figure 10.6 Iraq.

Shatt al Arab or the Khor Zubair. A further key element of location is that Iraq occupies most of the former Mesopotamia, a hydraulic culture and is, by Middle Eastern standards, well provided with water. However, Iraq controls neither the headwaters nor the major tributaries in the Tigris–Euphrates network. With regard to the other basic Middle Eastern resource, oil, Iraq is well placed with reserves bettered only by those of Saudi Arabia.

Recent history, since the accession to power of Saddam Hussein in 1979, has been dominated by conflict. First was the prolonged Iran–Iraq war from 1980 to 1988 and then the seizure of Kuwait and the subsequent Gulf War of 1991. Independent since 1932, the same year as Saudi Arabia and earlier than every other Middle Eastern state, except Egypt and Turkey, Iraq had before 1979 vied with Egypt as the most powerful Arab state.

With an area of 442,000 km², under half the size of Egypt, Iraq occupies 3.8 per cent of the area of the Middle East and is one of the larger states. The landscape comprises broad, fluvial plains with mountains along the crucial borders with Iran and Turkey. The population is 22.2 million, 6.7 per cent of the Middle Eastern total, making it the fifth largest in the Middle East after Egypt, Iran, Turkey and Sudan. The literacy rate of 58 per cent is higher than that of Egypt, but below that of every other Middle Eastern state except Sudan and Yemen.

Potentially, Iraq is economically strong with two mega-fields and vast reserves of oil, but the post-conflict rebuilding of the state together with UN sanctions have destroyed a large part of the economy. The GDP is difficult to assess, but has been estimated at $18.3 billion, or well under a third that of Iran. This represents only 2.6 per cent of the total Middle Eastern GDP at present. The current *per capita* GDP is lower than every Middle Eastern state except Sudan and Yemen. While having a relatively large agricultural sector, Iraq has also developed industrially, but the present situation cannot be assessed. The other significant figure is the external debt which is estimated to be $42 billion and, apart from that of Turkey, is by far the highest in the Middle East.

Judged by the number of active personnel, 383,000, Iraq has the fifth largest armed forces in the Middle East after Turkey, Iran, Egypt and Syria. The army remains large and relatively well equipped, but the navy is tiny and the medium-size air force suffers from the lack of serviceable equipment. The Iraqi military was potentially the most powerful in the Arab world and was viewed by the country as the bastion of the Arabs against Iran. Apart from the Republican Guard Forces, the armed forces have been eviscerated. However, the defence budget is listed as $2.7 billion, placing Iraq fifth in the Middle East, marginally ahead of Iran and Egypt. Iraq expends 5.8 per cent of the total Middle Eastern budget.

Throughout the war with Iran, Iraq was supported by the USA and the West in general and also by the Arab oil states. Neither the protection of terrorist Palestinians nor the gassing of the Kurdish population appeared to deter this support. However, since the invasion of Kuwait, the Gulf War and

UN sanctions, there has been no consistent external support for Iraq. In attempts to lift sanctions, aid has been forthcoming from France and Russia, and the Arab world in general is increasingly sympathetic towards the plight of the Iraqi people, but the only overt, unswerving sympathy for Saddam and the cause has been expressed by sections of the general populace in principally Jordan, Sudan and Yemen. The most influential figure in the recent past has been Saddam Hussein, seen by some in the Arab world as a successor to Nasser as the leader of the Arab people. His personal imprint appears on many of the events which have engulfed Iraq.

Iraq, by Middle Eastern standards a relatively powerful and developed state, has been decimated by almost two decades of first war and then sanctions. However, given its resources and location, Iraq remains potentially powerful.

Like Iran, Iraq has been involved in a wide range of geopolitical issues. Most obvious have been the two major wars, the continuing conflict with the Kurds and threats of terrorist activity beyond the boundaries of Iraq. While agreements have been reached with Iran and Kuwait, final long-term settlement is awaited. Major boundary issues have concerned the Shatt al Arab and both the land and maritime boundary with Kuwait. While the UN-imposed settlement of the Kuwait issue appears to have been successful, doubts remain as to the finality of the *thalweg* boundary in the Shatt al Arab. In the case of Kuwait, it is difficult to believe that the issue of Warbah and Bubiyan Islands will not at some stage re-emerge as the remaining maritime boundaries at the head of the Gulf are delimited.

Geopolitics is also highly significant, both internally and externally. Internally, the Third River project has resulted in the displacement of large numbers of Marsh Arabs, a sizeable population of Shi'a Muslims. Developments upstream on both the Tigris and the Euphrates, particularly on the former in Turkey, have raised many classic issues of hydropolitics. Other transboundary issues include the movement of refugees and the import of arms such as the celebrated supergun and of the prerequisites for the manufacture of chemical, biological and nuclear weapons of mass destruction.

More obviously than in anywhere else in the Middle East in recent history, the geopolitical issue of food security has arisen in Iraq. The effect of UN sanctions during the middle 1990s was so severe that malnutrition and undernutrition became prevalent. As a result, UN Security Council Resolution 986 and a series of successors have been passed to allow the sale of oil by Iraq sufficient to import food and other humanitarian requirements. Also related to the Gulf War has been the other geopolitical issue, that of Pan-Arabism. In an attempt to rationalize his assault on Kuwait, Saddam sought to establish linkage with the Israeli occupation of territory, a breach of UN Security Council Resolutions which had at that time lasted 23 years. It was hoped in vain that such rhetoric would rouse the Arab nations.

Israel

In that it was founded upon a specific religion, race and culture, Israel (Figure 10.7) is unique in the world and with the potential exception of Cyprus, the only non-Islamic state in the Middle East. The implanting of Israel following the Balfour Declaration (1917) and the partitioning of Palestine by a narrow vote in the UN General Assembly (1947) produced the one geographical break in the otherwise continuous Arab world. The area awarded constituted a significant part of the Land of the Bible, but within the original borders, barely one-third of the population was Jewish. Since 1948, the boundaries have expanded to take in the Golan Heights, the West Bank, the old city of Jerusalem and the Gaza Strip. To these has been added effective control of southern Lebanon. Thus, Israel is now considerably bigger than the intended state and the four neighbouring states: Lebanon, Syria, Jordan and Egypt have all at some stage lost territory. However, the major loss has been sustained by the Palestinians, originally the dominant population of the area.

Israel is located adjacent to the bridge linking Asia and Africa, with access to both the Mediterranean and the Red Sea. By capturing the entire coastline, the Arabs to the east became effectively landlocked. However, it is more for political, economic, social and cultural reasons than any geopolitical factors of location that Israel has been the focus of global attention in the Middle East throughout the period since the Second World War.

The history of the state of Israel, independent as an entity since 1948, has been dominated by war. The state was effectively born out of conflict with the Arab world and, with few exceptions, this enmity has continued until the present time. Following the wars of 1948–49, 1956, 1967 and 1973, the watershed event was the visit of President Sadat of Egypt to Israel to address the Knesset. After some setbacks, the result was the Camp David Accords, signed in 1978, and a formal peace treaty between Egypt and Israel which followed in 1979. This effectively removed Egypt from the military equation, but the subsequent invasion of Lebanon in 1982 proved in the long run disastrous. The situation with the Palestinians deteriorated and in 1987 the *intifada*, a well-augmented show of defiance began. This focused the world's attention on the plight of the Palestinians and, even in the USA, led to a marked decline in support for Israel. Subsequently, the debate over whether to leave the Occupied Territories in return for guaranteed peace has continued to concern the Israeli government, enhanced in 1988 by the decision of King Hussein to withdraw Jordan's claims to the West Bank.

The Gulf War in 1991 highlighted the fact that Israel had been in breach of Security Council Resolution 242 for 23 years in its retention of the Occupied Territories. Despite coming under attack from Scud missiles, Israel did not join the war, which left Israel militarily unrivalled in the Middle East. Following secret negotiations, the Oslo Accords of 1993 brought agreement between Israel and the PLO and a declaration of principles was signed. This

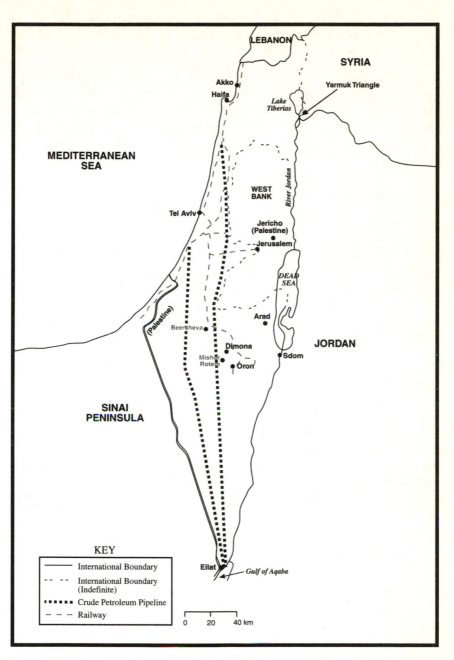

Figure 10.7 Israel, Palestine and the West Bank.

was followed in 1994 by the Gaza–Jericho Agreement giving self-rule to those areas and one year later Israel and the PLO signed the full interim agreement, which called for the redeployment of the Israeli defence forces away from most Palestinian cities. During the course of these events, Israel elected a new Prime Minister, Netanyahu, a hard liner, and progress towards the development of the state of Palestine has been deliberately slowed. The selection of Barak in May 1999 appears to offer a return to the serious move towards long-term peace initiated by the assassinated Prime Minister Rabin.

With an area of 22,000 km², Israel is a micro-state. Its area amounts to 0.2 per cent of the Middle East and only Bahrain, Cyprus, Lebanon, Qatar and Kuwait are smaller. The landscape comprises a desert in the south, a low coastal plain and a central upland ridge. The population, 5.5 million, represents 1.7 per cent of the total Middle Eastern population and, while small, is substantial for the size of territory. The literacy rate of 95 per cent is almost 10 per cent higher than any other Middle Eastern state except Cyprus and Lebanon.

The GDP of Israel is $78 billion, the third highest in the Middle East after Turkey and Saudi Arabia. This figure represents 11.2 per cent of the total Middle Eastern GDP and is completely out of phase with the size of the territory and population. It results partly from external sources of revenue from world Jewry and official US aid. Until the mid-1960s, the economy had also been helped by Holocaust reparations from West Germany. Both agriculture and industry have been intensively developed and can be characterized as high technology, and this is a further reason why, despite the lack of resources, the economy has flourished. Furthermore, valuable expertise has been continuously brought to the state by immigration.

In line with its economic strength, Israel is also militarily powerful. The defence budget totals $7.2 billion, or 15.4 per cent of the total Middle Eastern defence budget. This places Israel second only to Saudi Arabia in defence spending. With a personnel of 75,000, the armed forces are small, but are backed by reserves of some 430,000. The army is relatively large, the air force middle ranking and the navy small, but all have the most modern equipment. In terms of personnel and materiel, the Israeli armed forces are the most powerful in the Middle East. Furthermore, they are backed by a nuclear capability of up to possibly 100 warheads.

Crucial to the economic and military development of Israel has been the external support it has received. The duty of all Jews to support the national cause, long one of the main ideals of political Zionism, has been realized. In addition, Israel has received the general backing from the West and the specific and extensive support of the USA. Without these external factors, the micro-state of Israel could never have become a regional superpower to outstrip in most ways the far larger and more populous states of Iran, Turkey and Egypt. From Herzl, a key thinker in the development of the Zionist movement, onwards, there have been many influential individuals in the history of Israel. With the high public profile of Israel, its leaders have all appeared significant on the world stage. Particular mention might be made of Ben-Gurion, Begin, Perez, Rabin and especially because of his intransigence, Netanyahu.

Despite its small area, small population and lack of resources, as a result of world-wide and particularly US support, Israel has developed into arguably the most powerful state in the Middle East. Political and economic support from the USA has allowed Israel to flout UN Security Council Resolutions and avoid signing the Nuclear Non-Proliferation Treaty (NPT). As the ultimate safeguard, Israel has the only known stockpile of nuclear weapons in the Middle East.

The major geopolitical issue with which Israel has been concerned is continuing conflict with the Arab states and terrorism, both external and internal. The conflicts have varied from full-scale war to low-intensity conflict, typified by the *intifada*. Many Islamic terrorist groups have been involved in the fight against Israel and Israel itself has, through the security service Mossad, been involved in state-sponsored terrorism outside its boundaries.

The history of Israel has been the history of boundary disputes and present boundaries are, in many instances, armistice lines. Boundaries will need to be negotiated with the Palestinians on the West Bank and in the city of Jerusalem. Boundaries have also been adjusted, most notably following the Taba award to Egypt in 1989. Tension over water has also been a recurrent theme since the draining of the Huleh Marshes and the construction of the National Water Carrier. In such a hostile environment, it is difficult to identify an Israeli hydrological imperative, but any agreement on land for peace must give water supplies a major consideration.

Apart from water, other transboundary issues include the large-scale movement of refugees from Israel to the surrounding Arab states and the even larger immigration of Jews into the state of Israel. A reminder of the exodus is provided by the many Palestinian refugee camps still in existence. One legacy of the mass immigration has been the increase in crime, including drugs smuggling and prostitution. Arms have also featured strongly in covert cross-boundary activities. As a result, Israel has been able to work closely alongside the burgeoning arms industry of the Republic of South Africa and also to develop its own nuclear capability. Covert activity was further encouraged by the long-term embargo by Arab states which at times has included oil. Of particular geopolitical significance has been the transboundary intelligence which Israel has enjoyed. The latest result of this has been in the co-operation with Turkey over military matters and future water supplies and developing links with Azerbaijan and Afghanistan.

Since the end of the Cold War, nationalism and Fundamentalism have been cited as key geopolitical issues. As Israel is clearly so different in almost every sense from the other countries of the Middle East, it has generated an extreme form of nationalism. Fundamentalism is usually considered in the context of Islam, but throughout the course of this century, the most successful Fundamentalist movement in the Middle East has been that of the Zionists.

Jordan

Having boundaries with Syria, Saudi Arabia, Iraq, Israel and the West Bank, Jordan (Figure 10.8) is in the cockpit of the Middle East and is virtually certain to be affected by almost every geopolitical event in the region. The exact boundary in the West Bank will be determined by further negotiations following the Israeli–Palestinian Interim Agreement. Jordan is almost land-locked with only 26 km of coastline, the shortest length of any Middle Eastern country. However, Jordan has been able to benefit from transit trade, both to the Mediterranean and the Gulf of Aqaba. Located where the desert meets the edge of the coastal plain, Jordan has developed as a mixture of Palestinian and Bedouin culture and it is locked into the affairs of both. In particular, its position makes it critical in the water problems of the Jordan River.

The entire political history of Jordan as a separate entity is modern. After the First World War, the Emirate of Transjordan was carved out of what had been formerly known as Syria and in 1946 this was granted formal independence. The ruler, who claimed descent from the family of the Prophet, proclaimed himself king. After the war of Israeli Independence (1948–49), the West Bank was attached and the state became known as the Hashemite Kingdom of Jordan. The family of the Ruler of the Holy Places of the Hijaz became the Ruler of the Holy Places in Jerusalem, Bethlehem and Hebron. However, the West Bank also provided Jordan with a large, relatively sophisticated and politically active population, in sharp contrast to that which lived to the east of the river.

The 1967 war resulted in the occupation of the West Bank by Israel and the loss to Jordan of nearly half its population and much of its economy. Furthermore, 400,000 more Palestinians fled to Jordan to join those already in the refugee camps established in 1949. These camps became centres for guerrilla groups linked to the PLO. In 1970, with heavy loss of life, the guerrillas were crushed and the PLO operations moved to Lebanon. Jordan avoided direct confrontation with Israel in the 1973 war. Since the Camp David Accords suggested autonomy for the West Bank rather than reintegration, they were opposed by Jordan which continued to press for the full implementation of Security Council Resolution 242. During the 1980s, Jordan received strong support from Saudi Arabia and other Gulf oil states, but this was cut off following the Gulf War in 1991 with Jordan's official neutrality and apparent support for Saddam Hussein. Relations with Saudi Arabia and Kuwait have since improved.

Jordan has an area of 89,000 km^2, which is 0.8 per cent of the total Middle Eastern area and four times the size of Israel. The territory stretches from the Jordan rift valley eastwards and comprises, other than the valley itself, desert plateau. The population of 4.3 million, 1.3 per cent of the Middle Eastern population, is considerably smaller than that of Israel, but large for the resources available. The literacy rate of 86.6 per cent puts Jordan ahead of all the Arab oil states and behind only Israel, Cyprus and Lebanon.

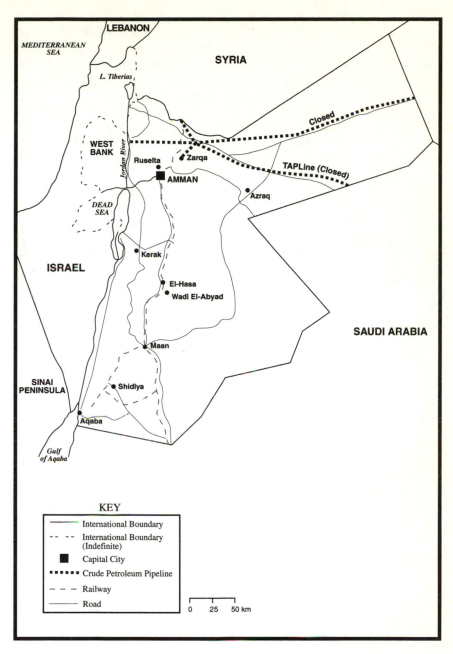

Figure 10.8 Jordan.

The GDP of Jordan, $6.6 billion, accounts for only 0.9 per cent of the total Middle Eastern GDP, a figure lower than any state except Bahrain and Palestine. Jordan has an educated population, but very few natural resources and depends upon remittances, transit trade and aid from the Gulf oil states. It has an external debt of $7.3 billion and, with Iraq and Sudan is the only Arab state in which the external debt exceeds the GDP.

The armed forces of Jordan comprise 99,000 active personnel and there is a heavy accent upon the army which is small, but relatively well equipped. The defence budget of $0.44 billion is 0.9 per cent of the total Middle Eastern defence budget, placing Jordan equal with Lebanon and ahead of six other states in the region.

With a literate population and virtually no resources, the situation in Jordan mirrors that in Israel. The state depends very much upon support from the Gulf states, the EU and the USA. However, the maintenance of the state is crucial for the peace of the Middle East and such support, other than following the Gulf War, is likely to remain forthcoming. This allows Jordan to not only develop its economy, but to maintain a relatively high-quality military. The history of Jordan has been dominated by one influential individual, King Hussein, who showed himself to be an extremely astute leader. His importance was indicated when his funeral in 1999 brought together not only Israelis and Arabs but the greatest gathering of world heads of state ever seen in the Middle East.

With an extremely limited cultivable area and a small population, Jordan is in many senses a micro-state, but owing to external support which results from its nodal position in the region, Jordan has been able to develop both economically and militarily. It cannot be considered a powerful state, but through the offices of King Hussein, it has wielded great influence and should never be discounted.

Despite its location, Jordan has been able to avoid many of the geopolitical issues which have confronted its neighbours. Following the catastrophic losses in the 1967 war, Jordan has become practised in conflict and terrorism avoidance. There have been no obvious transboundary issues other than the influx of refugees from the Israeli–Arab wars and the returnees from Kuwait. There remain a number of boundary issues to be settled, but these depend essentially upon negotiations between Israel and the Palestinians.

The most obvious geopolitical problem has concerned water and the sharing of the flows of the Yarmuk and particularly the Jordan. With effectively no other sources of surface water, the use of these rivers is critical to Jordan. Irrigation in the rift valley depends greatly upon the East Ghor Canal which was bombed by Israel in 1969.

Kuwait

Kuwait (Figure 10.9) is strategically situated at the head of the Gulf between Saudi Arabia and Iraq with Iran as a near neighbour. It is a micro-state with

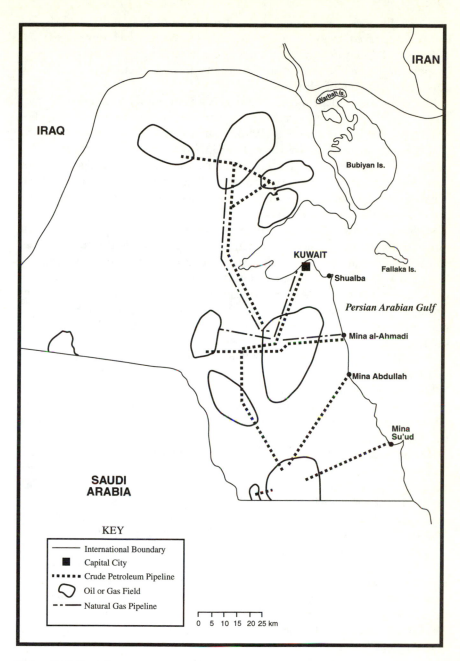

Figure 10.9 Kuwait.

massive oil reserves approximately equal to those of Iraq and the UAE and is therefore locked firmly into oil geopolitics. Its territory includes two islands, Warbah and Bubiyan, which lie off the coast of Iraq in the approaches to its port of Umm Qasr. Maritime transport from Kuwait must transit the potential flashpoint of Hormuz.

Kuwait became independent in 1961 and was promptly threatened by Iraq. Such threats, covert and overt, have continued intermittently, finding full fruition in the Iraqi invasion of 1990. Prior to that, Kuwait had supported Iraq throughout the Iran–Iraq war. The Gulf War of 1991 proved disastrous to Kuwait, particularly in the firing of the oilfields. It was also disastrous for the many foreign workers, predominantly Palestinians, who were expelled. Throughout this time, close relations have been maintained with its other powerful neighbour, Saudi Arabia.

With an area of 18,000 km², 0.15 per cent of the Middle East, Kuwait must be considered a micro-state, although it approaches the size of Israel and is larger than five other states in the region. It comprises essentially a large bay surrounded by desert plains. The population of two million is 0.6 per cent of that of the Middle East and is only approximately 45 per cent Kuwaiti. It is also balanced 30 per cent Shi'a and 45 per cent Sunni. The rate of literacy is 78.6 per cent, very similar to that of the other Gulf sheikhdoms.

The GDP of $26.7 billion is 3.8 per cent of the total Middle Eastern GDP which, in *per capita* terms, places Kuwait alongside Israel and only behind the UAE in the region. This high figure results basically from oil, but also from investments and industrial diversification. The external debt is $8 billion, accounted for in part by the reconstruction necessitated by the Gulf War.

The defence budget of $3.1 billion is 6.6 per cent of the total Middle Eastern defence budget, a figure exceeded only by Saudi Arabia, Israel and Turkey. *Per capita* Kuwait's defence expenditure is the largest in the Middle East. With only 15,300 personnel, the armed forces are small, but extremely well equipped. They are supported by the UN Observer Mission in Iraq/Kuwait (UNIKOM) forces and prepositioned army and air force equipment.

Continuing external support for Kuwait has come from Saudi Arabia, but the critical support occurred during the Gulf War when the USA organized a multinational coalition force, comprising predominantly Western and Arab states. For this endeavour, financial support was forthcoming on a generous scale from Japan and Germany. With regard to individuals, the state has been run since its inception by the Al-Sabah family, which has exerted little influence beyond Kuwait.

As a micro-state with large-scale oil reserves, Kuwait has long been viewed enviously by Iraq and is capable only of reacting to events. It depends essentially upon its strong economy and external military support. The major geopolitical issues concern oil and conflict. The Gulf War was in effect the first conflict in the era of resource geopolitics. The chief transboundary issues which resulted were the influx of coalition forces and the expulsion of foreign workers.

Warbah and Bubiyan Islands remain as potential geopolitical flashpoints, although the maritime boundary with Iraq has been agreed. The land boundary has also been agreed, but since this leaves part of the Rumaila oilfield in Kuwait, there could still be longer-term tensions. The most important boundary which remains to be settled is the maritime boundary with Saudi Arabia which results from the division of the former Neutral Zone.

Lebanon

Located centrally in the Levant, Lebanon (Figure 10.10) comprises a mixture of Christian and Muslim communities, based originally upon Mount Lebanon, but extended on separation from Syria in 1926 to include the Anti-Lebanon mountains. Since its official independence in 1943, the history of Lebanon has been dominated by the roles adopted by its two powerful neighbours, Israel and Syria, and by the need to balance the changing populations of the confessional communities. With its mountainous terrain in a region dominated by plains and plateaux, Lebanon is the one Middle Eastern state other than Turkey and parts of Iran which receives adequate rainfall. Lebanon acts almost as a buffer between Syria and Israel, but unlike Jordan, has been handicapped by a divided leadership. The religious community comprises five recognized Islamic groups, making up 70 per cent of the population and eleven recognized Christian groups which total the other 30 per cent.

Recent history has been dominated by the 16-year-long civil war which began in 1975. There were Israeli invasions in 1978 and 1982, after the first of which the UN Interim Force in Lebanon (UNIFIL), was emplaced. In 1987, there was violence from Beirut to the Israeli border during which there was internal fighting between the Palestinians and with and between Amal, the Shi'a militia, and Hezbollah, an extreme Shi'a Fundamentalist group. In the meantime, the delicate confessional balance had been disturbed by the influx of refugees and also by the addition of the PLO, between 1970 and 1982. The Kuwait crisis of 1990 allowed the government to seize the initiative to destroy many of the divisions erected in Lebanon. One group not attacked by the army was Hezbollah which was allowed to continue its attack upon the South Lebanon army and its allies, the Israeli military, in the Israeli Security Zone.

With an area of only 10,500 km², Lebanon occupies 0.09 per cent of the Middle East and is the smallest mainland state. It comprises a narrow coastal plain, and the mountains of Lebanon and Anti-Lebanon, separated by the Bekaa valley. Other than Cyprus, Lebanon is the only Middle Eastern country with no desert landscape. The population is 3.5 million, 1.1 per cent of the Middle East population and its literacy rate of 92.4 per cent is marginally behind those of Israel and Cyprus.

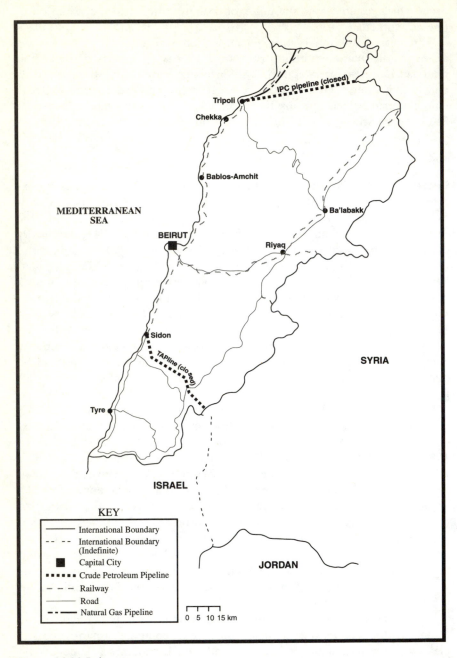

MEDITERRANEAN
SEA

Tripoli

Chekka

Bablos-Amchit

Ba'labakk

BEIRUT

Riyaq

Sidon

TAPline (clo sed)

Tyre

IPC pipeline (closed)

SYRIA

ISRAEL

JORDAN

KEY

	International Boundary
	International Boundary (Indefinite)
■	Capital City
	Crude Petroleum Pipeline
	Railway
	Road
	Natural Gas Pipeline

0 5 10 15 km

Figure 10.10 Lebanon.

The civil war from 1975 to 1991 almost removed Lebanon as a political and economic entity. However, in the recent past there have been impressive improvements and the current GDP is $7.7 billion, or 1.1 per cent of the total Middle Eastern GDP. This is the same as Cyprus and Qatar, both of which have considerably lower populations. As the infrastructure is rebuilt, so Lebanon may regain its position as the financial hub of the eastern Mediterranean. Its external debt is $3 billion, similar to that of Bahrain.

The armed forces, which number 49,000 personnel, are concentrated almost exclusively in the army, which remains relatively well equipped for the internal maintenance of peace. The defence budget is $0.41 billion which comprises, like that of Jordan, 0.9 per cent of the total Middle Eastern budget. Militarily, the requirements of the two countries are similar.

External support has come largely from Syria and Israel, both of which are still effectively in occupation. American involvement ceased abruptly in the mid-1980s, following the suicide bomb attack which killed 241 US marines. Israel maintains troops in the south and has done so since 1982, to support the proxy militia, the army of South Lebanon, along the self-declared 20 km wide Security Zone northwards from the border of Israel. Although Syria has had some troops in the south since 1976, there are still 30,000 troops in Lebanon, based mainly in the north, the Bekaa valley and Beirut. This deployment was legitimized by the Arab League and also according to the Ta'if Accord. Many temporarily influential individuals from the different communities have flitted momentarily across the world stage, but the only long-lasting contribution has been that of the Prime Minister, Hariri, who on the foundations established by President Harrawi, has largely rebuilt the country.

Since its economy was shattered, Lebanon has been among the least powerful states of the Middle East. However, its status is fast being restored as a key financial and entrepreneurial centre. Stability in Lebanon would represent a vital step towards lasting peace in the Middle East.

Geopolitically, the chief involvement of Lebanon has been in conflict. There has been more continuous violence than in any other Middle Eastern state, ranging from high-technology warfare to low-intensity conflict among and between a range of militias and terrorist groups. As Beirut is rebuilt, sporadic violence continues. Given the virtual disappearance of any state control throughout much of the 1970s and 1980s, there has been little border control. There has been illicit cross-boundary movements of terrorists, militias, arms, drugs and particularly refugees. Lebanon has become a key focus of cocaine processing and trafficking and an important trans-shipment point for hashish.

Having kept itself out of direct involvement in the Arab–Israeli wars, Lebanon has also remained aloof from hydropolitics. Much speculation has surrounded the Litani River and the possible diversion of water to the Jordan system. This is the only sizeable river wholly within Lebanon, and its lower reaches are within the Security Zone defended by Israel. The Orontes River and the Hasbani, one of the three tributaries of the Jordan, both rise in Lebanon and the state may therefore be implicated in future water issues.

Libya

Libya (Figure 10.11) is located centrally along the North African coastline between Egypt and the Maghreb proper and extends southwards well into the Sahara. It has the longest Mediterranean coastline of any Middle Eastern country and strategically is well placed to influence central Mediterranean traffic. Libya provides a link between central Africa and the Middle East. The Maghreb to the west is largely cut off by mountains from African influences and has far closer ties with Europe.

Libya was colonized by Italy which renounced control in 1947 when the area was placed under a joint British and French trusteeship. Independence was declared in 1951 under a monarchy. In 1969, there was a coup by junior officers, the most influential of whom was Qadhafi. His views fashioned by Nasserism and nationalism gave rise to his doctrine, known as *The Third Universal Theory* and set out in *The Green Book* published in 1976. The result in foreign policy has been implacable opposition to Israel and continuous confrontation with the West. For reasons of security, Libya has sought allies among all its neighbours and since 1989 has been a member of the Arab–Maghreb Union. With Syria, Iraq and Algeria, it formed part of the Confrontation Front in opposing the Camp David Accords.

The constant perception of Libya has been its involvement with terrorism. In 1980, 1982 and 1984, Libyans were allegedly involved in assassinations and assassination attempts. The long-term alienation of the USA resulted in two air attacks in 1986. Since then, Libyan foreign policy has moderated, but following alleged complicity in the bombing of flight PA 103 over Lockerbie, Scotland in 1988 and UTA 772 over Niger in 1989, sanctions were imposed. In mid-1999, when Libya handed over for trial two suspects for the Lockerbie case, sanctions were lifted.

Libya has an area of 1,759,500 km², 15.1 per cent of the Middle East, a figure exceeded only by Sudan and Saudi Arabia. It comprises desert plains and plateaux with a narrow coastal strip. The population of only 5.6 million, 1.7 per cent of the total Middle Eastern population, is concentrated along the coastal strip. The literacy rate of 76.2 per cent is around the average for the Middle East. Territorially, Libya is a major state, but it is almost a micro-state in terms of population.

The GDP of $25 billion, 3.6 per cent of the total Middle Eastern GDP, places Libya just below Kuwait, but significantly below Turkey, Saudi Arabia, Israel, Iran and Egypt. However, the GDP which is among the highest in Africa, has fluctuated wildly since it depends very largely upon oil, the commodity most hit by sanctions. Libya has an external debt of only $2.6 billion, one of the lowest in the Middle East in comparison to its GDP.

With the personnel total of 65,000, the armed forces are in the middle range by Middle Eastern standards, but their equipment is now relatively poor and they are considerably less powerful than many smaller militaries. The defence budget totals $1.4 billion, or 3 per cent of the total Middle

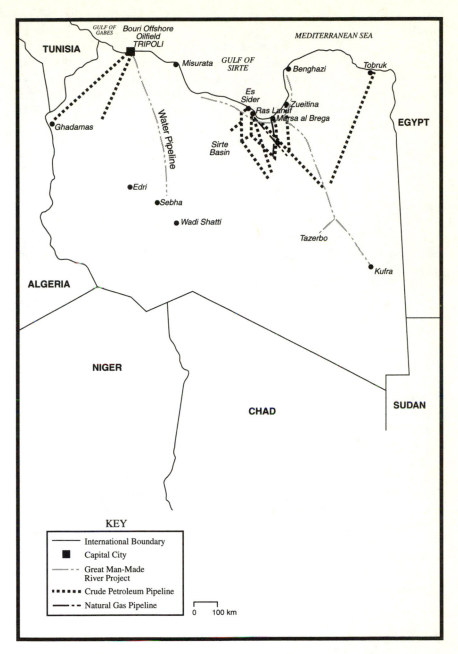

Figure 10.11 Libya.

Eastern defence budget. This places Libya centrally in Middle Eastern defence spending.

Libya has mounted a continuing quest for external support within the region, but globally, allied itself with the Soviet Union. Regarded by even its neighbours with suspicion, Libya is considered a pariah state by the USA. However, there is mounting evidence that Libya was not involved in several of the incidents, including the Lockerbie bombing in which it has been allegedly implicated. The entire period since the coup in 1969 has been dominated by Qadhafi and no Middle Eastern leader has been in power as long. Apart from King Hussein, only Saddam Hussein, Asad and Khomeini can be said to have been as influential in the development of their states.

With its oil resources, Libya remains potentially economically powerful, but is handicapped by the size of its population and, at present, by its virtual exclusion from world affairs.

Several of the geopolitical issues with which Libya has been associated remain to be proved, but there is strong evidence that terrorism has been generated and there has been illicit cross-boundary transfer of arms and possibly material for weapons of mass destruction. As the one major Middle Eastern oil producer west of Suez, Libya has been in a particularly advantageous position to use the oil weapon geopolitically.

More than any Middle Eastern state other than Saudi Arabia, Libya has been involved in boundary negotiations. The Aozou Strip over which a small-scale war was fought, was awarded to Chad in 1987, but Libya received favourable judgements from the ICJ with regard to its maritime boundaries with Tunisia (1982) and Malta (1985). Part of both of those maritime boundaries remains to be settled, as do the land boundaries with Niger and south-eastern Algeria. The other long-running boundary issue concerns the Gulf of Sirte since Libya claims that this very large water body constitutes internal waters.

As a largely arid country, Libya has developed the Great Man-Made River Project, the largest water development scheme in the world, to transport water from the Fezzan, notably the oasis of Kufra, to the Coastal Zone. This is an internal transfer, but it involves the mining of water from deep aquifers, the major one of which overlaps the boundaries of both Egypt and Sudan.

Qadhafi took power as a disciple of Nasser and has pursued the path of Arab nationalism ever since. However, apart from membership of the Arab–Maghreb Union, other offers of unification have been rebuffed.

Oman

Oman (Figure 10.12) is located at the entrance to the Gulf, but east of Hormuz. This has allowed an independence of action, giving it a certain detachment from the rest of the Arabian Peninsula. With long coastlines on the Gulf of Oman and the Arabian Sea, Oman developed maritime traditions and was the only non-European state to have an African colony,

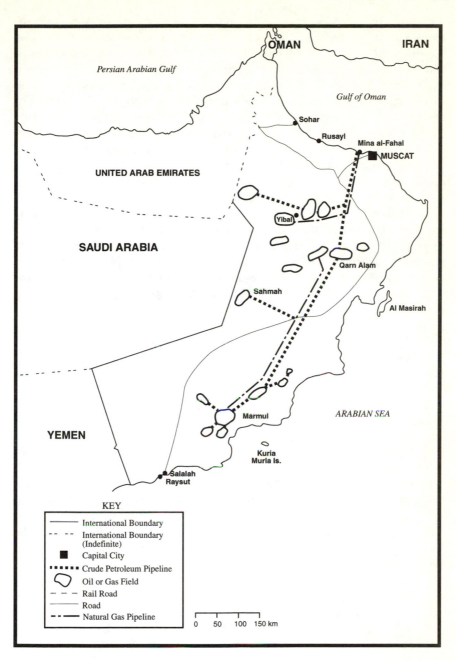

Figure 10.12 Oman.

Zanzibar. It also developed close links with the Indian subcontinent. Ironically, while itself being free of the constraints of the Strait of Hormuz, the tip of the Musandam peninsula is within Omani territory and thus Oman with Iran exercises control over the choke point. Its independence has been demonstrated by a lack of involvement in many Arabian issues, continuing close links with Iran and its willingness to host the US Rapid Deployment Force. There has been conflict with its neighbours, particularly South Yemen in the Dhofar and the UAE over Buraimi oasis. Regionally therefore, Oman occupies a highly strategic position. Internally, the country is effectively split into two or even three units by the terrain. The coastal strip and Muscat are divided from Nizwa by a high mountain range and Nizwa is separated from Salalah and the Dhofar in the south by desert.

The history of the area goes back a very long way, but the official date for the independence of the state is 1650. In the recent past there have been three key disputes. The first two, over Buraimi oasis (1952) and over the possible independent state claimed by the Imam in Nizwa (1957–59), occurred before the *coup* which began the reign of Sultan Qabus. Until that time, the state was effectively medieval, but Qabus immediately began modernization. The name of the state was changed from Muscat and Oman to Oman and the country was admitted to the Arab League and the UN in 1971. However, until 1976, a third dispute, that with the People's Democratic Republic of Yemen (South Yemen), dominated events. Since then, Oman has been politically stable and has grown significantly economically.

The area of Oman is 300,000 km², 2.6 per cent of the Middle East and is a medium-sized state. It comprises coastal plains in the north and south, separated from the interior desert by high mountain ranges. In terms of population, with 2.2 million, or 0.6 per cent of the total Middle Eastern population, Oman is a micro-state. The literacy rate of 80 per cent is relatively high.

The economic strength of Oman depends upon the oil industry which accounts for 75 per cent of export earnings. The agricultural sector is small and industry is developing, but is limited in scope. The GDP is $12.2 billion, 1.8 per cent of total Middle Eastern GDP and the external debt is $2.7 billion, much the same as Libya, but in an economy which is only half the size. The underlying weakness will be decreasing oil revenues as resources are very limited.

With a personnel total of only 43,500, the armed forces are relatively small, but they are well balanced and extremely well trained, with high-quality equipment. The defence budget is $1.8 billion, 3.8 per cent of the total Middle Eastern defence budget. This figure is similar to that for the UAE and is considerably in excess of that for other very small states in the region, apart from Kuwait.

As a result of its partial detachment from Arabian affairs, Oman has enjoyed support from most important sources, regionally and globally. There have been traditional close links with the UK, which was responsible for training the military, and relations with the USA are close. At the same time, Oman has maintained friendly contact with Saudi Arabia and other members of the GCC, together with Iran. The significance of its location has been enhanced by the diplomacy of Sultan Qabus, who has been highly influential in moulding the progress of the state.

For a variety of reasons, particularly location and foreign policy, Oman has been an increasingly significant actor in Middle Eastern affairs. It is not particularly strong economically, but has a small and powerful military.

Oman's level of detachment is illustrated by the few political issues with which it is involved. Apart from the dispute over Buraimi oasis, where questions may still be raised, all the boundaries have now been settled, most recently with Yemen (1992) and Saudi Arabia (1990). The other issues are merely potential in that Oman controls the southern shore of the Strait of Hormuz and shares a major aquifer with the UAE.

Palestine

Although not yet a state, Palestine (Figure 10.7) has territory alongside and within the state of Israel. The present two units are the Gaza Strip, which provides a coastline, and a small landlocked area around the city of Jericho. With completion of negotiations, the West Bank should also come under Palestinian control and a key geopolitical issue then will focus upon the provision of a permanent link between it and the Gaza Strip. Palestine has been the focus of four wars between Israel and the Arab states and must therefore qualify as perhaps the most critical geopolitical flashpoint in the world.

The history of the embryonic Palestinian state provides a mirror image of the history of Israel. After its independence in 1948, the state of Israel was extended at the expense of the Palestinians in 1949 while in 1967 the West Bank and Gaza Strip were occupied. The PLO was founded in 1964, and, despite many vicissitudes in its fortunes, has retained the leadership role under Yasser Arafat in the current negotiations for statehood. The PLO–Israel Declaration of Principles on Interim Self-Government Arrangements was signed in 1993 and provided for a transitional period of up to five years of Palestinian interim self-government in the Gaza Strip and West Bank. Permanent status negotiations began in 1996 and under the Declaration of Principles Israel agreed to transfer certain powers to the Palestinian Authority. The Palestinian Authority includes a Palestinian Legislative Council, elected in 1996 as part of the arrangements for the West Bank and Gaza Strip. In 1994, the Cairo Agreement on the Gaza Strip and the Jericho Area came into effect and power and responsibilities were transferred to the Palestinians, who had thereby achieved a territorial base, the key requisite for statehood. In addition, certain areas of the West Bank were transferred, but Israel retains overall responsibility during the transition period for security, both external and internal.

The area of Palestine, should it include the entire West Bank, would be approximately 6,200 km^2, apart from Bahrain, the smallest state in the Middle

East. The region comprises the West Bank, which is predominantly dry upland, and a section of coastal strip. The population of about 1.5 million constitutes 0.46 per cent of the Middle Eastern population and the figure for literacy is not available. However, as with the Jewish Diaspora before it, the Palestinian Diaspora has resulted in the highest value being placed upon education. Therefore, the level of literacy among returnees is likely to be high.

The GDP, in that it can be calculated, is $3.2 billion, which is the lowest in the Middle East and represents only 0.5 per cent of the total. Within the new Palestinian state, there will be few resources and heavy reliance must be placed upon the skills of immigrants and the levels of aid available. The external debt is said to be $0.8 billion, the lowest in the Middle East.

The security forces number 16,500 personnel and are essentially paramilitary. The PLO has a number of linked militias including the Palestine National Liberation Army, but is opposed by an equal number of militant groups. Palestine spends $0.09 billion on security, or 0.2 per cent of the total Middle Eastern defence budget.

Continuously since 1948, the Palestinians have been accorded spiritual support throughout the Arab world, but material support has varied. Most significantly, funds were cut off following the Gulf War when Palestinian support was given to Saddam Hussein. However, progress with the Peace Process will ensure the financial backing of the EU and the USA. The assistance of the Arab world has been crucial in the establishment of what will become a Palestinian state and, without it, the slow progress made both internally and externally by the PLO could never have been sustained. Nevertheless, there has also been unwanted assistance from a number of terrorist organizations and this has interrupted the progress towards statehood. Despite changes in fortune, Yasser Arafat has remained the consistent and influential figure throughout the protracted negotiations which have led to the current position.

When permanent status has been achieved, assuming that most of the West Bank is by then Palestinian territory, the state, despite having few resources and no real military, will be in a strong position geopolitically because of its intimate geographical relations with Israel.

In their progress towards statehood, the Palestinians have been implicated in a wide variety of geopolitical issues. These have concerned chiefly conflict and terrorism with the large-scale transboundary and indeed world-wide movement of refugees. Other transboundary movements have included arms, intelligence and funding from many sources.

Boundary issues, primarily over the West Bank, have yet to surface, but are likely to be extremely complex. They will also involve considerations of water as a result of the presence of key aquifers and the Palestinians will continue to be involved in hydropolitics.

Nationalism has also loomed large as a factor in the development of the state. For fifty years, Palestinians have had to accept that they were a nation without a state and nationhood has been fostered from both within the region and from the Diaspora.

Qatar

Qatar (Figure 10.13) comprises a peninsula on the Arabian side of the Gulf, approximately halfway between the Strait of Hormuz and Kuwait. It is therefore, like its neighbour Bahrain, centrally placed in oil geopolitical events. It is contiguous only with Saudi Arabia. Immediately to the north-west is the major Saudi oil export terminal of Ras Tanura, but the feasibility of a new exporting point in the Saudi corridor between Qatar and the UAE, immediately east of the peninsula, is being considered.

From 1968, Qatar discussed terms for federation with Bahrain and the seven sheikhdoms which comprised the Trucial Oman, but did not obtain satisfactory terms. It became independent in 1971 and a member of the Arab League and UN. Since 1986 there has been a territorial dispute with Bahrain and Qatar has shown increasing independence in its foreign policy. In particular, it has tried to maintain good relations with both Iraq and Iran.

With an area of only 11,500 km^2, Qatar constitutes 0.1 per cent of the Middle Eastern area, making it slightly larger than Lebanon. It comprises a low, flat desert platform. The population is 0.7 million, 0.21 per cent of the Middle Eastern total, and is only 40 per cent Arab, the expatriate workers outnumbering the indigenous population. The literacy rate is 79.4 per cent, ranking closely with Kuwait, Oman and the UAE.

The economy depends upon the petroleum industry. Oil provides approximately 80 per cent of export earnings and Qatar's proved reserves of natural gas are the third largest in the world. The GDP is $7.4 billion, 1.1 per cent of the total Middle Eastern GDP. External debt is $5.7 billion, relatively high in proportion to GDP for an Arab oil state.

With a total personnel of just under 12,000, the armed forces of Qatar are very small, only marginally larger than those of Bahrain and Cyprus. The defence budget is $0.33 billion, 0.7 per cent of the total Middle Eastern defence budget, and marginally smaller than that of Sudan. Despite its limited size, the military cannot be totally discounted as it possesses modern equipment.

Since it is economically strong and has not been attacked, Qatar has required little in the way of external support. However, it has maintained good relations with Iraq, Iran and its Arab neighbours. Recently, there have been problems with the GCC, but these have been of a temporary nature. The Al-Thani family has held power since independence, but its significance has been national.

Of the truly micro-states in the region, Qatar is by far the strongest economically, but it, like Kuwait, will never be in a position to do other than react to events. The only significant geopolitical issue concerns the continuing territorial dispute with Bahrain over the Hawar Islands but there is also a maritime boundary to be settled with Bahrain. In 1996, agreement was reached with Saudi Arabia to demarcate the land boundary according to the original 1992 Accord. The only other potential issues concern the proportion of expatriate labour in the population and the limited water supply. If,

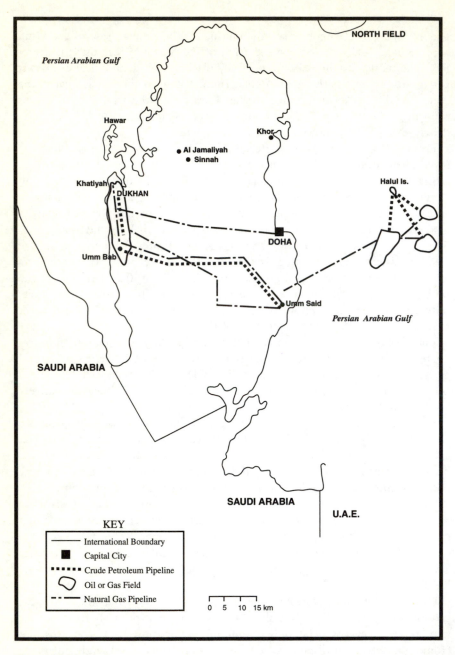

Figure 10.13 Qatar.

as planned, Iran supplies water through a pipeline, then this will result in dependence on a foreign source for a key resource.

Saudi Arabia

Saudi Arabia (Figure 10.14) occupies the greater proportion of the Arabian Peninsula, but has no direct access to the Gulf of Oman or the Arabian Sea. Within the Peninsula are the two holy mosques of Mecca and Medina and Saudi Arabia is thus the spiritual heart of Islam. With its Gulf and Red Sea coastlines, a total of 2,640 km in length, Saudi Arabia has the longest coastline in the Middle East, but because of the restricted nature of the two water bodies, it controls only a very limited offshore area. Only Oman and Yemen in the Middle East can claim their full Exclusive Economic Zone (EEZ). The territory also includes three of the world's mega-oilfields and Saudi Arabia has the world's largest oil reserves (26 per cent of the proved total). However, with no direct access to the global ocean, oil exports need to transit either the Strait of Hormuz, the Suez Canal or the Strait of Bab el Mandeb. Saudi Arabia has land boundaries with seven neighbouring states and maritime boundaries are either agreed or need to be negotiated with a further five states.

The Kingdom of Saudi Arabia was finally unified by Ibn Saud, King Abdul Aziz, and proclaimed as an independent state in 1932. In 1938 came the momentous discovery of oil in the Eastern Province, al-Hasa. Following Ibn Saud's death in 1953, he has been succeeded as king by four of his sons, the most notable of whom, in that he laid the foundations for the modern state, was Faisal. Under Faisal, Saudi Arabia was anti-Communist and a strong supporter of the Palestinian Arabs. He also came to terms with Iran, Yemen and Abu Dhabi. During the 1973 Arab–Israeli war, Saudi Arabia embargoed oil shipments to the USA and accompanied this by a 10 per cent cut in oil exports, thereby paving the way for the OPEC 1973–74 price rises.

Since that time, Saudi Arabia has co-operated closely with the USA. The seizure of the Grand Mosque in Mecca in 1979, the year of the Iranian revolution, was a stunning blow for the Kingdom. During the Iran–Iraq war, large-scale military purchases were made from the USA, France and the UK, but military dependence has been chiefly upon the USA.

Using its key position in the Middle East, Saudi Arabian diplomacy has been instrumental in the Lebanese settlement and the re-admittance of Egypt to the Arab League.

With the invasion of Kuwait by Iraq in 1990, Saudi Arabia felt itself threatened and requested US support. This has resulted in continuing disagreement between those who favour increased Islamic morality and those who support greater democracy. Relations with Yemen, following the 1994 Yemeni civil war, have been largely repaired, but agreement still has to be reached on the boundaries.

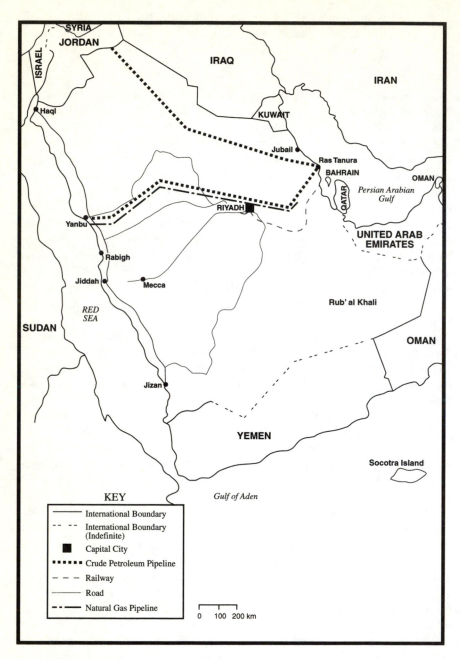

Figure 10.14 Saudi Arabia.

With an area of 2,240 million km², Saudi Arabia occupies 19.2 per cent of the Middle East, a figure only exceeded by Sudan. The landscape is predominantly desert, and includes the Rub' al Khali, the largest sand sea in the world. The latest estimate of the population is 20 million, 6.1 per cent of the total Middle Eastern population, but the literacy rate is low at 62.8 per cent.

Saudi Arabia is economically strong as a result of oil which accounts for 90 per cent of export earnings. There is a modest agricultural sector, dominated economically by agribusiness and developing high-technology industry, following a series of offset agreements. The GDP is $125 billion, 18 per cent of the Middle Eastern GDP and second only to that of Turkey. The external debt is $16.6 billion, slightly greater than that of Sudan.

The armed forces comprise 105,500 active personnel and there is a strong army and air force and a developing navy. All are equipped with the latest technology. There is also a Peninsula Shield Force of 7,000, supplied by the USA, France and the UK. This raises the question of when, with its powerful military, Saudi Arabia will act independently and when it will require Western assistance. The defence budget is $13.2 billion, 28 per cent of the total Middle Eastern defence budget and almost twice that of Israel which is the next largest.

As a result of its position, geographically and within Islam, and particularly its possession of the world's major oilfields, Saudi Arabia has been assured of external support, both regionally and globally. In fact, certain of the internal problems result directly from the overt nature of US support. Given the authority with which he can speak, each king is an actor upon the world stage, but undoubtedly the most influential individual in the recent history of Saudi Arabia was King Abdul Aziz. He continues as an heroic figure who personally fashioned the state and has left as his legacy the House of Al-Saud.

Despite the relatively small population for its area, Saudi Arabia is economically strong and, by Middle Eastern standards, militarily powerful. With the legitimacy and authority provided by Islam and King Abdul Aziz, Saudi Arabia must always remain a crucial actor in Middle Eastern affairs.

In fact, Saudi Arabia has remained an island of relative calm with conflicts of both high and low intensity occurring all around it. Nonetheless, as a partisan and a mediator, the Kingdom has been involved in the geopolitics of conflict. More importantly, Saudi Arabia is always likely to be implicated in resource geopolitics and any use of the oil weapon.

The other major geopolitical issues to concern Saudi Arabia have been international boundaries, both land and maritime. In boundary settlement, Saudi Arabia has consistently shown flexibility in agreeing to shared areas or porous boundaries. The major unresolved issue is the maritime and most of the land boundary with Yemen which remain undelimited. The one section of the boundary which is agreed, is the demarcated section from the Red Sea coast across the mountain chain to the western edge of the interior (Treaty of Taif, 1934). With the UAE, there is a *de facto* boundary which reflects the 1974 agreement and the maritime boundary with Kuwait, including the ownership of Qaruh and Umm al Maradim Islands, has yet to be settled.

Sudan

Located to the south of Egypt and abutting on only one other Middle Eastern state, Libya, Sudan (Figure 10.15) is geographically in the Middle East but, like Turkey and Cyprus, is greatly influenced by its neighbouring region. Indeed, although the country is 70 per cent Muslim, it is only 39 per cent Arab. It is the largest state in the Middle East and has land boundaries with nine other states and a maritime boundary yet to be delimited with Saudi Arabia. It is also the largest state in Africa and is dominated by the continent's largest river, the Nile. Therefore, relations with Egypt are crucial and, although it is politically peripheral, Sudan has been involved in Middle Eastern foreign affairs through relationships with Iran and Iraq in particular.

The country itself tends to be split between the Muslim north and the animist and Christian south, where there has been a crisis since 1983. Owing to its relationships, particularly its alignment with Iraq during the Gulf War, Sudan has been cited by the USA as a pariah state. As a result of its human rights record, UN sanctions have been mooted. Since independence in 1956, Sudan has alternated between civilian and military rule. Between 1972 and 1983 the south was autonomous, but this situation ended when Numeri, the President, re-divided the region into three and implemented *Shari'a*. This resulted in rebellion led by the Sudan People's Liberation Movement and Army (SPLM/SPLA), a formidable military–political force. The military was again succeeded by a civilian rule before the military once more took power in 1989. With the support of other Middle Eastern states including Iran, the new military regime, led essentially by the doctrines of the Muslim Brotherhood, sought to crush the southern rebels. However, despite a split in their ranks and the loss of support from Ethiopia, the war continues.

In the Middle East, Sudan supported Iraq in the Gulf War and also aligned itself with Iran, thereby antagonizing many Arab states. It was also allegedly implicated in the World Trade Center bombing (1994), thereby incurring the grave displeasure of the USA. Relations with Egypt to the north and Kenya and Uganda to the south are also poor. Ethnic, social and religious differences appear unreconcilable. With an area of 2.506 million km^2, Sudan occupies 21.5 per cent of the Middle East. There are mountains in the east and west, but the major part of the terrain is flat plain. The population is 32.6 million, 9.9 per cent of the total Middle Eastern population, but the literacy rate of 46.1 per cent is, with the exception of Yemen, the lowest in the Middle East.

With continuing political instability, civil war, hyper-inflation and a reduction in remittances from abroad, the economy of Sudan is in a parlous

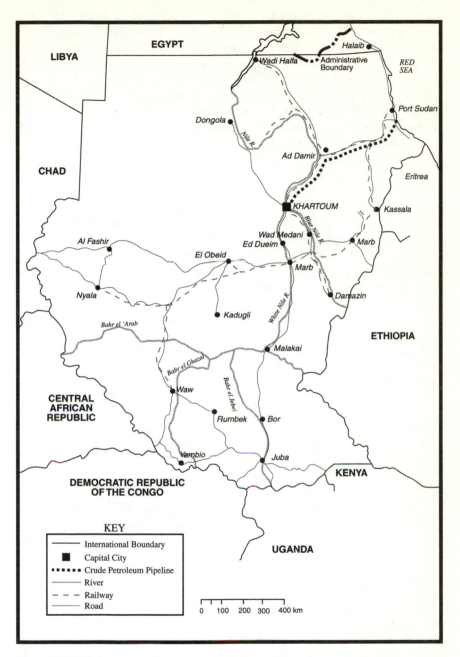

Figure 10.15 Sudan.

position. The agricultural sector is large and industry is relatively little developed. The GDP is $9.1 billion, 1.3 per cent of the total Middle Eastern GDP, a figure similar to that of Yemen which has a population of only 13.9 million. The external debt is $18.5 billion, or twice the GDP, a situation only exceeded by Iraq.

Membership of the armed forces stands at 89,000, 96 per cent of whom are in the army. Equipment is very modest in both quality and scale and, despite its size, the army of Sudan cannot be considered potentially powerful in anything approaching modern warfare. The defence budget is $0.39 billion, or 0.8 per cent that of the Middle East, putting it marginally ahead of Yemen and Cyprus, but behind Jordan and Lebanon.

Given the political ineptitude since independence, it is hardly surprising that Sudan has virtually no consistent external support. Furthermore, in its brief history there have been no influential individuals, unless Numeri, as the leader who finally set the civil war on course for the current chaos, is counted.

Considering its size and population, together with its potential resources, Sudan could be a global actor of significance. Furthermore, it exercises a controlling influence over a large part of the Nile Valley and therefore hydropolitically is in a strong position. However, long-term mismanagement, allied to its indigenous differences, have resulted in little other than continuous civil war.

The key geopolitical issue which involves Sudan is conflict and terrorism. This has led to large-scale movements of arms, militias and refugees across boundaries. In fact, in the south in particular, boundaries mean relatively little. Nonetheless, there are official international boundary disputes with Kenya and with Egypt over the Hala'ib Triangle. Internal disputes have centred upon religious differences and there is now a major question of food security. The other potential geopolitical concern must be with water, since Egypt is already, by agreement, using part of Sudan's share of the Nile.

Syria

Located in the core of the Middle East and abutting on to Turkey, Jordan, Iraq, Israel and Lebanon, Syria (Figure 10.16) is likely to be involved in most of the key geopolitical issues of the region. All its neighbours, other than Jordan, are at present involved in some form of conflict. Syria shares the Turkish question with Turkey and Iraq, is itself in occupation of parts of Lebanon and is in part occupied by Israel. Syria is a riparian of the Tigris, the Euphrates, the Orontes and parts of the upper Jordan. It has a coastline on the Mediterranean and shares Ba'athist philosophy, but not practice, with Iraq. The boundary with Turkey represents the northern limit of the Arab world and it shares three of the major Middle Eastern landscapes: the Levantine coastal plain, Mesopotamia and the desert.

Independence occurred in 1946 when, through the efforts of the UN, the USA, the Soviet Union and Britain, France was forced to withdraw. Over the next 21 years there were more than a dozen military *coups*, but following

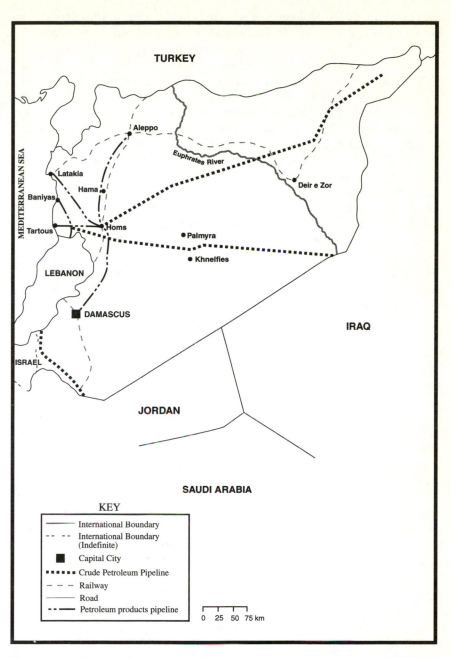

TURKEY

Aleppo

Euphrates River

Deir e Zor

MEDITERRANEAN SEA

Latakia

Hama

Baniyas

Tartous

Homs

● Palmyra

● Khnelfies

LEBANON

DAMASCUS

IRAQ

ISRAEL

JORDAN

SAUDI ARABIA

KEY

—————— International Boundary

– – – – International Boundary (Indefinite)

■ Capital City

▪▪▪▪▪ Crude Petroleum Pipeline

– – – Railway

——— Road

–▪–▪– Petroleum products pipeline

0 25 50 75 km

Figure 10.16 Syria.

the loss of the Golan Heights to the Israelis in the 1967 war, a new Ba'ath party military regime under Al Asad seized power. Asad was elected President in 1971 and in 1973 Syria again fought against the Israelis and recovered a small strip of territory. In 1975, attention turned to Lebanon where Syria, in the interests of its own security, became a military peacekeeper. The Arab League created the Arab Deterrent Force, largely Syrian, to restrict both the Palestinians and the Maronite militias. With Egypt neutralized following the Camp David Accords, Syria became the key front-line state and Israel formally annexed its part of the Golan Heights in 1981.

In 1987, Syria again entered Lebanon, but by early 1989 had become isolated from Western support as a result of alleged terrorism and within the Middle East, because of ideology, disagreement with the Camp David Accords and quarrels with the Palestinians and the Arab League proposal for peace in Lebanon. However, Asad managed to reach agreement at Taif that Syria should have the right to station troops in Lebanon indefinitely. Syria's position was undermined by the demise of the Soviet Union, but strengthened later by its support for the coalition during the Gulf War. As a result, EU sanctions were dropped and financial assistance was received from the Arab oil states, Europe and Japan. Negotiations with Israel over Golan continued in the early 1990s, but resulted in stalemate.

Syria is 185,000 km² in area, 1.6 per cent of the Middle East, and comprises a coastal plain, desert plateau and mountains in the west. The population is 16.1 million, 4.9 per cent of the total Middle Eastern population, and the rate of literacy at 70.8 per cent is just below the average for the Middle East.

The economy of Syria is balanced in that there remains a well-developed agricultural sector, but oil production is modest and may have peaked and there is a relatively heavy external debt of $22 billion. The GDP is $30 billion, 4.3 per cent of the total Middle Eastern GDP, putting Syria just above the average for the region. This accords well with its population which is the seventh largest in the region.

With a total of 421,000 personnel in the armed forces, Syria has a very large army, a large air force and a relatively modest navy. All three services are well equipped, the air force and navy mainly by equipment from the FSU, while the army has drawn on a number of sources. The defence budget is $2 billion, 4.3 per cent of the total Middle Eastern defence budget, the eighth largest in the region.

Until 1989, Syria could count upon the support of the Soviet Union, but was frequently isolated otherwise. Ideologically, it disagreed with most of its neighbours and its alleged association with terrorism for a long time alienated the West in general. It is only since the Gulf War that Syria has received both political and economic support from the West. After a long period of chaotic mismanagement, Asad has, by his diplomatic skills, made Syria a serious actor upon the Middle Eastern stage.

In area, population, GDP and defence spending Syria is consistently in the upper middle rank of Middle Eastern countries. It has a balanced but somewhat volatile economy and is relatively strong militarily. It is likely to remain, while not a regional superpower, an important and relatively powerful actor in the Middle East.

Geopolitically, Syria has been involved in rather fewer issues than many of its neighbours. Most obviously, it was active in the Arab–Israeli wars, losing the Golan Heights in 1967. It has been associated with terrorism, both in and beyond the region, but has escaped the notoriety accorded to Iran and Libya. It is heavily involved in hydropolitics, particularly with regard to the Euphrates, but the question of the Orontes is closely allied to its boundary dispute with Turkey over Hatay. The Golan Heights remain occupied by Israel and Syrian troops have been in northern, central and eastern Lebanon since 1976.

With regard to transboundary movements, Syria is a transit point for cocaine, heroin and hashish, bound for Western markets. Following Ba'athist ideology, Syria has long been a proponent of Pan-Arabism and for three years was a partner with Egypt in a joint state, United Arab Republic (1958–61).

Turkey

Turkey (Figure 10.17) occupies the north-west of the Middle East and is in a sense a counterpoise to Iran which occupies the north-east. Both are vestiges of great former empires, both are Islamic, but neither is Arab. The key difference is that Iran is a theocracy, while Turkey is secular. Since part of its territory is in Europe, Turkey provides the ideal bridge between the Middle East and Europe. As a long-serving member of NATO, it has closer military ties with the West than with any other Middle Eastern state. However, it has been refused EU membership, while Cyprus is in the throes of negotiation.

Like Iran, Turkey also provides a link to the Trans-Caucasus, the Caspian basin and Central Asia. Indeed, as Central Asia opened following the demise of the Soviet Union, it was surmised that Turkey and Iran would compete for influence.

Turkey holds perhaps the most strategic location in the whole of the Middle East, which is further enhanced by its control of the Turkish Strait and therefore Black Sea traffic. Its location also makes it critical in the debate over pipeline geopolitics concerned with exit points for Caspian oil. A further highly significant aspect of its location is that Turkey controls the headwaters of both the Tigris and the Euphrates, and indeed has sufficient surplus water that it was able to propose the Peace Pipeline and to implement Operation Medusa.

In the Gulf War, Turkey's position was critical in that it, with Saudi Arabia, was able to cut off the oil exports of Iraq by closing the pipelines. In addition, the NATO air bases were used for Operation Desert Storm. Immediately to the west, its relations with Greece are fraught, as a result of history but most visibly in the dispute over Cyprus and the long-standing problem of the Aegean maritime boundary. Most (95 per cent) of Turkey is in Asia,

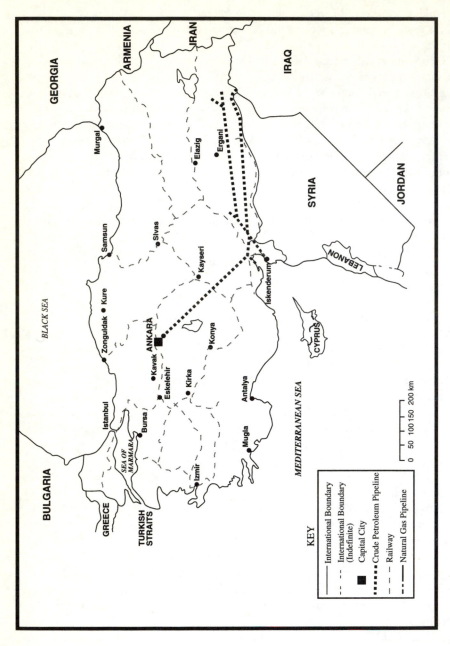

Figure 10.17 Turkey.

where it abuts on to five countries, and 5 per cent of it is in Europe, bordering Bulgaria and Greece. With Syria, Iraq and Iran, it is heavily implicated in the Kurdish issue.

As the successor to the Ottoman Empire defeated in the First World War, the state was rebuilt by Mustafa Kemal, later known as Ataturk, and by the Treaty of Lausanne became independent in 1923. Building upon Western models, Ataturk secularized and modernized the state and when he died in 1938 he was acknowledged as the 'hero of the nation'. In the Second World War, fears of the Soviet Union forced what had been a neutral Turkey into the Western camp and in 1951 it joined NATO. Since then, there has been a mixture of civil and military rule.

In 1974, Turkey invaded Cyprus in response to what was taken as Greek provocation. During the Iran–Iraq war Turkey remained neutral and indeed increased trade with both. In 1987, Ozal was elected Prime Minister and, both in that role and as President, he brought order to what had been chaotic political and economic development. In the Gulf War, Turkey was a critical component of the coalition without engaging in the fighting. Meanwhile, throughout the 1990s there has been fierce fighting in the east of the country with the PKK and this continues. Apart from the Kurdish uprising, Islamic militancy also threatens the stability of Turkey, but this appears to be contained.

Turkey has an area of 779,500 km² or 7.7 per cent of the Middle East. The territory is large, but is dwarfed by those of Sudan, Saudi Arabia and Iran. It comprises mostly upland with a high central plateau and a narrow coastal plain. The population of 63.5 million is 19.3 per cent of the Middle Eastern population total, placing Turkey marginally behind Iran and Egypt as one of the three most populous states. The next largest population is that of Sudan which has approximately half the total of the three. The literacy rate of 82.3 per cent is relatively high, placing Turkey above all of the Gulf states, except Bahrain.

The economy is dynamic; a complex mix of modern industry and commerce, alongside traditional production. There is a large agricultural sector and remittances are received from approximately 1.5 million Turks who work abroad. The GDP is $167 billion, 24 per cent of the total Middle Eastern GDP and by far the highest in the region. However, the debt stands at $75.8 billion, more than twice that of any state in the Middle East, other than Iraq.

With 639,000 personnel in its armed forces, Turkey has by far the biggest military in the Middle East. All three services are large and all are equipped with relatively modern material. Turkey provides the headquarters for the NATO Allied Land Forces South East Europe and also for Operation Provide Comfort (directed at Iraq), which is supported by France, the UK, the USA and Israel. Indeed, military links with Israel have increased to the great discomfort of the Arab Middle Eastern states. Turkey therefore provides a formidable military, easily the most powerful in the region and indeed of significant power by global standards. The defence budget is $6 billion,

12.8 per cent of the total Middle Eastern defence budget, putting Turkey third behind Saudi Arabia and Israel.

During the Cold War period, as the only NATO state having an important strategic boundary with the Soviet Union, Turkey was fully supported by the West. Support has, however, never been unequivocal in that problems have arisen over its relationship with Greece. Its ability to link the Middle East with Europe has been important in raising support on both sides and its membership of the coalition during the Gulf War ensured continuing US and EU support. However, developing relations with Israel have led to some alienation in the Arab world, while its human rights record, particularly in the context of the Kurdish uprising, has been generally criticized. Like King Abdul Aziz, Ataturk is the heroic figure who provides the authority and legitimacy for the current secular state. Any government attempting to dispose of his legacy would face problems, particularly with the military. Of recent leaders, Ozal is the most notable for his consistency in bringing both political and economic order.

Turkey is a powerful state with a large, well-educated population, a history of close links with the West, continuing external support from both East and West, a relatively strong economy and a very powerful military. In many ways Turkey is the regional superpower in the Middle East and major geopolitical events in which it is concerned will require its agreement.

With probably the most strategically important location in the Middle East, it would be expected that Turkey would be involved in a range of geopolitical issues. The focus must be upon boundaries and particularly with the complex maritime, air and territorial disputes with Greece in the Aegean Sea and over the question of Cyprus. There is also the issue of Hatay which is disputed with Syria and this involves the hydropolitics of the Orontes. However, the major hydropolitical issues concern the Tigris and particularly the Euphrates rivers. When the flow of the Euphrates was stopped for one month in January 1990, this was interpreted by both Syria and Iraq as a major geopolitical signal. Agreement still has to be reached on the management of the Tigris–Euphrates basin.

The location of Turkey was critical in the Cold War and in the subsequent Gulf War. Nevertheless, the only conflict involving the Turkish military has been with the PKK, both inside Turkey and in northern Iraq. As a result of the Kurdish issue in general, there have been transboundary flows of refugees. A further cross-border movement has been the supply of expatriate labour to Western Europe. Turkey is a major transit route for heroin and hashish from south-west Asia to the West by land, sea and air. It is known that major Turkish, Iranian and other international trafficking organizations operate in Istanbul and there are laboratories in Turkey for producing heroin.

Apart from boundaries, the major geopolitical issue of the future concerns the use of the Turkish Straits and the development of a pipeline network from the Caspian oilfields. While not an oil producer, Turkey will therefore be heavily implicated in oil geopolitics. Both of these transit problems will be highly significant. They will require Turkish control and will provide an important source of transit dues.

The reconstruction of Turkey after the demise of the Ottoman Empire generated feelings of nationalism which have been sustained. These have been enhanced recently by close ties with the other Turkic-speaking peoples of Central Asia. Nationalism has also been illustrated by Turkish support for the Turkish Republic of Northern Cyprus and has been demonstrated by the establishment of a water supply to the island from Turkey.

United Arab Emirates (UAE)

The UAE (Figure 10.18) forms all but the tip of the peninsula which divides the Persian–Arabian Gulf from the Gulf of Oman and forms the Strait of Hormuz. The tip of the peninsula is the territory of Oman. The UAE comprises seven emirates, but the area of Abu Dhabi is approximately three times the size of the other six combined. Fujairah, together with very small detached remnants of Ajman and Sharjah, form the east-facing Gulf of Oman coastline, while the major areas of Ajman and Sharjah, together with Ras al Khaymah, Umm al-Qaywayn, Dubai and Abu Dhabi, form the Gulf coast, which extends from the territory of Oman at Hormuz to the corridor of Saudi Arabia just short of the Qatar peninsula. The UAE therefore defines the limit of the Gulf geographically and is itself a Gulf state and a Gulf of Oman state. This excellent trading position has been utilized by the development of free trade zones in Jebal Ali (Dubai) and in Fujairah. The nodality of the state is also indicated by the presence of three international airports within just over 100 km. Proximity to Iran has resulted in close relations with that state, although there are three disputed islands. The other significant factor of location is that it coincides with the end of the major oilfield zone which extends from Iraq and the neighbouring parts of Iran through Kuwait and Al Hasa province of Saudi Arabia. Abu Dhabi has approximately the same reserves as Kuwait, while Dubai and Sharjah also have oil, but in very limited quantities.

Oil production began in Abu Dhabi in 1962 and, on the British withdrawal from the Gulf in 1971, six of the seven emirates formed the UAE. Ras al Khaymah joined in 1972. Each sheikh runs the internal affairs of his own emirate, while at the federal level, authority lies with the Supreme Council of Rulers, with the membership comprising all seven sheikhs. Abu Dhabi, the most powerful because of its vast oil resources, and the contributor of most of the national budget, provides the president, while Dubai, the second most powerful emirate, provides the vice president, who is also prime minister. During the 1980s, the UAE suffered, partly as a result of the Iran–Iraq war and partly through the drop in oil prices and, like Kuwait, sold more than its OPEC quota. This drew the condemnation of Iraq in 1990 and the UAE joined the coalition which won the Gulf War. In 1992, the dispute resurfaced with Iran over three islands in the approaches to Hormuz: Abu Musa, Greater Tunb and Lesser Tunb.

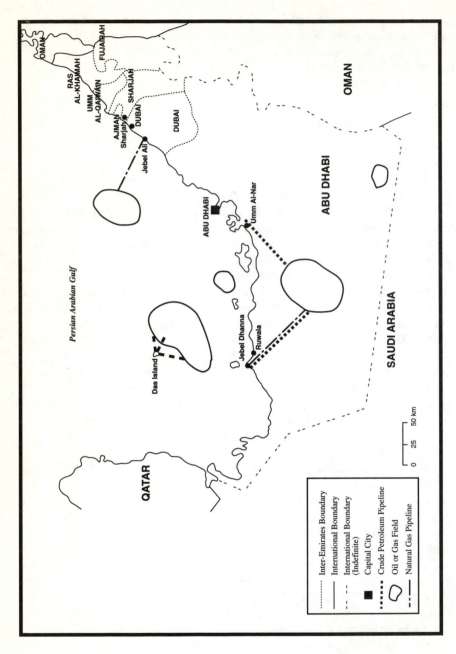

Figure 10.18 United Arab Emirates.

With an area of 78,000 km², the UAE comprises 0.67 per cent of the land area of the Middle East and is therefore not a micro-state in territory. From east to west, the landscape consists of a mountain chain, an area of sand sea and a coastal plain. In terms of its population, 2.2 million (of which less than 20 per cent are UAE citizens), which accounts for 0.67 per cent of the total Middle Eastern population, the UAE is clearly a micro-state. The literacy rate of 79.2 per cent is very similar to that of the other Gulf sheikhdoms.

The UAE is economically very strong as a result of its huge oil reserves, located primarily in Abu Dhabi. There are also important gas reserves, but at current rates of extraction, the oil should last for over 100 years. Upon the strength of this income, the UAE has diversified its industry and services. It has also become a major trading centre, particularly through Jebal Ali. The GDP is $39 billion, 5.6 per cent of the total Middle Eastern GDP, a figure exceeded only by Saudi Arabia, Israel and the three major states of Turkey, Iran and Egypt. The external debt of $14 billion is relatively low in terms of GDP by Middle Eastern standards.

The armed forces are composed of 64,500 personnel, 30 per cent of whom are expatriates. The army predominates, but all three services have modern and effective equipment. The defence budget is $1.9 billion, 4 per cent of the Middle Eastern budget, marginally less than that of Syria.

Like Saudi Arabia and the other Gulf sheikhdoms, the UAE has been able to count upon the support of the West. Within the region, the UAE is a key member of the GCC, but also retains good relations with Iran. President from independence and the architect of the UAE is Sheikh Zayed.

The UAE is a smallish state territorially and a micro-state with regard to population, but it has immense wealth and a small, but potentially powerful military force. Through his effectiveness, the ruler has retained a significant voice in Middle Eastern affairs. The only geopolitical issues of note in which the UAE is involved concern islands and boundaries. At present, Iran occupies Greater Tunb and Lesser Tunb and shares the administration of Abu Musa with the UAE. Since 1992, Iran has attempted to exert unilateral control over Abu Musa, but the UAE still disputes the ownership of all three islands. Their significance lies primarily in their strategic location in the eastern Gulf at the approaches to the Strait of Hormuz. Although the tip of the Musandam peninsula is not within the territory of the UAE, since it has coastlines on both sides, the Hormuz remains a key issue.

The boundary with Saudi Arabia is at present *de facto*, reflecting the 1974 agreement, and there is no defined boundary with most of Oman but in the far north there is an Administrative Line. The boundaries, land and maritime, associated with the Saudi Arabian corridor immediately to the east of the Qatar peninsula also require delimitation.

Owing to its relationship with Iran and south-west Asia, together with its development as a Hong Kong-style trading centre, the UAE has become a focus for heroin trans-shipment and money laundering.

Yemen

The location of Yemen (Figure 10.19) is at the mouth of the Red Sea, a virtual mirror image of that of Oman at the entrance to the Gulf of Oman. Both have extensive lengths of Indian Ocean coastline and considerably shorter lengths, respectively, on the Red Sea and the Gulf of Oman. In addition, both control the major strait. In the case of Yemen, this is Bab el Mandeb in which is Perim Island, part of Yemeni territory. In the case of Oman, a detached part of the territory shares control of Hormuz with Iran. Yemen also has a detached part of its territory which is of great strategic importance, the island of Socotra. Socotra lies in the approaches to the Gulf of Aden and in the centre of the main tanker route from the Gulf to the Cape of Good Hope. During the early 1980s, when very large oil tankers which could not transit Suez followed the Cape Route, Socotra was a staging post for the Soviet Indian Ocean fleet. At that time, it was considered a key potential flashpoint in the global oil infrastructure. Like Oman, Yemen comprises internally two or possibly three distinct areas. There are North and South Yemen, but the latter can be subdivided into the Aden area and the Hadhramaut, which are quite distinctive. Yemen's strategic location on the Red Sea is highlighted by the dispute with Eritrea over the median line Hanish islands, recently arbitrated in favour of Yemen.

North Yemen, or the former Yemen Arab Republic, is the Arabian Felix of the ancient world, which gained its independence from the Ottoman Empire in 1918. South Yemen, the former People's Democratic Republic of Yemen, was granted independence from Britain in 1967 when it comprised the Federation of South Arabia. The two joined to form the Republic of Yemen in 1990.

In North Yemen, the Imam was removed in 1962 and the country became a republic. Civil war followed in which Egypt provided support for the republic. However, a Zaidi religious leader was installed in 1967 and the civil war ended in 1970. In South Yemen, Britain had developed Aden as a Crown Colony and by 1966 had persuaded all the local rulers to join the Federation of South Arabia. Nevertheless, unrest ensued, the Federation collapsed and in 1967 Britain ended control in Aden, handing over to the National Liberation Front. This was succeeded two years later by a Marxist government. There followed during 1972 a war with North Yemen.

In the North, Salih came to power in 1978 and established a stable government during the 1980s. With the demise of the Soviet Union, the major support for South Yemen ended and union with the North occurred in 1990. During the Gulf War, Yemen was the only Arab member of the UN Security Council and its abstention was taken as support for Iraq. The consequence was that approximately one million Yemeni workers were expelled from Saudi Arabia and the economy was in chaos.

In 1992, significant oil discoveries were made and the first parliamentary elections were held. However, antagonism between the former North and

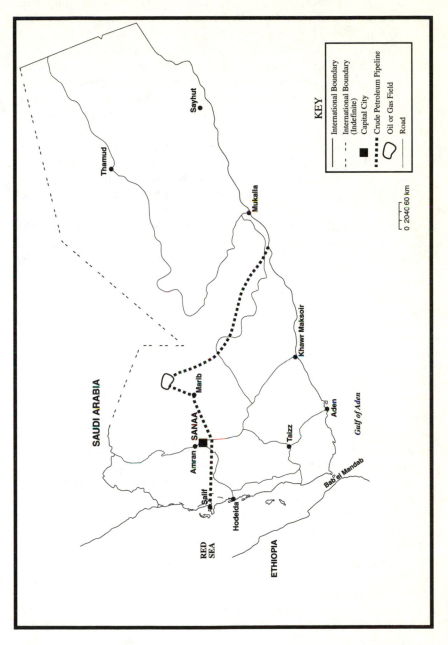

Figure 10.19 Yemen.

South continued and in 1994 a civil war ensued. As peace returned, negotiations with Saudi Arabia over the settlement of boundary issues were resumed. In 1995, Eritrean forces occupied the Hanish Islands, sovereignty over which had not been claimed since 1923 and in 1996 the two sides accepted arbitration which found in favour of Yemen.

With an area of 532,000 km², Yemen is almost twice the size of Oman and accounts for 4.5 per cent of the Middle Eastern land area. It is therefore the seventh largest Middle Eastern state. It comprises a narrow coastal plain, backed by mountains and upland, leading to an interior desert. The population is estimated at 13.9 million, 4.2 per cent of the Middle Eastern population, although estimates as high as 16.6 million have been made. This population is of a different order from Oman and the Gulf sheikhdoms and within the Arabian Peninsula, rivals Saudi Arabia. However, the literacy rate of 38 per cent is by 8 per cent the lowest in the Middle East.

Aden is the economic and commercial centre of Yemen and plans to redevelop the port, both commercially and as a base for the US Navy, are under consideration. Other sources of revenue are oil and remittances, the latter having dropped catastrophically following the Gulf War. The agricultural sector is unable to sustain self-sufficiency and is increasingly dominated by *qat* production. The GDP is $9 billion or 1.3 per cent of the total Middle Eastern GDP which is exactly the same as Sudan and marginally more than Qatar. The deficit is $8 billion, a little less than the GDP.

Yemen has modestly sized armed forces with 42,000 personnel, mostly members of the army. All three services have a mixture of Eastern and Western equipment, most of it rather dated. The defence budget is $0.35 billion, or 0.7 per cent of the total Middle Eastern defence budget, the same figure as Qatar and Cyprus.

During the various political and economic vicissitudes, the two Yemens have enjoyed support from different sources. North Yemen republicans have been supported by Egypt and royalists by Saudi Arabia, while in the South, the main source of aid was the Soviet Union. Following the Gulf War, Yemen lost considerable support, both regionally and internationally. The only influential individual over the recent past has been Salih who has, despite many setbacks, brought a certain stability to Yemeni affairs.

With its long history and strategic location, Yemen is potentially a significant actor in Middle Eastern affairs. However, mismanagement and civil wars have brought a period of political and economic disruption and the recently unified state is only now recovering. The major geopolitical issue over recent history has been internal conflict and terrorism. Potentially, the most significant issue is control over the Gulf of Aden and Bab el Mandeb which Yemen can exert as a result of its mainland location and its possession of Socotra. Nevertheless, there are boundary disputes to be settled with Saudi Arabia over both the maritime and the land boundary. The Yemen–Saudi boundary problem is undoubtedly the most complex of its kind in the world.

11 Geopolitics

Introduction

The relationship between geography and politics varies from country to country throughout the Middle East as each location and, to a lesser degree, each political system is unique. To these differences and the possibilities of tension engendered by them, must be added a number of issues which affect more than one country and are, in most cases, obviously transboundary. These transnational problems include security, political, economic and social concerns. They are all either transnational or possess the potential toward transnational dimensions but they vary considerably in importance.

Of major significance are the proliferation of weapons of mass destruction, closely followed by drug trafficking, conventional arms smuggling and terrorism. All these are on a global scale and each threatens the stability of the international system. The movements of refugees are a major social problem particularly in the Middle East. Food security is of specific concern in the region particularly among the GCC states but is only transnational in its concomitants. Oil and water problems have obvious transboundary implications and both are key components of what has become characterized as resource geopolitics. Pan-Arabism and Fundamentalism pose problems for relations between states but only the latter would appear to be of international importance. International boundaries are in some way associated with approximately 70 per cent of conflicts. Boundary concerns may be the source of tension or may be the resultant of a totally different disagreement. Although some of these issues are obviously more potent than others, any could potentially result in conflict.

Pervading all transboundary issues is the movement towards one global economic system known as globalization. With the integration of Eastern Europe, the states of the FSU and China into the market economy, there are effectively no states which remain entirely isolated from the international economic system. With the development of communications, the world is for the first time economically interdependent. This may well be as significant a moment in history as was the final exploration of the globe approximately a century ago. In that environment Mackinder (1904) pioneered geopolitics.

For the first time it was possible to examine the inter-relationships of the globe in its entirety. Now, it is almost impossible for any state to remain immune from the effects of global economic forces. Any change may reverberate throughout the system. Furthermore, global communications now ensure that knowledge of any change is available world-wide almost instantaneously.

Globalization is seen, by some economists at least, as bringing benefits to all participants. However, the same forces which generate benefits also render the possibility of global economic crises or even collapse that much more likely. Economic turmoil can cross state boundaries with the devastating consequence of unrestricted transboundary capital flows. The ease of capital mobility can transform the possibility of crisis into a self-fulfilling prophecy. The market economy depends upon the sum total of individual decisions but while each decision may appear in itself rational, the collective outcome may be irrational. Thus, theoretically governments possess the capabilities to resolve such problems but on a global scale, the issues are beyond the scope of any individual government. What is clear but ironical is the fact that it is essential for there to be government intervention into the workings of the free market if capitalism itself is to survive.

At the international level, the role of the IMF, although it has not been very active in the Middle East, is hindered by lack of resources and credibility.

> The idea that IMF 'bailouts' distort the orderly functioning of the markets and perpetuate the inefficiencies of recipient economies is a criticism that comes from the right of the political spectrum. From the left, complaints focus on the extent to which the IMF interventions are geared towards the interests of the markets rather than the recipient country or that the IMF is little more than a servant of its dominant member, the US.
>
> (Anderson *et al.* 1999)

In the Middle East, the countries with the higher GNPs are either heavily dependent upon oil or were effectively created as a result of oil. Fluctuations in oil prices have brought economic reality to the region resulting in deficits and cash-flow problems for even the major producers. Nonetheless, these economies represent the more robust within the region. The increasing fragility of the Middle Eastern economies generally means that they are likely to be particularly susceptible to transnational economic forces.

International boundaries

In his seminal Romanes Lecture (1907), Lord Curzon stated that: 'Frontiers are indeed the razor's edge on which hang suspended the modern issues of war or peace ...'. Almost a century later, following two World Wars and a prolonged Cold War, boundaries remain a major concern of international relations. However, it is possible to discern a change in emphasis. In the earlier part of the twentieth century thinking on such issues remained basically Darwinian. Large and powerful states grew at the expense of their weaker neighbours and therefore the establishment of internally recognized boundaries was crucial in terms of sovereignty, legitimacy and ultimately, survival. In the late twentieth century, the development of international norms through such bodies as the UN and the ICJ has ensured that the forced territorial integration of lesser states by their greater neighbours is far less of a risk. Although there were other key issues such as the guarantee of oil supplies, the Gulf Conflict (1991) provides a recent example of an international force safeguarding the existence of a small state.

However, the maintenance of territorial integrity depends importantly upon the possession of clearly defined territorial boundaries. Boundaries and security are therefore still closely linked. Indeed, the concept of territoriality presupposes the existence of boundaries. At one level, given the development of the global market place, boundaries may be seen as increasingly redundant, but at another level, the functioning of the state as an entity depends upon boundaries. Elements of statehood, ranging from the resource base to national law, security and iconography are dependent upon national boundaries. If there were any doubts about the continuing importance of territory to national sovereignty, then these should be dispelled by events in Palestine. By accepting even the minimal areas of the Gaza Strip and the environs of Jericho as the basis for their state, the Palestinians have demonstrated how vital they consider the possession of defined areas of land to be.

On a global scale, the parallel processes of fission and fusion (Waterman 1994) can be witnessed. While in Western Europe there is increasing integration, the former states of the Soviet Union and Yugoslavia have broken down into their constituent parts. In the Middle East, events have been less dramatic, but examples of fission can be seen in Palestine/Israel and fusion in Yemen.

In the post-Cold War era, boundaries have become particularly significant as the major focus of conflict. At least a quarter of the world's land boundaries are at best unstable, while two-thirds of the maritime boundaries have yet to be settled. In the Middle East, for example, several boundaries in the Arabian Peninsula have yet to be delimited and very few have been demarcated, while in the Red Sea, no maritime boundaries have been agreed. The dispute between Eritrea and Yemen over the ownership of the Hanish and adjacent islands was arbitrated in favour of Yemen in 1998 but has yet to be delimited.

'Boundary', since it implies 'limit' is considered the most appropriate term for the line which divides the jurisdiction of states. Both 'frontier' and, to a lesser extent, 'border' include an element of width. The frontier is commonly taken to be a zone, while borderlands have become a focus of academic study in their own right. Indeed, it is apparent that the inhabitants of borderlands on either side of a boundary are likely to have more in common with each other with regard to lifestyle than either group has with the remaining population of its own state.

This theme has been developed by Martinez (1994) who recognizes four models of borderland interaction. In alienated borderlands, the boundary is closed and transborder interaction is negligible. In this case, given the lack of interaction, the inhabitants on either side may have relatively little in common. With the development of very limited transboundary movement, Martinez recognizes co-existent borderlands in which there is a narrow but discernible border zone of interaction. Increasing interaction produces first interdependent borderlands and then as movements across the boundary become unrestricted and the border zone increases in size, integrated border-lands. In the Middle East, at different times in the period since Second World War and for different reasons, all four models can be recognized. For much of the time, the boundaries between Israel and its neighbours have been either alienated borderlands or, under favourable circumstances, co-existent borderlands. The boundaries of Lebanon have been increasingly interdependent or effectively integrated borderlands.

Following the definition set out by Lord Curzon, who distinguished 'natural' from 'artificial', boundaries may be classified as:

1   physiographic;
2   geometric;
3   anthropomorphic; and
4   compound.

Physiographic are what were formerly designated 'natural' boundaries and comprise features of the landscape ranging from rivers, watersheds, coastlines and deserts to marshes and forests. Unless there are offshore islands, the coastline everywhere provides an obvious end to the state, but with the advent of maritime claims from the three-mile wide territorial sea to the more expansive claim lines of today, it does not represent the limit of the state. Linear features such as rivers provide ideal lines on maps, but they divide drainage basins which are perhaps the basic landscape unit of human occupants. Deserts provide a barrier of movement, but in most cases there is no definite line. In the Middle East, the most obvious physiographic boundaries are wadis and watersheds.

Geometric boundaries are generally based upon latitude and longitude, but, as in the case of Kuwait, may be produced by other geometrical construc-tions. In the absence of suitable landscape characteristics, geometric boundaries predominate in the Middle East. If they are to be features of the landscape, these boundaries are best demarcated.

Tribal and other traditional territorial divisions are characterized as anthro-pomorphic boundaries and these are particularly relevant in the Middle East. For example, the eastward continuation of the demarcated Treaty of Taif boundary between Saudi Arabia and Yemen was defined by tribal occupance. Compound boundaries comprise any combinations of the other three cate-gories. The western boundary of Kuwait follows the Wadi al-Batin which marks the division of territory between tribes. It is therefore a compound boundary.

Boundaries may also be classified according to whether delimitation was before or after the settlement and development of the area. Frequently in the Middle East, delimitation was by the colonial powers and development referred to the development by those external powers. Furthermore, the majority of the boundaries were geometric and were constructed in ignorance of underlying resources of petroleum, natural gas or water. In many cases, such as that of the Iraq–Kuwait boundary, this delimitation has raised issues of resource geopolitics.

The landscape may also contain the scars of former or relict boundaries. For example, the former international boundary between the Yemen Arab Republic and the People's Democratic Republic of South Yemen is now a relict within the Republic of Yemen, which is in places discernible in the landscape.

Under International Law, since it has the right of exclusive sovereign jurisdiction over the territory, control of the state is essentially territorial. According to Islamic constitutional law, a different model for the state within the Islamic World has been defined. Since the primary purpose of government is to defend and protect the faith rather than the state, the basis of the Islamic state has been ideological rather than territorial, political or ethnic. Furthermore, it was envisaged that the entire Islamic World should form a single unit, the *Umma* in which the precepts of the *Shar'ia*, Islamic law, could be fulfilled. Thus, the Islamic state was concerned with community rather than territory and therefore there could be no question of sovereignty over unoccupied land. Therefore, while political units according to the Western model obviously exist and were accepted as providing a structure for Islamic society, there was relatively little concern with boundaries. According to Blake (1992), in the Middle East boundaries do not normally represent a source of international tension and formal treaties exist for 80 per cent of the land boundaries. Thus, in a sense, boundaries and boundary conflicts are issues which are alien to the Middle East.

Of greater relevance is the fact that the region as a whole has witnessed, over thousands of years, the rise and fall of disparate empires and states, which have each influenced the political landscape. In the context of current problems, the most important influences were those of the Ottomans before the First World War and France and Great Britain afterwards. Following the Sykes-Picot Treaty (1916) (Figure 6.4) and subsequent treaties, France and Britain redrew the political map, delimiting new boundaries, which often cut across existing social and economic divisions as new states were created. This process resulted in the legitimacy of the boundaries so constructed and therefore the states enclosed by them open to question (Joffé 1994). Indeed, it is possible to argue that the divisions imposed were for the convenience of the delimiting powers, which set up and supported states ruled by local allies to maintain them. For example, according to Joffé (1994):

The modern sovereign political structures of the Arab states of the Gulf region are, in virtually every respect, a testimony to British imperial policy, spurred on by the interest in oil and in commercial control.

With the loss of influence of the colonial powers during the post Second World War period, the region, as a result of its position and resources, was divided by the superpowers along bipolar lines. Iraq and Egypt both effectively changed sides, while Syria and the People's Democratic Republic of Yemen became Soviet allies.

Following the departure from the Gulf by Britain in the early 1970s, US policy was seen in the twin pillar approach as a result of which, specific support was given to Iran and Saudi Arabia. This ended with the Iranian Revolution in 1979, after which US policy became more pragmatic.

From the end of the Cold War, the world political map has become more malleable and echoes of this process are to be seen in the Middle East today. The two Yemens united in 1991, Palestine finally emerged as an independent entity in 1995, but, in contrast, questions must be raised about the future shape of Lebanon.

From the legal standpoint, the issue of *uti possidetis* is of interest. The concept was accepted in the Cairo Declaration of the Organization of African Unity (OAU) (1964) to apply to colonial boundaries in Africa. In the Asian part of the Middle East, there remain colonial boundaries, but their acceptance is dependent upon the full panoply of diplomatic and legal procedures, ranging from negotiation to arbitration.

Since this is the driest of all the major world regions, problems associated with Middle Eastern boundaries are different in a number of respects from those experienced elsewhere. For most countries of the region, the population distribution is extremely uneven, being concentrated either along the coast or inland at sources of water. The result is that large areas of most of the countries are either uninhabited or inhabited only temporarily by nomadic people. Thus, for much of their length, boundaries appear to have little relationship to the human occupants of the landscape. As a result, there are problems of boundary management and, in particular, control.

Another characteristic, best exemplified in the Middle East, is the importance of subterranean resources. The basis for water management is, for most of the countries, the aquifer and in some places, the deep aquifer. The key resources of the region are petroleum and natural gas, both occurring in underground fields. In the case of all three resources, the boundaries of the fields are unlikely to be exactly located, while the political boundaries on the surface were defined, in most cases, before the extent of the underground resources was known. Therefore, there is permanently a mismatch between state boundaries and the limits of underground resource supplies. Cases which, in the past, have resulted in tension, include the water supplies at Buraimi oasis and oil in the Rumeila field on the Iraq–Kuwait boundary.

Four stages are discernible in the establishment of an international boundary:

1   *allocation* in which the need for the boundary and its basic course is agreed;
2   *delimitation* in which the boundary line is agreed and drawn on a map;
3   *demarcation* in which the position of the boundary line is identified on the ground by boundary markers; and
4   *management*, as a result of which the form and function of the boundary is maintained.

Different problems are presented at each stage. For land boundaries in the Middle East, these concern particularly the scale of maps available and the general lack of linear features, physiographic or anthropomorphic.

In such environments, a wide range of variables may be employed to reach a boundary settlement. Historically, these would include considerations of sovereignty, national allegiance and occupance. However, it is essential that any boundary must be considered viable under today's conditions and therefore current anthropology, human and physical geography, geology and security issues all need to be taken into consideration. As far as possible, the boundary should separate, rather than cut across, local, political, economic and social systems. Oil and mineral concessions must be considered, while geomorphological features may provide secure boundary markers.

Modern boundary delimitation is greatly facilitated by the use of aerial photography and satellite imagery. Nonetheless, fieldwork is still a necessity to provide the most accurate, precise and equitable line. Once agreement has been reached, in the absence of obvious geomorphological features, demarcation is essential. Formerly, as in the case of the Treaty of Taif (1934), the boundary between Saudi Arabia and Yemen was demarcated by heaps of stones, tree trunks and other such ephemeral markers. Today, as exemplified by the Kuwait–Iraq boundary, markers are monumented, usually by ferro-concrete pillars, topped by reflectors, so that the boundary line can be identified on aerial photographs and satellite images.

Maritime boundaries present a range of different problems. In the Middle East, one of the major difficulties is that the seas tend to be enclosed or semi-enclosed, with very few countries having access to a full 200 nml EEZ. A feature of boundary settlement in the region has therefore been negotiation for delimitation between opposite coasts. Furthermore, United Nations Conference on the Law of the Sea (UNCLOS), which came into force one year after ratification by the sixtieth state in November 1994, provides only general guidelines for defining boundaries. For example, the equitable principle is advised as a guide for overlapping resource claims, but no definition of 'equitable' is given. Added complications in the Middle East include the presence of strategic islands and the offshore location of many oil and natural gas fields.

In the Middle East today there remain several land boundaries which are unsettled and could be characterized as disputed (Figure 11.1). Predominant are the boundaries of Israel, initially with Lebanon, Syria and Jordan, and most significantly, with Palestine. With Jordan there has been settlement

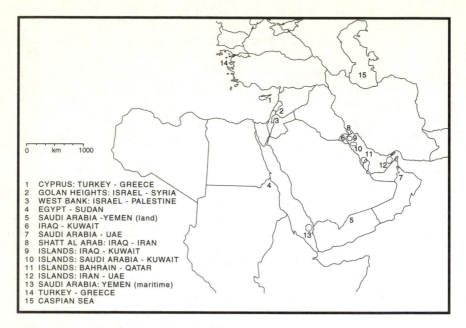

1   CYPRUS: TURKEY - GREECE
2   GOLAN HEIGHTS: ISRAEL - SYRIA
3   WEST BANK: ISRAEL - PALESTINE
4   EGYPT - SUDAN
5   SAUDI ARABIA -YEMEN (land)
6   IRAQ - KUWAIT
7   SAUDI ARABIA - UAE
8   SHATT AL ARAB: IRAQ - IRAN
9   ISLANDS: IRAQ - KUWAIT
10 ISLANDS: SAUDI ARABIA - KUWAIT
11 ISLANDS: BAHRAIN - QATAR
12 ISLANDS: IRAN - UAE
13 SAUDI ARABIA: YEMEN (maritime)
14 TURKEY - GREECE
15 CASPIAN SEA

Figure 11.1 Middle East: boundary issues.

and, while Israel still occupies part of their territory, the actual boundary lines of Lebanon and Syria are unlikely to be adjusted. However, the final pattern of boundaries will depend upon the outcome of negotiations with Palestine which began with the Madrid Conference in 1991.

The land boundary between Saudi Arabia and Yemen, with the exception of a western section demarcated following the Treaty of Taif (1934), will be over 800 km in length and, if viewed optimistically, has reached the stage of allocation. Potential boundary lines, according generally with the claims of Yemen, appear commonly on maps but this sector remains the largest gap in the global land boundary network.

Given longer-term difficulties which are likely to affect water use within the Nile basin, boundary disagreements between Egypt and Sudan retain the potential to become disputes. The geometrical boundary of the 22°N latitude was established as a result of the Anglo-Egyptian Agreement (1899) but later in the same year, a small area known as the Wadi Halfa salient was transferred to Sudan. Approximately midway between this area and the Red Sea coast the straight line boundary was modified (1902) to regularize the administration of nomadic tribes in the area. Egypt does not accept the long-term loss of territory which has resulted from both adjustments. Furthermore, the terminal point of the land boundary at the Red Sea coast will obviously exercise a crucial influence on subsequent maritime boundary settlement between the two states.

The boundary about which there continues to be most tension is that between Iraq and Kuwait, formally agreed and delimited following the Gulf War of 1991. Provision for the demarcation of the boundary was included in UN Security Council Resolution 687 (April 1991) and this was completed by early 1992. However, an assessment of the demarcation indicates that there are still potential problems particularly with regard to the Rumeila oilfield, some 2 km of which remains within Kuwait. The boundary follows the line of the Wadi al Batin, an alignment made on physical and ethnic grounds, a geometrical boundary positioned according to a known location south of Safwan and then a straight line boundary to a predetermined place on the Khor Zubair. The maritime section follows the Low Water (Springs) line of Khor Zubair and then a median line down the Khor Abdulla between Warbah Island and the mainland of Iraq. Thus, Iraq gained better access to the port of Umm Qasr in that the previous boundary in the Khor Zubair had followed the median line (Figure 11.2). However, the question of the Rumeila oilfield remains a live issue and in fact in early 1999, Iraq has again begun disputing the location of the entire northern section of the boundary. The other land boundary dispute in the Middle East concerns an area known as the Ilemi Triangle over which there remains disagreement between Sudan and Kenya.

Two definitions of Middle Eastern land boundaries have involved two high profile examples of judicial arbitration. In the case of the dispute between Libya and Chad over the Aouzou Strip, the ICJ found in favour of Chad

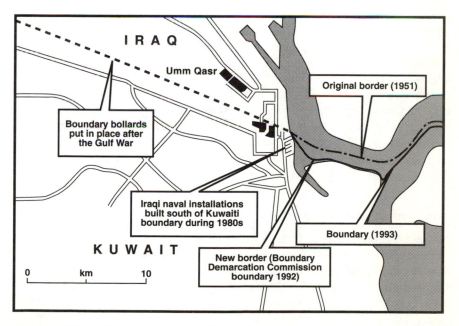

Figure 11.2 Iraq/Kuwait boundary at Umm Qasr.

using the principle of *uti possidetis*. The location of the final pillar on the Gulf of Aqaba of the Israel–Egypt boundary was also judicially settled. The issue was known as the Taba Case and historical evidence was used to decide in favour of Egypt.

Maritime disputes in the Middle East and its environs are of concern in the Caspian Sea, Mediterranean, the Red Sea and the Persian–Arabian Gulf. In the Caspian, boundary delimitation will crucially affect oil extraction and the boundaries between Iran and its neighbours have still to be settled. In the Mediterranean, several unsettled boundaries remain but the most intractable would appear to be that in the Aegean between Greece and Turkey. If sovereignty is taken to be the only factor of significance, Greece can claim almost the entire Sea. However, if equitable use is considered, then it may be necessary to set aside the question of sovereignty before settlement proceedings are initiated. A precedent for such a procedure has been established between the UK and Argentina in the context of the Falkland Islands (Malvinas).

Maritime boundary settlement is relatively far advanced in the Persian–Arabian Gulf but a particularly difficult dispute between Bahrain and Qatar can be identified. The issue is ostensibly over the ownership of the Hawar Islands but in reality is focused upon seabed resources, notably natural gas. In the Red Sea there has been no maritime boundary delimitation. A sovereignty dispute over the Hanish Islands was arbitrated in favour of Yemen (1998) and delimitation will follow. Otherwise, there is a shared resource zone agreed between Saudi Arabia and Sudan.

Although in the Middle East the number of disputes over land boundaries is limited, those which remain include some of the most problematical in the world. Furthermore, there have been recent examples of conflict directly related to boundary location. The most obvious is the Gulf War (1991) after which there was a UN-sponsored delimitation of the Kuwait–Iraq boundary. A key factor in the Iran–Iraq War (1980–88) was the alignment of the boundary with regard to the Shatt al Arab. Following a short conflict, the boundary between Libya and Chad was arbitrated in Chad's favour in 1987.

Delimitation of the adjacent water bodies has, with the exception of the Red Sea, not lagged and indeed, the Libya–Tunisia and Libya–Malta Cases, decided by the ICJ, provide important precedents.

Weapons of mass destruction

Nuclear, chemical and biological weapons are all characterized as weapons of mass destruction and it can be assumed that there are stockpiles of all three in the Middle East. Chemical and biological weapons are outlawed by international treaty and the control of nuclear weapons is targeted by the NPT (1968). Thus the development or procurement of such weapons takes place in conditions of the utmost secrecy and reliable data are almost impossible to obtain. However, owing to the requirement for high technology, materials and components, it is somewhat easier to monitor nuclear development than anything in the chemical or biological field where the basis for weapons can be relatively small amounts of easily obtained substances. Furthermore, delivery systems for nuclear weapons need to be far more sophisticated and for any country to be a nuclear power requires both the weapons and the means of power projection.

Within the Middle East, both nuclear and chemical weapons are known to have been used. During the Gulf War (1991), the Allied Forces led by the USA employed weapons containing depleted uranium resulting in the subsequent disastrous spread of cancers in Iraq. In 1988, during the Iran–Iraq War, Iran with Kurdish assistance captured the city Halabja. The next day, Iraq used gas on Halabja causing some 5,000 deaths. Later in the year, the Iraqi military again used chemical weapons to recapture Fao and Mehran from Iranian occupation. Both these instances were on a relatively limited scale but both have revealed the potency of the weapons.

As so little is known of chemical or biological weapons capability in the region, the focus of this section is upon known or suspected nuclear capacity. According to Spector *et al.* (1995) in a survey by the Nuclear Proliferation Project of the Carnegie Endowment, within the Middle East, Israel can be classified as an undeclared nuclear weapons state; Iran and Libya as having suspected nuclear weapons programmes and Iraq as a country which has recently renounced nuclear weapons (Figure 11.3). The allegedly extensive nuclear programme of Iraq was dismantled by UN inspectors following the Gulf War and the country remains subject to special UN mandated monitoring.

Israel is a *de facto* nuclear-weapon state and its probable capability together with the means of delivery is set out in *The Military Balance 1997/8* (The IISS 1998). However, officially Israel denies that it has nuclear weapons and it is not a party to the NPT. Original estimates of Israel's nuclear capacity were based upon an analysis of the evidence provided by the former nuclear technician Mordechai Vanunu (*The Sunday Times* 5 October 1986). Subsequently, estimates have been reduced from some 200 nuclear devices to approximately 100. Whether Israel has conducted nuclear tests is not clear but the detection of a signal by a US VELA monitoring satellite on 22 September 1979 over the South Atlantic has been interpreted as indicating a low-yield Israeli nuclear test. Certainly there is no doubt that Israel deploys two nuclear-capable ballistic missile systems: the Jericho I and the Jericho II. The latter has a range of approximately 1,500 km. Furthermore, it is believed that Israel's base launch vehicle can be modified to become an intercontinental ballistic missile (ICBM).

Thus Israel has the delivery systems and there is very strong evidence that it also has the weapons. Indeed, Hersh (1991) considers that all current estimates are far too low and that Israel has a far larger and more advanced arsenal. Whatever the case, Israel is as a result of its nuclear infrastructure a superpower in the Middle East. Even the lower estimates mean that it has the sixth largest nuclear arsenal in the world.

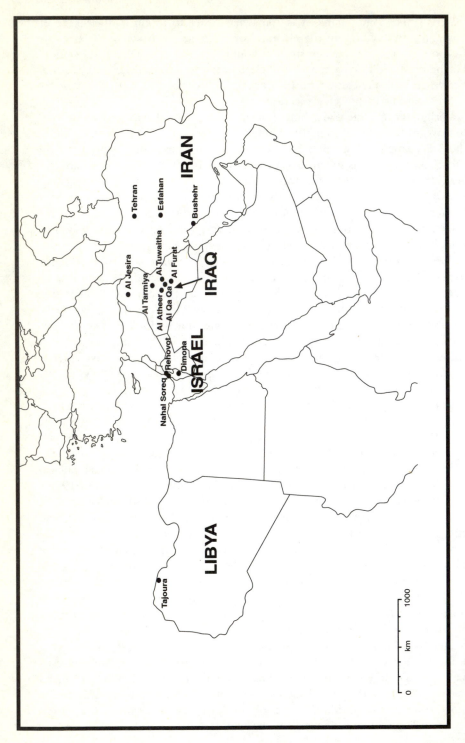

Figure 11.3 Middle East: key elements of nuclear infrastructure.

The key location for the nuclear weapons programme and therefore among the most sensitive geopolitical flashpoints in the region is Dimona. At Dimona is an array of facilities including a plutonium production and associated plutonium extraction plant, uranium purification conversion and fuel fabrication facilities together with a pilot-scale laser and centrifuge enrichment programme. Other parts of the nuclear infrastructure include Nahal Soreq, a research reactor and pilot-scale plutonium extraction plant. It is here that research on nuclear weapons design is thought to take place. This was revealed in a report (Hough 1994) which also indicated the location of assembly facilities: the nuclear missile base and bunker for storing nuclear gravity bombs and the storage site for nuclear weapons. The other location indicated by Spector *et al.* (1995) is the pilot-scale heavy water production plant at Rehovot.

Israel has encouraged ambiguity about its nuclear status for a number of reasons. By law, the USA is unable to supply aid to any country which is producing a nuclear capability and Israel is sustained by large-scale US military and economic aid. Strategically, ambiguity has greater deterrence value while still allowing Israeli criticism of other states in the region which might be attempting to build nuclear capability. Indeed, apart from opposing the transfer of nuclear technology to other Middle Eastern states, Israel has used force to prevent nuclear development. In 1981, Israel destroyed the Osiraq nuclear facility in Iraq.

As the sole nuclear power in the Middle East, Israel can use its status as a bargaining ploy. In 1973 and again during the Gulf War, Israel deployed nuclear weapons (Aronson 1992).

The remaining three Middle Eastern countries with suspected or actual nuclear weapons programmes are all signatories of the NPT. The nuclear programme in Iran is thought to be at least five years away from the production of nuclear weapons but the time for their development is clearly dependent upon the transfer of nuclear-related technology from China. Recently, there have also been contacts with Russia with the possible acquisition of nuclear materials and technology.

However, Iran has been meticulous in accepting inspections from the International Atomic Energy Agency (IAEA) as required by the NPT. Visits were made in 1992 and 1993 and no violations were found. The main facilities are in Teheran at the University of Teheran and the Sharif University of Technology and the Nuclear Research Centre at Esfahan. Bushehr on the Persian–Arabian Gulf is the site being developed with Russian and Chinese assistance for the generation of nuclear power.

Following the Gulf War, UN Security Council Resolution 687 effectively removed Iraq's potential to develop nuclear weapons. Furthermore, a monitoring programme was established to prevent reconstruction. During inspections by the IAEA working under the UN Special Commission on Iraq (UNSCOM), the extent of Iraqi violation of the NPT was revealed. Specific violations included possession of significant amounts of unsafeguarded nuclear materials related to uranium enrichment, the undeclared construction of the facility at Al Tarmiya for the production of enriched uranium and the extraction of small amounts of plutonium from irradiated fuel at Al Tuwaitha.

By late 1994, the IAEA had completed the destruction or the removal of all known nuclear weapons-usable material facilities and equipment in Iraq. In the same year, continuous presence for monitoring was established in Iraq. Such monitoring continued until 1998 when the activities of UNSCOM were suspended following attacks on alleged facilities by missiles from aircraft and ships of the USA and the UK. Iraqi frustration has been fuelled by the refusal of the UN, in response to extreme pressure from the USA and the UK, to lift the economic sanctions imposed on the country in 1991 or even to indicate a time-scale for lifting them.

Prior to the Gulf War, Iraq had ballistic missile capabilities including Scud-Bs with a range of 300 km and derivatives of the Scud-B, the Al-Husayn with a 600 km range and the Al-Hijarah/Abbas with a range of 750 km. Under Security Council Resolution 687, all ballistic missiles with a range in excess of 150 km have been dismantled.

Among the large range of nuclear-related sites identified in Iraq, six are outstanding although their capability has been largely destroyed. Al Tuwaitha is the nuclear research centre and the key to the development of the entire programme. The main development and testing of nuclear weapons occurred at Al-Atheer and the high explosive and propellant facility at Al-Qaqaa is now under IAEA monitoring. Other facilities concerned with nuclear enrichment and processing and now all severely damaged are at Al-Tarmiya, Al-Furat and Al-Jesira.

Like Iran, Libya has long been suspected of attempting to obtain nuclear weapons but its only facility at present is a small research reactor at Tajoura. However, there is strong evidence that Libya has a substantial array of chemical weapons. With regard to delivery systems, it can deploy Scud-B missiles with a range of 300 km and is developing the al-Fateh which has a probable range of 950 km.

Of the other Middle Eastern countries, none has nuclear weapons but Egypt has suitable delivery systems and plans to promote a nuclear energy programme. Syria has no weapons and no reactors but several weapons-capable delivery systems including Scud-Bs and MiG 29 aircraft. Like Libya, it is thought to have a well-developed arsenal of chemical weapons. Turkey has no weapons but a research reactor in operation at Akkuyu and there are plans to build nuclear power plants. It has several nuclear weapons-compatible delivery systems.

In the Middle East, the only non-signatories of the NPT are Israel, the UAE and Oman. Geopolitically, the key to future developments in the Middle East on the banning of weapons of mass destruction lies principally with Israel. As the only nuclear weapons power in the region, Israel enjoys unique security advantages. Furthermore, the strenuous efforts made by the USA to eliminate any nuclear capacity in Iraq appear hypocritical in the light of the

Israeli arsenal. Given the potential military power resulting from its nuclear capability and the uncritical support it receives from the USA, Israel poses a permanent threat to stability in the region. Indeed, the successful development of nuclear technology in Pakistan (1999) and the resulting potential for the construction of an Islamic 'bomb' must be blamed in part upon Israel.

Conventional arms

As Peres (1993) observes, there are direct links between the conflict, the high levels of military spending and the economic problems confronting the Middle East. Not only do arms purchases affect the economy, but the question is raised in IR as to whether there is a relationship between arms races and conflict. The bipolar arms race during the Cold War did not end in catastrophe, but the basis of that was non-conventional weapons, the use of which could have produced cataclysmic global consequences. Certainly, the arms stockpiles of the Gulf region were laid bare during Operation Desert Storm (1991) and, in the immediate aftermath, great efforts were made politically to prevent a further round of the arms race.

In pursuing arms sales to the Middle East, the USA faces a continuing dilemma (Cordesman 1993). As a result of the close relations between the USA and Israel, the USA cannot sell arms to many Middle Eastern states. US Congressional policies preclude supply to any states which are actively hostile to and threaten the security of Israel. As a consequence, many key oil suppliers and other strategically placed states have to obtain arms and military advice elsewhere.

With the USA effectively sidelined, the counterbalance is provided by the UK, France, Germany and Italy. In fact, the total West European sales within the region are approximately one-quarter of those of the USA. A direct effect of this situation is that GCC states frequently buy from different sources and consequently their equipment is incompatible. In the present round of arms sales, Kuwait has favoured the USA, the UAE has bought from France and Oman has remained committed to the UK. A further effect is that the Western European countries are loath to work too closely with the USA in case this is interpreted as support for Israel, which might lead to difficulties over trade and, particularly, oil supplies.

When considering the arms race in the Middle East, both the quantitative and the qualitative aspects must be considered. Since 1967, the numbers of land battle tanks and combat aircraft have risen dramatically among the potential belligerents, but even more striking is the improvement in quality. The gap between the arms technology of the main NATO powers and that of the Middle Eastern countries has been significantly narrowed with deployments of supersonic jet aircraft and missiles, and with the introduction of electronic warfare. Weapons systems are now equivalent to those of the Great Powers (Maull 1990). Indeed, in some cases, the systems have been specifically developed or redesigned, supported by the Middle Eastern buyers and with their specifications in mind. Examples are the financial support given by Saudi Arabia to the French *Mirage* fighter aircraft and the *Maverick* missile (Saudi version).

The fact that, increasingly, arms supplies to the major Middle Eastern buyers comprise state-of-the-art equipment introduces a number of further problems. For the purchasers, the costs are greater the more sophisticated the equipment and therefore fewer items can be obtained for the arsenal. Given that many of the potential conflicts would be with opponents using far more basic equipment, a question mark must be raised against the rush towards arms elaboration. In the event that such highly priced sophisticated equipment really were needed, it is likely that Western allies would make a timely intervention. Furthermore, such weapons systems require sophisticated training and support and therefore dependence upon the original suppliers.

Among the more extreme examples are the purchase of the *Abrams* main battle tank by Kuwait. This equipment cannot be maintained in the field and any breakdown requires replacement parts directly from the US defence industrial base. This indicates a further problem in that the defence industrial bases of the Western countries are all, following the end of the Cold War, in decline. Therefore, whether the concern is with consumables such as ammunition or missiles, key parts or whole equipments, any guarantee of resupply within a reasonable time period is unlikely. The conclusion might be that for the states for whom it is an option, a decision needs to be made as to whether the latest high-technology military equipment should be purchased or whether they should settle for dependence, which is already apparent, upon Western allies.

After the most recent conflict in the region, the Gulf War, serious thought was given to the relationship between the arms race and security. On 29 May 1991, George Bush, then President of the USA, revealed details of an initiative to promote arms control in the Middle East (Sadowski 1993). Among a number of initiatives, he called for negotiations among major weapons-exporting countries, to reduce the flow of arms to the region.

While the majority of countries in the region applauded the initiative, few practical steps were taken to implement them, either by consumers or suppliers. Multilateral arms negotiations in the region were identified with and consequently became bogged down by the overall framework of Peace Talks between Israel and its Arab neighbours. The 'Big Five' suppliers: the USA, the UK, France, Russia (the FSU) and China, met three times, but reached no agreement. As a result, the debate turned from arms control to such nebulous concepts as 'balance of power'.

Certainly, the Middle East is a region in which real conflicts, rather than perceptions of potential conflict or even paranoia have driven the arms race. Consequently, it could be concluded that the pattern of overlapping and reinforcing antagonisms indicates that arms control methods which have proved successful elsewhere will not work in the Middle East (Sadowski 1993). There are no clear groupings of countries or blocs, there appears to be little desire for military parity and diplomatic relationships tend to be erratic.

An arms build-up is thought to be motivated by a number of factors, all of them present in the Middle East. The most compelling is perceived insecurity, as a result of potential or imagined threats. There is also the desire for international status. The possession of arms is thought to equate to military power, which is taken as a symbol of national independence (Maull 1990). However, military power comprises more than hardware and many states of the Middle East possess large sophisticated arsenals, but cannot be considered militarily powerful. If sophisticated weaponry can be used effectively, then it can enhance the power of the state in the pursuit of foreign policy. Nevertheless, in the Middle East, this applies normally to very local conflicts only in that the key underpinning for major foreign policy initiatives is provided by powerful backers outside the region. This can be seen in the changes of status within the region, which have occurred following the end of the Cold War and the demise of the Soviet Union.

Apart from external pressures, there are also internal factors which may generate an arms build-up. In particular, in several Middle Eastern countries there is an obvious need to enforce the power of the government. Furthermore, the military, particularly in countries with politically unstable regimes, form an influential pressure group. However, even internally induced arms purchases may influence neighbouring countries to indulge in an arms race in response to the perceived increased threat.

Taking a broader perspective, it must be remembered that during the Cold War, both superpowers considered the Middle East, and particularly its oil-rich states, as within the zone of their strategic interests. Therefore, there was a tendency for the consumer states to receive arms under financially favourable conditions. It was under such terms that the Soviet Union supplied Iraq during the Gulf War.

For the arms suppliers, there are both political and economic objectives. Politically, key arms suppliers gain influence in the consumer countries and, as a result, are able to pursue their own broader interests. Furthermore, the consumer becomes locked into the defence industrial base of the supplier and this allows the latter to exercise some control over the employment of military power. More importantly, modern sophisticated equipment is beyond the absorption capacity of the local military and, for its use, demands training and other forms of assistance.

Increasingly, given the problems of the defence industrial bases in the West, it has been economic considerations which have been paramount among producers. The costs of new weapons development can only be absorbed if offshore sales can be guaranteed. The size of the armed forces in many of the supplier countries provides a base which is insufficient to support the necessary research and development. Also, with the increasing levels of sophistication, the costs of key equipments are so large that they provide an important element of the balance of payments in the supplier countries. Given the potential significance of state-of-the-art equipment, governments in the supplier states normally retain some control over exports. However, as military supplies to Iraq prior to the Gulf War illustrate, government regulations can be circumvented and even internationally known companies may be involved in the illegal arms trade.

In contrast, government-to-government transfers of non-conventional weapons are rare but private Western companies and individuals have played an important role in helping Middle Eastern states develop chemical and nuclear weapons (Navias 1993). Indeed, since for Middle Eastern countries there is total dependence on external sources of technology and expertise, it is reasonable to assume that all major non-conventional weapons projects have at some stage received outside help. An added dimension for non-conventional but particularly conventional arms since the break-up of the Soviet Union has been the available supply of relatively cheap former Soviet weaponry. Other sources of growing importance have been the countries of Eastern Europe, China, North Korea and, to a lesser extent, Brazil, Argentina and South Africa. Within the region itself, the only indigenous arms industry of military significance is that of Israel, although various joint ventures between Arab states have been attempted.

Given the wide range of issues involved, it is hardly surprising that the Middle East remains the largest regional arms market (The IISS 1998). As it has been consistently over the past decade, Saudi Arabia remains the largest international buyer of arms (1995: $8.1 billion at constant 1994 prices). Other major purchasers are Egypt ($1.9 billion) and Kuwait ($1 billion). Egypt and Israel receive virtually all the US annual Foreign Military Financing (FMF) grants and Jordan is now to receive Foreign Military Assistance (FMA). Iran remains a major purchaser of Russian weaponry, but has also received equipment from the Far East (The IISS 1997).

In the Middle East, the arms race continues. Agreements between suppliers to control imports into the region appear unrealistic on a number of counts. The development of modern arms industries depends crucially upon a guaranteed export market, such as that of the Middle East, where there is demand and purchasing power. Furthermore, supply-side controls are not likely to reduce demand. The likely effect of an agreed reduction in arms supplies from the West would be an increase in prices and a proliferation of suppliers elsewhere. Control of arms imports is far more likely as a result of economic problems in the consumer states. There is great pressure to reduce military budgets and if this could be achieved across the region with equal reductions everywhere, security would not be placed in jeopardy. Indeed, arms control might then be viewed as a strategy towards achieving military security (Sadowski 1993).

Migration and refugees

During the last part of the twentieth century, world migration patterns have been dominated by the movements of refugees. The total involved is estimated by the UN to be well in excess of 20 million people. Political scientists have recognized such vast movements as being largely beyond the scope of any one state, however powerful, to manage. Such movements are, as a result, viewed alongside arms smuggling and drug trafficking, as part of the macro-political agenda.

However, statistics require detailed analysis in that clear differences need to be recognized between, for example, people displaced within their own country and those who become refugees beyond its borders. As a result of civil war or natural disasters, people may become temporary refugees who intend to return to their homeland when conditions are again judged to be normal. Many of these would not conform to the definitions laid down in the Geneva Convention (1951), but it must be remembered that the focus when the Convention was drafted was upon conditions in Europe after the Second World War.

The UN Convention on Refugees (1951), amended by the New York Protocol (1967) defines refugees as people who are:

> outside their own country, owing to a well-founded fear of persecution for reasons of race, religion, nationality, membership of a political social group or political opinion.

Such a definition allows a wide range of interpretation, exacerbated by the fact that the recognition of refugees is a matter for receiving states (Frelick 1992). Black (1993) points out that in Africa and Central America, signatories respectively to the OAU Convention on Refugees (1967) and the Organization of American States (OAS) Cartagena Declaration (1985) use different definitions. In the case of the OAU, those forced to leave their country due to external aggression, occupation or domination are included, while the OAS lists those who become refugees by internal conflict or massive violations of human rights.

There is also a major difference in approach between the developing countries and the developed world, particularly Western Europe. In the former, it is customary to grant asylum to whole groups of refugees displaced for the same basic reason. In the latter, the case of each individual is examined separately, so that there is a need to prove individual persecution.

In the Middle East, the two groups which obviously fall outside the 1951 Convention are those who are refugees within their own countries and those who might be classified as economic rather than political refugees. During the period after the Second World War, the Middle East has been the focus for the global refugee problem and distinctions between migrants and refugees, let alone different categories of refugee, become extremely blurred.

As Cohen (1995) points out, even the most significant migration into the Middle East, that of the Jews into Palestine in the late nineteenth and twentieth centuries, can be interpreted in two completely different ways. The conventional Zionist explanation is that the movement was a homecoming after exile. In contrast, a group of Israeli historians have laid stress on the similarity between this migration and European colonization and settlement generally, during the same period. In this second context, the departure of some 780,000 Palestinians can be equated with the ill treatment of native peoples in general. It was one of the effects of colonialism.

By 1992, Aldelman (1995) estimated that the number of Palestinian refugees, including their descendants, approached 5.5 million. To these two explanations, Cohen adds a third element, that of ethnic homogeneity towards which many states have tended during the period of nation-building. A classic case occurred with the division of India and Pakistan in 1948.

Following the oil price rises in the early 1970s, the boom economies of the Gulf states attracted many Palestinians. These were joined by some 3.5 million Asian migrant workers (Abella 1995) and workers from other Middle Eastern states. With the outbreak of the Gulf War in 1990, some 450,000 Asian workers were repatriated, causing massive economic problems in their countries of origin: Bangladesh, Pakistan, Philippines, Sri Lanka and Thailand (Amjad 1989). At the same time, as a result of their government's overt support for the Iraqi invasion of Kuwait, 750,000 Yemenis were repatriated from Saudi Arabia. Most of the Palestinians were deported, many to Jordan which had to accommodate some 300,000 refugees who had formerly been resident in the West Bank. According to Van Hear (1995), these refugees brought skills and capital to Jordan triggering something of an economic boom, despite the obvious hardships resulting from UN sanctions against Iraq.

Such movements have been put into context by Pikkert (1993) in his discussion on the three layers of social reality. At the base is the predominantly inert structure of geographical and climatological change. On that is superimposed the layer of demographic, economic and social change and then, lastly, comes the superficial turbulence of daily developments. The movements resulting from the Gulf War would be included in this last group, Pikkert's '*longue duree*' factors.

Certainly, while the numbers of refugees involved are very large, they must be seen in the context of the annual population growth rate of some 3 per cent in the Middle East, together with internal migrations. For example, one thousand people a day move into Cairo, which, at the millennium numbers something like 18 million.

The mass movements of people as a result of the Gulf War (1991) and other Middle Eastern crises emphasize two key aspects of the relationship between states and people in the Middle East (Humphrey 1993). Many states rely on foreign labour for their development, the numbers of foreigners in the work-force commonly outnumbering numbers of indigenous workers. However, in the context of refugees, the second point is the more important, that national integration is very fragile. Citizenship is strictly controlled and few foreign workers, however long they have resided in their countries of work, are accorded full citizenship.

As a result of these two factors, conflicts in the Middle East have produced the greatest refugee flows of any region in the world. By the end of 1991,

two-thirds of the world's total refugee population of 16.7 million had originated in the Middle East.

As boundary controls have tightened, so there is a danger that refugees
will be reclassified as illegal migrants. For the greater part of history, the
Arab world has been relatively free of boundaries and it has been predominantly in the present century that the system of nation-state delimitation has
resulted in stateless refugees. However, despite denationalization and dispos-
session, refugees are not excluded from labour markets (Humphrey 1993)
but, lacking political and legal rights, are highly vulnerable in their host societies. Thus, they are likely to be removed at any time and this has fuelled
fears in the developing world of unstoppable mass migrations. One result
has been a UN-supported change in the 'international rules'. States from the
North, predominantly the USA and Europe, can now apparently legitimately
intervene in the affairs of states of the South, to keep populations within
borders. Examples have been seen in Somalia and Iraq.

Thus, as a result of the increasing globalization of the labour market, the
Middle East has become highlighted as the major exporter and importer.
For this new indentured labour, social conditions have been constructed in
the interests of the importing states. As a result, the workers are not only
insecure, but in every sense, temporary. Therefore, it is likely that the mass
movements of people and the creation of refugees will continue to be a characteristic of the Middle East.

Drugs and drug trafficking

Drug trafficking is a key element in the macro-political agenda. In its control
of all operations from production through to street sales, the narcotics industry
is truly multinational. Furthermore, the profits involved are so phenomenal
that the drugs trade exercises a major influence on guerrilla warfare, terrorism,
crime and corruption generally (Clutterbuck 1994).

The Middle East has various connections with drugs, ranging from the
most damaging and addictive such as heroin to *qat*, a relatively mild stimulant. The fresh shoots and leaves of *qat* contain a psychoactive substance and
when they are chewed or drunk as tea stimulant effects similar to that of
amphetamines are engendered. *Qat* rarely, if ever, produces toxic psychosis
or aggressive behaviour (Inciardi 1992). Nonetheless, it is highly significant
in that increasingly in the northern part of Yemen land once used for export
crops such as cotton, fruit and vegetables has been converted to *qat* growing.

However, it is through narcotics, in the form of opium–heroin, that the
Middle East has become implicated in the global transboundary drugs trade.
Within the region, opium is grown, processed and transported. These facts
are known but the detail of the trade is virtually impossible to obtain. Since
drug trafficking is a criminal activity, a high proportion goes undetected
and host governments are understandably loath to provide any publicity.
Consequently any data produced are unlikely to be reliable.

Heroin, which is more addictive and damaging than cocaine, is produced by processing the seedpods of opium poppies. The opium produced is concentrated in two main areas: the Golden Triangle which includes parts of Myanmar, Laos and Thailand and the Golden Crescent which stretches through Pakistan, Afghanistan and into Iran. These two areas have vied for world leadership in production and it is reasonable to assume that both contribute equally to a total output from the two which is in excess of 3,000 metric tonnes of opium annually. To this must be added production from Lebanon and Latin America to produce a global total of between 4,000 and 5,000 metric tonnes.

Within the Middle East, the major opium producer is Iran followed by Lebanon and probably Turkey (CIA 1998). Iran is an illicit producer of the opium poppy for both the domestic and the international drug trade while Lebanon is a small illicit producer of hashish and heroin. In Turkey, the government maintains strict controls over the areas of legal opium poppy cultivation and the output of poppy straw concentrate.

Processing involves the initial conversion locally of opium into morphine and later the transformation of morphine into heroin in relatively sophisticated laboratories. Such laboratories are located near Istanbul and also in the more remote regions of Turkey. However, it is as a result of transit points that the Middle East is most implicated in the global drugs infrastructure.

Apart from the producing countries, Egypt, Syria, Cyprus and UAE are all involved. Egypt is a transit point for both south-west Asian and south-east Asian heroin and opium being transported to the West. It is also a meeting point for supplies from West Africa, notably Nigeria. Iran is a similarly important trans-shipment point, while Lebanon combines this function with cocaine processing and trafficking. Syria is a transit point for both refined cocaine and heroin together with hashish for Western markets. Cyprus, via air routes and sea-bound container traffic to Europe, is a transit point for heroin, hashish and cocaine while the UAE is a growing heroin trans-shipment and money laundering centre (CIA 1998).

While the various facets of the drugs' trade are at present highly localized within the Middle East, the location of the region between the major production areas and the key markets of Western Europe and the USA means that the attendant problems can only increase. In particular, governments are becoming greatly exercised about internal consumption and Saudi Arabia has taken the lead by introducing the death penalty for drug trafficking.

Petroleum

Energy geopolitics, and probably the idea of resource geopolitics, began with the oil era. When the only industrial energy source was coal, the issue did not arise. In most cases, countries had become industrialized largely as a result of their own or nearby coal resources. Global trade relied entirely upon shipping and ships were fuelled by coal. Bunkering ports such as Aden sprang up on all the major sea-lines but fuel vulnerability was never considered a threat. Furthermore, the world order was far more tightly controlled by the Great Powers and the global distribution of coal is far more even than that of oil.

As, from the early part of the twentieth century, dependence upon oil grew, first for shipping and then for industry, the situation changed radically. The major sources of oil have always been relatively few and therefore oil trade routes tend to have become fixed. This is even more obvious in the case of pipelines, the world network of which remains largely devoted to oil. Key sources of oil tend to be developing countries, many of them unstable, although it must be admitted that instability is frequently related to the possession of oil. Oil therefore presents a complex of vulnerabilities from the sources to the pipelines and the sea-lines with their attendant choke points.

The focus of this complex is the Middle East and, in particular, the Persian–Arabian Gulf. To this core region of countries with major oil reserves can be added to the west Libya and Algeria and to the north the oil-rich states of Central Asia. The Arab world accounts for 60 per cent of the proved global oil reserves (1997) and the Muslim world as a whole 74 per cent. Only thirteen countries in the world have 1 per cent or more of the proved reserves and five of those countries border the Persian–Arabian Gulf. When Libya is added, almost half the countries with over 1 per cent of the reserves are Middle Eastern.

The demand for oil is concentrated in the three economic core zones of North America, European and Asia-Pacific which subsumes both China and Japan. The reserves within those areas are listed in Table 11.1.

These three areas between them account for well over 80 per cent of world trade. The states of the FSU developed their economies largely on internal trade but they are now gradually becoming part of the global pattern. The FSU accounts for 6.4 per cent of proved reserves with an R/P ratio of 24.7 which provides an indication of the longevity of the reserves. The R/P ratio is greatly affected by new discoveries and the development of new extraction technology.

In sharp contrast to these statistics, 65.2 per cent of the world proved reserves are located in the countries of the Persian–Arabian Gulf and the overall R/P ratio is 87.7. Among these, Saudi Arabia is outstanding with

Table 11.1 Proved oil reserves 1997

	Percentage of world reserves	*R/P ratio*
North America	7.4	16.0
Total Europe	1.9	8.2
Asia-Pacific	3.5	15.5

R/P ratio = reserves to production ratio (number of years which the reserves will last at current rates of production).

25.2 per cent reserves and an R/P ratio of 79.5. For Iraq (10.8 per cent), Kuwait (9.3 per cent) and the UAE (9.4 per cent), the R/P ratio is over a hundred years. The other major source is Iran with 9.0 per cent of the reserves and an R/P ratio of 69. These figures illustrate the extreme dominance of the region and the very uneven distribution of global resources. Furthermore, there are other sources in the Middle East, the major one being Libya with 2.8 per cent of the world reserves and an R/P ratio of 55.6. There are few alternatives to Middle Eastern sources. The major ones are shown in Table 11.2.

These three are all, of course, developing countries and are likely to require an increasingly large proportion of indigenous production for their own industrialization. Only two other countries possess significant global reserves: China has 2.3 per cent with an R/P ratio of 20.5 and Norway has 1.0 per cent with an R/P ratio of 8.6.

In cases in which there are few alternatives to a particular resource, it is often possible to escape vulnerability by substitution. However, oil as an easily transportable liquid source of energy has a number of end uses for which substitution is currently not possible. In particular, in transport and especially air transport there is no substitute. For certain uses, particularly in industry and energy generation natural gas can replace oil. However, total natural gas production, measured in terms of oil equivalent, amounts to only 58 per cent of oil production. Furthermore, the Persian–Arabian Gulf region accounts for 33.7 per cent of natural gas proved reserves with an R/P ratio well in excess of 100 years. To this can be added Algeria (2.6 per cent with an R/P ratio of 54.8), Egypt (0.5 per cent with an R/P ratio of 66.5) and Libya (0.9 per cent with an R/P ratio of over 100 years). Thus, in total the Middle East region accounts for well over a third of global natural gas reserves. The other major source is the countries of the FSU which account for 39.2 per cent with an R/P ratio of 86.2. Two countries dominate natural gas reserves: the Russian Federation with 33.2 per cent and an R/P ratio of 85.9 and Iran with 15.8 per cent and an R/P ratio of over 100 years. The other significant reserves are found in Qatar (5.9 per cent), UAE (4.0 per cent) and Saudi Arabia (3.7 per cent), all with an R/P ratio of over 100 years. Thus, sources of natural gas are even more concentrated than those of oil and the use of natural gas does relatively little to alleviate oil vulnerability.

Table 11.2 Major alternative proved oil reserves

	Percentage of world reserves	*R/P ratio*
Mexico	3.8	33.6
Venezuela	6.9	59.5
Nigeria	1.6	20.2

R/P ratio = reserves to production ratio.

The other source considered particularly important for the future is nuclear energy but, for a variety of reasons, development has largely stalled. After a rapid initial increase in production, a number of nuclear energy station alarms followed by the partial meltdown of the station at Chernobyl, there is little public confidence in nuclear energy as a source of power. Consumption is dominated by three countries: the USA (27.7 per cent), France (16.5 per cent) and Japan (13.5 per cent). The only other countries with significant consumption are Germany (7.1 per cent), the Russian Federation (4.5 per cent), the UK (4.1 per cent), Ukraine (3.3 per cent) and South Korea (3.2 per cent). In terms of equivalence, nuclear energy consumption is some 18 per cent of oil consumption.

The only other sources of primary energy are coal with an oil equivalent of 31 per cent and hydro-electricity with an oil equivalent of 7 per cent. Thus, a brief summary of global primary energy reserves illustrates clearly the predominance of oil. Furthermore, when an analysis is made of sources of proved oil reserves, the Persian–Arabian Gulf, with almost two-thirds of global reserves, is pre-eminent. With the large oil requirement in the developed world and the marked concentration of reserves in the developing world, oil can be considered the most strategic commodity. As a result, there is obvious potential for the geopolitical use of oil in which access or denial, usually the latter, depend upon political factors rather than normal commercial transactions. As a result of the laws of supply and demand, the Persian–Arabian Gulf states have leverage internationally, a factor which has been designated the oil weapon. Oil is a vital resource for modern life, it is the major commodity traded and it is the one truly global commodity. However, distinction must be made between the possession of a potential weapon and its use.

Since the second round of oil price hikes in 1979, the conservation programmes established following the first major series of price rises in 1973–74 have produced an effect. Also, many new suppliers have come on stream. Although many of these have relatively small reserves, they can nonetheless be important in the short term. To be truly effective in the use of the oil weapon, major producers would need to operate as a cartel. OPEC is such a potential cartel but its weakness is that there are several significant producers who are not members. Thus the potency of OPEC has been diminished over the past twenty years with a result that oil prices have at times sunk below $12 a barrel which, for most producers, is little above the cost of production.

For a variety of reasons, therefore, the oil geopolitics has proved less effective than would seem likely in theory. Nonetheless, there remains the psychological aspect which is still potent. At the onset of the Gulf War (1991) oil prices rose significantly, not because demand increased or supplies declined but as a result of Wall Street neurosis. The sophisticated financial instruments which had been developed for the sale of oil were thought to be in danger and, although supplies of oil remained virtually unchanged, there was a minor global shock. This enhanced the theory that the interest of the West was more in the security of oil supplies than in any violation of the boundaries of Kuwait.

The potential for oil geopolitics appears obvious but the reality is less clear. Sudden price rises can cause severe disruption to the global economy and aspects of this rebound upon the oil producers. In particular, the major oil producers among the GCC countries have invested heavily in Western Europe and North America. If economies in those regions decline, then losses are felt in the Arabian Peninsula. Increasing economic globalization means that any major change is likely to have repercussions world-wide and the knowledge of this greatly reduces any temptation to employ the oil weapon.

Water

From the global perspective, resource geopolitics in the Middle East is generally equated with the potential for the denial of international access to oil. Within the Middle East the liquid which most influences the relationship between states is undoubtedly water. The concept of strategic resources, those which were critical but for which there was vulnerability, has been implicit throughout history. However, it was only after the oil price rises of 1973–74 and 1979 and the fivefold rise in the price of cobalt after the invasion of Shaba Province in the then Zaire in 1979 that the concept moved into the realm of high politics. By the mid-1980s, with the relatively low prices of oil and most strategic minerals, the focus of attention turned to water and particularly water in the Middle East. Instead of a given, water became increasingly regarded as a resource, the value of which needed to be quantified so that water codes involving the regulation and pricing of the resource could be introduced.

Important as the introduction of commercial considerations has been, the emergence of water as an element of high politics has been even more profound. It remains the one resource to be characterized by its own terminology: 'hydropolitics'. Furthermore, the term encapsulates much of the dilemma of the subject. 'Hydro' indicates a number of related sciences such as hydrology which are based upon quantification. In contrast, 'politics' implies discussion and decision-making which may be factually related but is primarily based upon perceptions. As a consequence, when critical water issues such as transboundary flows are being negotiated, hydrologists and politicians are likely to share little common ground. In regions of water scarcity, planning and policy development must include both hydrological and geopolitical considerations and therefore water scientists and politicians need to co-operate.

The potential for misperceptions was well illustrated when on 13 January 1990, to allow solidification, the diversion channel beneath the Ataturk Dam was closed and the flow of the Euphrates was effectively terminated for one month. Water scientists from all three riparian states understood the need for such a measure but so potent was the geopolitical symbolism that momentarily, Baghdad and Damascus were united in their opposition. Indeed, the word 'rival', derived from the Latin *rivus* – a stream, originated from the concept of using the same stream.

The Middle East and in particular the Arabian Peninsula, Egypt and the coastal Levant other than Lebanon, is the global region with the greatest actual and potential water deficiency. It is therefore not surprising that concepts such as water diplomacy, water conspiracy and water wars are normally discussed in the context of the Middle East. Apart from the mountainous areas to the north and in Lebanon, rainfall is extremely low and potential evapotranspiration is high. Indeed, it is only the seasonal differences between the two, combined with the extreme variability of the precipitation, that allow surplus surface water and any recharge. Throughout most of the Middle East there is no permanent surface flow and all the perennial rivers, with the exception of the Nile system, are to the north of latitude 30°N. The Nile itself receives almost all of its discharge from rainfall outside the region.

With water supplies throughout most of the region under severe strain, the situation is likely to be further exacerbated by the extremely high rates of population growth in the Middle East. The population of the region appears set to double within the next twenty-five years. With population increase has come urban migration and in many of the Middle Eastern countries the urban proportion of the population already exceeds 75 per cent. The significance of this is that, under urban conditions, water use is virtually exponentially enhanced. A nomadic family may consume between 10 and 30 litres per person per day but urban life will result in anything from a tenfold to a twenty-fold increase in that figure. As the population has increased, so the results of social and economic activity have resulted in greater pollution problems. However, of more concern is the requirement for food and therefore for irrigation. Although in the more developed parts of the region techniques for water application have become more sophisticated, the use of water for agriculture in the Middle East is still profligate and in most of the countries accounts for between 60 and 70 per cent of the water use.

The minimum annual requirement per person is considered to be 1,000 cubic metres and, with an average below that figure, a country is designated water scarce. In the region, Israel, Jordan, Palestine, Syria and the whole of the Arabian Peninsula are already water scarce. For Israel, Jordan and Palestine the situation is even more extreme and absolute water scarcity is said to exist.

As with any other strategic resource, when demand exceeds local supply, there are a number of options. These can be summarized as:

1   stockpiling or storage;
2   research and development in relation to substitution, recycling and conservation; and
3   importation.

In the three main river basins of the region, storage is in naturally occurring or artificial lakes which allow the regulation flow. Lake Kinneret provides a natural regulator for the Jordan system while Lake Nasser has been impounded to provide Egypt with control over supplies. However, throughout most of the Middle East in which surface flow is ephemeral, storage must be in aquifers. Natural recharge can be augmented by the construction of recharge dams and there are more than sixty of these in Saudi Arabia alone. The major problem inherent in such a system is that the detailed geology of the aquifers, particularly the deeper aquifers, is not well known and, given the slow rate of water transmission through rock, the effects of artificial recharge on downstream wells are almost impossible to evaluate.

So far, the chief output of research and development has been desalination and more funding has been spent on installations in the Middle East than in any other part of the world. Over 60 per cent of the global installed capacity is in the Arabian Peninsula. MSF predominate but increasingly other methods such as reverse osmosis (RO), electro-dialysis (ED) and vapour compression desalination (VCD) are being introduced either separately or in concert. The major drawback is cost which varies according to normal economic factors but also crucially the salinity of the water. To provide half the annual flow of the Jordan or 1 per cent of that of the Nile, in the case of Israel would cost between $1.2 and 1.8 billion per year (Hewedy 1989). In the smaller Gulf countries which rely on one or two major facilities, there is the added consideration of supply vulnerability.

A second method of supply enhancement is through recycling of wastewater chiefly from sewage disposal. Few cities in the Middle East have adequate sewage systems but recycling is receiving increasing attention. Israel, Jordan, Qatar, Kuwait and Saudi Arabia all have significant capacity. The major early scheme was the Dan Wastewater Reclamation Project which was built to treat the sewage from Tel Aviv (Shuval 1980). In 1997, the leading countries in the UN ESCWA area were for desalination, Saudi Arabia with a production of 795 bcm; and for recycling, Syria with a total of 1.45 bcm (*The Yemen Observer* 1999).

Conservation procedures can be applied throughout the water supply system: the effective collection of precipitation to the most efficient use in irrigation. In the rural areas throughout the region, water-harvesting techniques are evident and the Middle East, particularly Iran, is celebrated for the development of *qanats*. Leakages from the infrastructure tend to account for anything from 30 to 50 per cent of supply and these pose major problems. However, the key approach related to conservation is increasingly seen as allocation. Allocation would result in restriction of water to those end uses which bring the greatest return. The requirement for potable water is unlikely to exceed 10 per cent of the available supply and, for most of the other uses, a lesser quality of water will suffice. Furthermore, although the idea runs counter to that of food security, the import of low-cost foods effectively amounts to using the water of the exporting country. This has been characterized as the use of 'virtual water'.

Water importation may be from indigenous or external sources. The most significant example of internal trans-catchment water transport is the Great Man-Made River Project in Libya which is intended to irrigate some 18,000 hectares at an annual cost of $500,000,000 (Agnew and Anderson 1992). Groundwater from deep aquifers is transported some 15,000 km from Kufra in the southern Fezzan to the coast to facilitate development around the Gulf of Sirte. Potentially the most spectacular international project is the 'Peace Pipeline' proposed by Turkey in 1986. Some 6 million cubic metres per day from the catchments of the Seyhan and Ceyhan rivers would be transported to either side of the Arabian Peninsula. The proposed western pipeline would supply 3.5 million cubic metres per day to Syria, Jordan, Palestine and Saudi Arabia while the eastern pipeline would transport 2.5 million cubic metres per day to Kuwait, Bahrain, Qatar, the UAE, Oman, if required, and other parts of Saudi Arabia. Feasible routes have already been surveyed but there is no decision on the final alignment of the pipelines. The problems are costs and vulnerability. The costs of desalination are declining and the vulnerability of pipelines to politically inspired interruption have been amply demonstrated by the restrictions placed on Iraq's oil exports following the Gulf War. If the intention is to supply an extra volume beyond that which is basic, then the costs appear excessive.

Turkey has also pioneered, using water from the same catchments, a sea transportation system known as Operation Medusa. Already supplies have been transported to northern Cyprus and the major intended beneficiary is Israel. Vast balloon-like structures are filled with water and towed by tugs to the reception points. One advantage is that a balloon might supply the needs of a number of settlements along a particular coastline. However, it is difficult to see that the costs involved will not preclude such a system from anything other than emergency use.

The Middle East is the most water-scarce region in the world and the situation is likely to deteriorate. Not only is the precipitation throughout most of the region very low but it is also highly variable while potential evapotranspiration rates are universally high. Many countries in the Middle East, particularly those in the Arabian Peninsula, have population growth rates which are among the highest in the world and the increasing urbanization only exacerbates the water problems. Most of the region is essentially agrarian and irrigation techniques tend to be profligate. Supply enhancement is expensive and potentially wasteful in the case of storage and highly expensive in the case of desalination or importation. Therefore, the water security of the Middle East as a region can only be characterized as fragile. The potential for hydropolitics is greater than in any other area of the world.

Hydropolitics is not scale-specific and may involve issues on any scale from the local to the international. Furthermore, hydropolitical problems may arise from atmospheric water, surface flow, groundwater and even temporary flow. However, owing to their wider implications, it is the potential problems existing between states in shared international basins that have dominated hydropolitical thinking. International basins are those shared between two or more states and it is calculated that such basins accommodate over 40 per cent of the world's population. Given the importance of water and its range of potential uses, the possibilities for dispute over sharing supplies are immense. Although they are not the only shared basins in the region, the catchments of the Nile, the Tigris–Euphrates and the Jordan exhibit the greatest potential for future conflict (Figure 11.4).

The total catchment of the Nile drains 10 per cent of Africa and has a more complex hydrological regime than that of any other river. It comprises the White Nile and two major tributaries, the Blue Nile and the Atbara, and from its conference with the Atbara some 1,800 km from the Mediterranean, there are no perennial tributaries downstream. Total flow shows wide variation and discharge figures range from 78 to 85 bcm, although the lower figure provides the more realistic guide. What is clear is that the key river is the Blue Nile and Ethiopia supplies approximately 80 per cent of the water of the entire catchment.

Egypt, with a population of approximately 65 million (1997), is the main user and also in every way the most powerful state in the basin. Sudan has a fast-growing population which at present comprises 35 million but does not utilize its full share of water, the surplus being consumed by Egypt. In 1929, the Nile Water Agreement was signed between the two giving Egypt 48 bcm and Sudan 4 bcm. As a result of the Aswan High Dam project, there was heightened tension in the 1950s and military confrontation in 1958. However, the Agreement for the Full Utilization of the Nile Waters was signed in 1959 and remains in operation. The result was that the 1929 apportionment was ratified but the additional water from Lake Nasser was shared between Sudan and Egypt in the ratio of approximately 2 : 1.

Rather than between Sudan and Egypt, more likely confrontation is between Ethiopia and Egypt. The suggestion of Blue Nile development projects in Ethiopia was met by the threat of force in 1979 and all indications are that the situation has not changed.

In the Tigris–Euphrates basin, there is at present a surplus of water but, with the completion of current developments, particularly the GAP, it is thought that Iraq may suffer water problems in terms of both quantity and quality. While statistics vary, the flow of the Tigris can be put at about 49 bcm and that of the Euphrates at 35 bcm, giving a total which is probably somewhat in excess of the Nile flow. Turkey provides 89 per cent of the annual flow of the Euphrates and 51 per cent of the Tigris. Thus, the basin resembles that of the Nile in that a non-Arab upstream state is the major source of flow. However, whereas Turkey is the most powerful state in the Tigris–Euphrates basin, Ethiopia is militarily weak compared with Egypt.

The focus of geopolitical attention is upon the Ataturk Dam and in particular the GAP which will, on completion, provide some 50 per cent of the irrigated area and 50 per cent of the power generation of Turkey. The area to be irrigated by water impounded by the Ataturk Dam is estimated at over

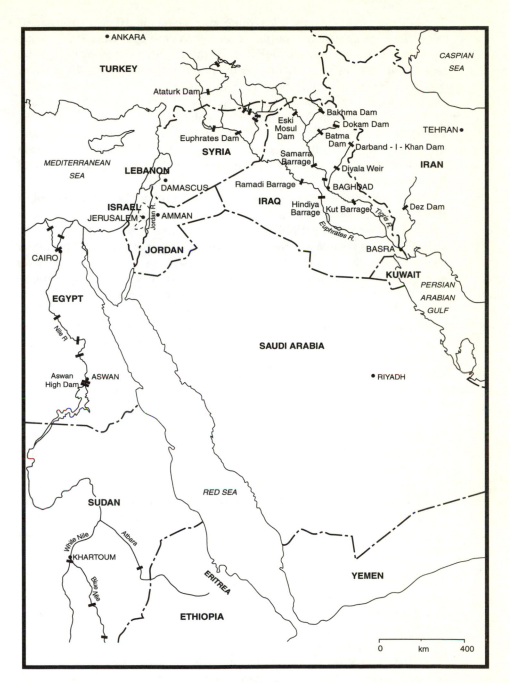

Figure 11.4 Middle East: major drainage basins.

7 million hectares and this will require annually some 10 bcm of water. If this is added to the volumes projected for Syrian development and allowances are made for evaporation, the discharge of the Euphrates entering Iraq could be reduced to as little as 11 bcm, representing something like a 20 per cent deficit on Iraq's projected minimum requirements.

Geopolitical activity over water occurred in 1974 when the filling of lakes behind two dams in the upper Euphrates coincided. The result was a significant drop in the discharge of the river and confrontation between Syria and Iraq. Again in 1990, the construction of the Ataturk Dam resulted in a marked reduction of the flow of the Euphrates and this created alarm in both Syria and Iraq.

The catchment of the Jordan is by far the smallest of the three but it remains the only one in which politically motivated water diversion has been seriously planned. Since 1948, international relationships within the catchment have been volatile and there has never been agreement on co-operation over water supplies. Most crucially, the flow at about 1.5 bcm approximates to only 2 per cent of that of the Nile. The situation has been further complicated by the international acceptance of Palestine as a separate political entity so that the number of riparian states has increased to five.

The political complexity is indicated by the geography. There are three main headwaters of the Jordan: the Hasbani, rising in Syria and Lebanon; the Banyas, from the springs in Syria; and Dan, wholly within Israel. The major tributary, the Yarmuk, provides the boundary successively between Jordan and Syria and Jordan and Israel while the River Jordan itself forms the boundary between Israel and Jordan and to the south, the West Bank and Jordan (Anderson and Rashidian 1991). Thus, bearing in mind the geopolitical importance of location, it can be seen that in the Jordan Valley Syria and Lebanon are upstream to Israel, but Israel is upstream to Jordan and Palestine.

Unbiased hydrological data are extremely difficult to obtain but it can be stated that only 3 per cent of the Jordan basin lies within the pre-1967 boundaries of Israel. Indeed, occupation of the West Bank has been significant hydrologically for Israel in that this is now the source of some 80 per cent of its water. Additional water comes from the Golan Heights and the occupation of southern Lebanon allows access to the Litani River.

There is no firm evidence that there has been a hydrological imperative in Israel's wars with the Arabs. Water has been one component in a long-running confrontation. Nonetheless, the construction of the National Water Carrier from Lake Kinneret met universal Arab disapproval and led to the first military action by al-Fatah, the Palestine National Liberation Movement. The 1964 Arab Summit resolved to divert the headwaters of the Jordan and work started in 1965. However, Israel used its military power to destroy the Syrian construction sites and skirmishes continued directly to the war of 1967. It was their victory in that war that allowed Israel to control the key geopolitical locations in the Jordan catchment.

While there has been no outright war over water, there has been confrontation or conflict in all three basins and the potential for further aggression must be considered high (Figure 11.5). Research indicates a number of factors against which the likelihood of conflict can be assessed. Problems are likely if there is disagreement over data, a marked asymmetry of power and clear ideological differences. All three factors obtain in all three catchments. Geographical position is also crucial together with the degree of interest in water problems and the potential, internal or external, for power projection. The level of interest in water is obvious among the riparian of the Tigris–Euphrates and the Jordan basins and the majority of those in the Nile catchment. However, power varies considerably in each basin, the catchment superpowers being Turkey, Israel and Egypt.

When negotiations fail, before recourse to conflict, international law may be consulted. However, there are four major traditional alternative principles concerning the sovereignty of states over water resources. Upstream states can quote the principle of absolute territorial sovereignty and downstream states the principle of absolute territorial integrity. The principle of community and the principle of equitable utilization or limited territorial sovereignty suggest co-operation. According to the last, a state can use the water insofar as it does not interfere with utilization by other co-riparians. It was from this principle that the International Law Association (ILA) developed the 'Helsinki Rules'. Further legal developments have followed from the thirty-two articles produced by the UN International Law Commission (1992).

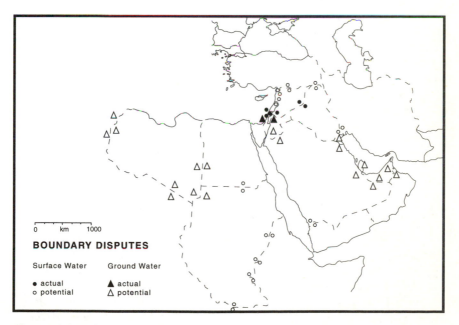

Figure 11.5 Middle East: hydropolitics.

In many ways the most effective form of coercion has been financial since for most major projects international funding, usually from the World Bank, is needed. The Bank has been able to develop a policy of refusing support where there are conflicting viewpoints about a project. Examples in the region are the Unity Dam on the Yarmuk and the Ataturk Dam on the Euphrates.

Food security

Petroleum, water, certain minerals and food are all considered to be strategic commodities in that they are critical in their use and supplies could be at risk. Food resembles the others in that supplies can be safeguarded through stockpiling, research and development or imports. In the case of food research and development, it might concern, for example, the development of higher-yielding crop varieties. The major distinction is the potential for substitution. Compared with the others, food is almost infinitely substitutable. The required number of calories and the necessary dietary balance can be obtained from a very wide variety of sources. Therefore food geopolitics, the interference with food supplies for political purposes, is less likely to be successful than in the other commodities. However, in the last resort, food is more vital than either minerals or petroleum and thus a threat against supplies is particularly psychologically damaging.

For most countries in the world, including many in the Middle East, food security implies the maintenance of the quantity and quality of food uptake per inhabitant. If supplies cannot be maintained then malnutrition, under-nutrition or starvation may result. Of the Middle Eastern states, Egypt, with a proportionately small cultivable area and a burgeoning population, is the most at risk. Local food supplies are inadequate and there is insufficient purchasing power locally, regionally or nationally to make up the deficit. However, starving people are a risk to the integrity rather than the security of a country if they migrate across borders. Their requirements may over-stretch the local market in the receiving state leading to tension with indigenous inhabitants, but large-scale conflict is unlikely.

Egypt has taken a number of measures to improve its food security and other more imaginative approaches such as seawater irrigation are under consideration. In particular, as Cairo and the Delta cities expand, every effort has been made to conserve the fertile soils of the Nile Valley. However, it is realized that, even with maximum cropping, there will still be food shortage. Therefore, irrigation channels are being built to link Lake Nasser with a succession of oases which occur in the desert to the west. Crucial to all Egyptian plans is the water of the Nile and a likely result of this dependence will be water disputes with the upstream riparian states, most obviously Ethiopia. Indeed, since some 80 per cent of the Nile flow originates from the Ethiopian highlands, Ethiopia can mount a good case to develop more extensive irrigation.

It is more likely that conflict results in starvation, rather than starvation produces conflict. Thus, while the only major famines recorded in the Middle East have been in Egypt in 42 BC and between 1064 and 1070 (Sasson 1990), conflict has certainly produced local areas of malnutrition and perhaps starvation in Sudan and in Iraq. The long-term effects of Operation Desert Storm (1991), supplemented by the continuing UN sanctions, have had a particularly devastating effect upon nutritional levels in Iraq. Normal Iraqi trade is highly constrained, the main element being the sale of oil for food allowed by UN Security Council Resolution 986. The passage of this Resolution illustrates the general acceptance by the global community that food security must be assured.

Field (1993) argues that nutrition is a weak political issue for four principle reasons. It lacks salience and nutritional concerns are difficult to activate or sustain at governmental level. Advocacy is generally poor and it is hard to define a policy with regard to nutrition.

Fourthly, nutrition is very susceptible to rhetoric and tokenism and it is therefore extremely difficult to determine the extent of serious political commitment. Nonetheless, in the peculiar situation of the Gulf states, food security has attained a high profile.

Following the first round of oil price rises in 1973, Henry Kissinger in various interviews, issued the threat of food geopolitics (Anderson and Rashidian 1991). The Gulf states are particularly susceptible to such threats in that they are materially advanced societies, heavily dependent upon food imports. Indeed, it has been advanced that one of the reasons why there have been relatively few restrictions on oil flow has been the dependence on Western food markets. Food could obviously be obtained from elsewhere, but the range might well be restricted.

One result was that in Saudi Arabia the government publicly stated that the security of food supplies was the foundation for their national independence. As a result, in an effort to achieve self-sufficiency in basic food commodities, huge investment was made in agricultural development. Indeed, agriculture was made the Kingdom's second priority after petroleum and self-sufficiency was achieved in wheat, potatoes, dates, eggs and dairy produce (Sasson 1990). The increase in wheat production was particularly spectacular and resulted in a surplus for export. Production increased from 3,300 tonnes in 1978 to 1.7 million tonnes in 1985. The down side was that production was subsidized and placed an enormous stress on groundwater resources.

It can be seen that the concept of food security involves a variety of issues. While less obviously and directly geopolitical than most of the other issues considered, the subject is sufficiently emotive that political perceptions can override practical realities. Furthermore, food security can be related directly to the concept of strategic minerals. These are normally taken to relate primarily to steel alloy minerals but food production depends critically upon fertilizers. The Middle East, as an arid region is a source of evaporites such as potash and also of downstream products from oil. However, the most significant deposits are those of phosphates of which the greatest share of the world resource base is owned by Morocco.

Although the concept of food security may be variously interpreted, it is interesting that the efforts to ensure supplies can have a disastrous effect environmentally in developed and developing countries alike. Rees (1990) shows in her discussion of flow resource scarcity that technological progress and market forces have not acted to reduce pressures on renewable resources. In the more advanced countries, increased demands on the landscape have resulted in many agricultural changes which have been little less than disastrous with regard to the environment. Such pressure to intensify production must be questionable in some countries. In developing countries, the failure of technology and market forces is even more marked as agriculture has advanced into more marginal areas so output has become more precarious and potential productivity has been reduced. In the Middle East, desertification has been one highly publicized outcome. Reclamation of the desert, whether in low-income economies such as that of Egypt, or the relatively high-income economies of the Gulf states, has produced problems in both the short and the long term.

In a detailed analysis of food security, Alamgir and Arora (1991) trace the concept of food security from the level of the household to that of the nation and, finally, the globe. In particular, they stress that food security hinges as much on purchasing power as upon production and supply. Therefore, long-term solutions must involve socio-economic, environmental and political factors. Thus, variables ranging from the level of world resources to transport infrastructures, management and planning, must all be considered. A useful indicator of the food situation is the food security index (Brown 1984). This includes measures of both stockpiles and the potential for idle crop land and the two combined can be expressed as potential days of consumption. Stockpiles are held generally in the exporting countries which are, for grain: the USA, Canada, Australia, Argentina and France. However, idle crop land has become a feature of both US and EU policy. When grain stocks and grain equivalent of US crop land drop below 50 days of world consumption, grain prices rise and become unstable. It is just this sort of eventuality which led the government of Saudi Arabia to develop its policy of grain self-sufficiency.

However food security is interpreted, the Middle East provides a number of telling examples of the geopolitics of food supplies. Given the possibilities for substitution, the transportability of food and the global concern over malnourishment, food will remain a potential geopolitical issue within the region but not one to rival water.

Pan-Arabism

Pan-Arabism or Arab Nationalism has been displaced from the centre of geopolitics in the Middle East by Fundamentalism. Islamic Fundamentalism has increasingly flourished as nationalism, in relation to specific countries, has declined. However, in the future, conditions conducive to the greater Arab Nationalism, Pan-Arabism, may again arise. Pan-Arabism is a movement dedicated to the ideal of unifying all Arabs and, as such, has the potential to become a significant force within the macro-political agenda.

The populations of the Arab countries inhabit a region of the world which is physically, socio-economically and geopolitically highly distinctive. The continuous Arab territories are enclosed within formidable natural barriers which also confine it to a uniform climatic zone (Mansour 1992). The only break in this swathe of territory is, of course, afforded by Israel, which lies centrally astride the crossroads separating Asia from Africa. The Arab world stretches beyond the Middle East to Mauritania and the far reaches of the Maghreb.

Historically, the Arabs are thought to have a common source, emanating originally in migratory waves from Arabia. Thus, there is thought to have been a common group of religions and languages and this accounts for the ease with which Islam and Arabic were commonly accepted from 635 AD onwards (Mansour 1992). The strength of Arab culture survived 400 years of Ottoman rule over most of the Mashrek and over 150 years of French influence in the Maghreb.

A nation can be considered a relatively long-established and recognized stable community based on common territory and sharing other common characteristics particularly language, economic life and culture.

The Arab world has a common language, territory and culture, and therefore fulfils the greater part of the definition and could thus be considered a nation. What is, as yet, not present is a common economic life (Welty 1987). This would imply an economic life in which most members participate and are mutually dependent. As a result, according to the definition, the 'nation' could only truly emerge in a modern era, since it is only under modern economic conditions that the efforts of individuals have been welded together into a national structure. In the Arab world, despite the advent of such bodies as the Arab League and the GCC, it has not proved possible to operate a Pan-National economy.

Undoubtedly, the key factor in any consideration of Arab nationalism must be the fact that, with the exception of a few minorities, the populations of all the countries speak a common language, Arabic. In the east, the Mashrek, Arabic has been the tongue for at least 1,300 years, while in the centre, Egypt and Libya, the language has been spoken for a least a thousand years and in the west, the Maghreb, for eight centuries. Of course, there are many dialects but the distribution of these does not accord with political boundaries and therefore they do not reinforce the fragmentation introduced by the delimitation of the nation-state. To this strong bond of Pan-Arabism must be added the fact that written Arabic has an even greater uniformity over space and time. This can be attributed to the effect of the Koran which was revealed in Arabic, the language which remains the privileged mode for its communication.

The other key element in the common culture for the Arab world is that the great majority of people share a common religion, Islam. Therefore, it is possible to talk about an Islamo-Arabic civilization (Mansour 1992). This cultural homogeneity was greatly enhanced until the present period of international boundary restrictions, by the mobility of the Arabs. Territoriality was not the basis for allegiance and therefore boundaries, such as they existed, tended to be porous and, as a result, the Arabs were among the world's great travellers.

The other key factor in any discussion on Pan-Arabism is territory. With the noted exception caused by the establishment of the state of Israel, the Arab world is continuous from east to west and north to south. Furthermore, it is well defined geographically and enjoys, if not a homogeneous, at least a basically similar climate and landscape throughout. The basic landscape elements are flattish desert platforms and scattered water sources. With the notable exception of the Nile and the Tigris–Euphrates systems, both of which draw their water from outside the region, there is virtually no surface flow and, in large measure, the water source is the aquifer. This shared extreme environment has resulted in reactions by man which are similar throughout the Arab world. Thus, modern-day developments aside, each part presents a microcosm of the whole.

The modern period, particularly since the advent of the oil era, has witnessed differential economic development within the broad and long-established pattern, largely as a result of the presence or absence of petroleum resources. As a result, the modern Arab world can be seen, in economic terms, to be somewhat polarized between the oil states and the non-oil states. The modernization apparent in the GCC states is unequalled elsewhere. Thus, economic development has militated against Pan-Arabism.

However, the failure of the Arab world, despite the existence of so many unique common factors, to achieve unity must also be attributed to the vicissitudes of history. Shuraydi (1987) identified four fundamental landmarks in contemporary Arab history, vital in any consideration of Pan-Arabism. First, the fragmented Arab world was unified under the Islamic Empire. After this disintegrated, the second landmark was the Ottoman Empire, which dominated the region for over 400 years. Third, after the First World War and the break up of the Ottoman Empire, large parts of the Arab world were divided between Britain and France. The fourth landmark was the creating of the state of Israel.

Shuraydi (1987) reasons that while these four landmark events may give the appearance that external factors were mainly responsible for the continuing divisions of the Arab world, there were also internal factors of significance. Furthermore, the tension between Pan-Arabism and Pan-Islamism clearly continues to influence the secularization of the Arab world. Indeed, the chief supporters of Arab nationalism during the Ottoman period were Western-educated Christians (Tibi 1981). Moreover, the main protagonist of Arabism has been Egypt, which has remained, at the same time, the most obviously nationalist of the Arab countries.

In the period since the Second World War, any incipient Pan-Arabism has been shattered by three events. In 1948, the Arab armies were defeated and Palestine was lost. There appeared to be a second major chance with the merger of Egypt and Syria into the United Arab Republic in February 1958. However, three and a half years later, Syria seceded. Then in 1967 came the shattering blow of the further defeat of the Arab armies and the clear demonstration of the military weakness of the Arab world. The immediate result of this catastrophe was, in some quarters, a total indictment of the entire Arab way of life (Shuraydi 1987). The message was (Ajami 1981): 'We have to slay the past to liberate the present'.

Over the post Second World War period, the dominant figure, forever associated with Pan-Arabism but regarded with ambivalence throughout the Arab world, was Gamal Abdel Nasser. Following the removal of King Farouk and the proclamation of the Republic of Egypt in June 1953, Nassar emerged as the key figure becoming President in 1954. For the Arab world, he was a charismatic figure who concentrated power in his own hands to a marked degree and developed the ideas known generally as 'Nasserism'. To a basic Pan-Arabism were added republicanism, socialism, non-alignment and a strident anti-imperialism. The man and his philosophy received a timely boost as a result of the Suez crisis of 1956, during which the world supported the new state of Egypt against the old colonial powers of Britain and France. On the strength of the euphoria generated, it seemed possible that the entire Arab world might be reshaped. However, the only outcome was the short-lived United Arab Republic. Military defeat in 1967 dramatically reduced the authority of Nasser who died in 1970 leaving a chaotic political, economic and social legacy to his successor.

Nevertheless, despite the sentiments often expressed by Arab intellectuals, the feeling generally throughout the region remains Pan-Arab and the myth or otherwise of Pan-Arabism is too strong to perish. Also, the continuing question of Palestine has exerted a unifying force throughout the Arab world. With the success of Islamic, rather than purely Arab groups, it would appear that, at least in the medium term, Pan-Arabism will remain an alluring and compelling vision but as a force its value will be only psychological.

Fundamentalism

In the context of the Middle East, popular Western understanding is that Fundamentalism is synonymous with Islamic Fundamentalism. However, the term 'fundamentalist' can be applied to an adherent of any religion who espouses a return to the fundamental tenets of that religion. Since the Middle East is the birthplace of several world religions, most notably Islam, Christianity and Judaism, there is no logical reason why Islam should enjoy a monopoly among fundamentalists. Indeed, while there are many fundamentalist Christian groups within the region, the Zionists, a fundamentalist group based on Judaism, have been more effective politically this century than any Muslim group. The history of Israel is inextricably intertwined with Zionism and the event with the most far-reaching political consequences in the Middle East this century was not the discovery of oil, but the establishment of the state of Israel.

Despite these misgivings, since the perception is so strongly held in the West, the geopolitical threat in the Middle East will be examined with a focus upon Islamic Fundamentalism. When, at the end of the Cold War, Fukuyama (1989) wrote of the 'end of history', meaning that the global opposition had finally been vanquished and liberal democracy would prevail, he added, in defence of his ideas, that there would still be threats. In the absence of communism in its various forms, problems would arise as a result of forces generated by Fundamentalism and Nationalism. In discussing these points, he clearly identified the Fundamentalism as Islamic. Such ideas appeared particularly timely, given the demise of the Soviet Union and therefore the bipolar model of the world as Pfaff observed in *The New Yorker* (28 January 1991): 'There are a good many people who think that the war between Communism and the West is about to be replaced by a war between the West and Muslims'. That such thoughts should be committed to paper illustrates the strength of feeling about the potential threat from the Muslim world. Islam is a multinational force, with about 1.2 billion members, who form a majority in some 45 countries, stretching round the globe from West Africa to South-East Asia. It is also represented by significant and growing minorities throughout Western Europe, in the USA, Russia and China. India, the largest democracy in the world, includes a Muslim minority which approximates to the total population of Pakistan in size and therefore, as a national body, is second only in size to the Muslim population of Indonesia.

Given this geographical distribution, any global subdivision between Muslim and non-Muslim would be infinitely more complex than that between East and West. Moreover, the world-wide East–West divide resulted largely from coercion, whereas a division according to religion would separate peoples who are culturally different. Therefore, any second Cold War which might result would be unlikely to end in the same resounding fashion. The question must be posed as to how great a threat, if any, is posed by Fundamentalism. It seems unlikely that a religious or, at most, an ideological challenge should constitute a major geopolitical threat.

Islamic Fundamentalism simply means a belief in the precepts and commandments of Islam, as stated in the holy Koran and practised by The Prophet Muhammad, 'the Sunna' (Rashidian 1991). It therefore represents a return to the purest form of the religion and is a movement aimed at cleansing Islam from impurities and revisionism. Therefore, Fundamentalism could be used as a term to include all practising Muslims who accept the Koran as the literal word of God and the example of The Prophet as a model for life.

As Fundamentalists can be identified in a number of world religions, the term 'Fundamentalism' can be considered pejorative in a totally different context from Islam. In fact, in the West, the term when applied to Protestant Christianity tends to be considered derogatory, implying a mindless, reactionary approach. In the Middle East, in contrast, many Fundamentalist leaders are among the best educated people, making full use of all modern advances.

A further problem with the term 'Fundamentalism' is that, particularly in Western media, it tends to be equated with terrorism and fanaticism. Of course, it is true that certain terrorist groups do publicly espouse Islamic Fundamentalism, but to extrapolate from them to the entire Muslim world is, at the very least, unscientific. To illustrate this point, Esposito (1992) illustrates in detail how the word 'Fundamentalist' has been applied to the governments of Libya, Saudi Arabia, Pakistan and Iran, four very disparate regimes. The contrast between, for example, the foreign policies of Saudi Arabia and Libya could scarcely be greater. One factor which all four have in common is that government legitimacy has been based upon Islam.

Thus, as a label, Fundamentalism is virtually meaningless. It has been demonized by Western media so that Fundamentalist represents a well-identified stereotype. At the same time, any significant Arab reawakening has been characterized by the stereotype. As a result, to the undiscriminating, there appears to be a monolithic Fundamentalist threat, emanating actually or potentially from every Muslim country. A preferable and more precise term would be 'Islamic Revivalism'. Esposito (1992) identifies revivalism and activism as the key words.

Ideally, the legitimacy of the Islamic state depended not so much upon any relationships its system might have to democracy, but upon its ability to protect the *Umma* (the total community) and to institutionalize the Shari'a (the law). Over the centuries, the Shari'a has been developed and enhanced to cover every aspect of life so that Muslim society world-wide shows a remarkable homogeneity. However, despite the flexibility of the Shari'a, it was, over time, supplemented to cover such areas as commerce, taxation, public and criminal law (Ruthven 1992). Ruthven recognized the distinction between religion and politics, which became more acute with the increasing power of the state from the eighteenth century onwards. The changes which exacerbated this separation were commonly seen as necessary to counteract political threats from the West. Herein lay the key dilemma. Muslim civilization might survive either by accommodating to changes emanating from the West or through resisting such changes.

When it was realized that nationalist leaders who had adopted Western modifications to Islamic values could not, even after independence, initiate the new Islamic 'golden age' (Ruthven 1992), they were severely criticized. Possible Islamic solutions to current problems were developed and religiously inspired groups such as the Muslim Brotherhood were established.

The period following the Second World War saw the withdrawal of the colonial powers from the Middle East and, particularly in Egypt and Syria, the promulgation of Arab nationalism, aimed at Pan-Arabism. The optimism engendered in the Arab world and, to an extent the Islamic world in general, ended abruptly with the catastrophic defeat of the 1967 war. Pan-Arabism has yet to recover. On the other hand, while the loss of the shrine in Jerusalem was a stunning blow for the Muslims, a new generation has arisen, seeking a solution to its problems through Islam.

The Islamic revival has varied widely from country to country in its organization and political objectives. There are a number of common ideological, socio-economic and cultural characteristics. Overall, there is a reaction against secular materialistic Western society and its increasing cultural influence over Muslim countries. However, traditionalist approaches to the restoration of Islamic values have been intermingled with the requirements of modern society. As a result, it is argued that the Islamic ideal remains only a possibility. Despite these attempts, there exist contradictions with, for example, the concept of Islamic banking and the Koranic ban on *riba* or interest of capital. A particularly significant geopolitical departure has been the demand that the Shari'a be applied to territory, rather than community. Until relatively recently, territoriality was a term foreign to Islamic and, particularly, Arab society, in which traditionally allegiance was owed to a group of people. The major cause of the conflict in Sudan was the insistence of the government that the Shari'a should be applied not only to the north of the country, but also to the non-Muslim south.

At the same time as the Shari'a and other bastions of Islamic civilization have been rejuvenated, disillusionment with the state has increased. In most Muslim societies, the traditional balance between the state and society has gradually moved in favour of the state. Institutions may have been created with reference to the Koran, but many were seen as secular and so lacked legitimacy. The overall result was that nationalism either never fully developed or lost popularity.

While 'modernization' is broadly acceptable, 'Westernization' is not (Serjeant 1994). Indeed, Islamic revivalism can be seen as a direct response to Muslim state dependency in a global order, managed by the West (Khoury 1983). Internally, the situation has been exacerbated by the apparent inability of the state to solve basic problems such as the disparity between the rich and the poor. Islam can be seen in this context as an instrument through which political, economic and social demands might be met. In the face of mass disillusion, people tend to question authority and Fundamentalists claim that Islam offers a viable alternative.

While several Middle Eastern states link their legitimacy to Islam, Iran is the only true theocracy and following the Revolution of 1979, an Islamic government was established. The distinguishing features included in the new Constitution, were three new bodies. The Council of Guardians was established to examine parliamentary legislation in the light of Islamic Law. The Assembly of Experts was set up for the interpretation of the Constitution and the Council for Determining the Expediency of the Islam Republic was tasked with giving approval or otherwise to legislative and executive decisions.

After the death of Ayatollah Khomeini in 1989, there has been some pragmatic modification of Iran's international posture. When, during the Gulf War, Iran maintained a neutral stance, relations with the West greatly improved. This was seen as vital to facilitate the development of the country which, with a burgeoning population, has undergone recurrent economic crises. However, while Iran remains a target for US embargoes, the opportunities for sharing in the development of Central Asia and particularly its energy resources are likely to remain limited.

The re-emergence of Fundamentalist movements has resulted essentially from the inability of governments to provide adequately for the majority of their people, from the increasing repression of dissidents and from the encouragement of Western values in place of traditional culture. It has led to a questioning not only of policies, but also of government legitimacy. In the long term, governments will be supported as a result of their performance, particularly in maintaining peace and stability in the Islamic world.

However, the influence of Islam can no more be underestimated and without Islamic legitimization, no state government is likely to be secure. Fundamentalism, to be effective, depends upon a strong coherent structure and leadership and this can only probably be challenged by a well-founded democratic nationalist movement. Whether this implies some form of pluralism and democracy as recognized in the West or whether some indigenous form has yet to be seen.

Terrorism

Terrorism is the terrorizing of a nation, a part of a nation or even an individual in order to strike at a range of targets from the state itself to a particular regime or, in the extreme case, the idea of the state (Gal-Or 1985). As the targets vary, so do the aims and forms of terrorism.

The rationale behind a particular activity may be merely to gain publicity for a cause, which can itself be seen as legitimate. Thus, certain acts which could be defined as terrorist were perpetrated during the *intifada* to bring to global attention the plight of Palestinians living in the Gaza Strip. The aim may, or course, be more extreme and be focused on de-legitimizing and eliminating, as a political entity, an entire group of people, such as the Jews in Israel, the Alawites in Syria or the Sunni Muslims in Iraq (Rubin 1989). By attacking high-profile public targets, often at random, terrorists hope to reveal the impotency of the opposing state forces. To succeed, however, there needs to be in place an environment which is at least marginally conducive. There may be, for instance, some underlying support for the terrorist cause or, on the other hand, the regime or government may enjoy only modest backing from the population. If such an environment is lacking, the rationale may be to shock the ruling group into implementing security measures, which so negate the principles upon which the regime is based that popular support is significantly reduced and policy is changed. To avoid this possibility, democratic states are reluctant to over-react to terrorism.

As a variety of aims can be distinguished, so can a number of categories of terrorism. Terrorism may be directed internationally at foreign targets or may be entirely domestic. Sometimes, it may spill over from one to the other. In the Israeli–Arab conflict, all three possibilities have been seen. Terrorism may be state-sponsored or perpetrated by independent groups. In the absence of overt government support, it may well be difficult to distinguish between the two. However, in a number of Middle Eastern countries, notably Libya and Iran, inflammatory statements justifying or promising acts of terrorism can be attributed to the leadership. Certain terrorist acts carried out by the Secret Service arm of government, such as *Mossad* in Israel, also provide evidence of government sponsorship. With regard to independent groups, the problem lies in attempting to establish independence, since such terrorist groups are obviously highly secretive in their operations. In the Middle Eastern context, to provide a classification of terrorist organizations is difficult enough without the additional problem of attempting to establish financial links with governments.

Domestic terrorism can be subdivided between what might be characterized as national liberation and rebellion, although, again, the distinction may be blurred. The actions of the Kurds in northern Iraq are seen by many as the result of nationalist aspirations, directed towards the establishment of a Kurdish state. In Baghdad, the same activities are considered as rebellion against the central government.

Thus, while it is possible to identify various forms of terrorism, many of the strands are closely intertwined. The chief distinction of geographical relevance is whether the scene of operations is internal or external. Internal terrorism remains, for the most part, the concern of the relevant state government. External terrorism is an example of the many transboundary movements, which constitute part of the macro-political agenda and can be of global concern.

It is the essential political element which separates terrorism from conventional crime, but, unlike political violence, terrorism is illegal (Gal-Or 1985). However, it is the political aspect which lends terrorism legitimacy in the eyes of those who sympathize with the political aims at issue or support a specific campaign. It is a cliché that what to one person is a terrorist, to another is a freedom fighter, but this is refuted by Clutterbuck (1994) on the grounds that terror tactics are never justifiable. Indeed, in asserting the fact that terrorism is used by all sides, he argues succinctly that while the cause may be justified, the technique never is. Nevertheless, the fact that sections of the population, actively or passively, provide support, serves effectively to legitimize terrorist intimidation. There is, given the background against which much of it occurs, a general feeling that terrorism is an inevitable outcome of social and economic deprivation. In contrast to this view, Rubin (1989) provides evidence to show that terrorism may be a carefully chosen strategy, often initiated by powerful, sometimes governmental, backers.

Terrorist organizations are not legitimate political bodies in that they do not offer an alternative source of government. This distinguishes them from groups or national armies of liberation and those engaged in classical guerilla warfare. Indeed, the aim of terrorism is to undermine and corrupt the basis of government, to take advantage of the resulting instability to further their own causes. It must be doubted whether terrorists, although they often receive it, are basically concerned with winning popular support. As a result, even in the cases where their aims are articulated, it is almost impossible to assess whether terrorist groups have ever achieved them.

Despite the fact that terrorism can occur anywhere, it has a strong association in the public mind with the Middle East. In identifying the most likely areas of terrorist conflict in the world, Clutterbuck (1994) includes in his list: countries associated with the Arab–Israeli conflict, Cyprus, Turkey, Egypt, Jordan, Libya, Sudan, Iran, Iraq and the Gulf monarchies. Of particular relevance in the region is Islamic Fundamentalism, which Clutterbuck (1994) considers to have overtaken Marxism as the prime ideological generator of international terrorism. The relationship between terrorism and the Middle East is, of course, not a chance factor, but reveals much about the political situation over time in the region (Rubin 1989). Indeed, terrorism can be traced back at least to the *Sicarii* in Palestine during the period AD 66–73. The area also witnessed the withdrawal of British administrative influence, which has often been associated with violence (Wilkinson 1981). The last period of British rule in both Cyprus and Palestine witnessed widespread terrorism. Added to these factors peculiar to the Middle East must be the general world-wide growth of urban terrorism, particularly, from the 1960s onwards. However, it must be remembered that certain states, notably Iraq and Libya, have practised terrorism on their own people, both domestically and internationally.

Nonetheless, whether it could be claimed justifiably that parts of the Middle East produce conditions ideal for the rise of terrorism must remain doubtful. There is little in the way of government which would be considered democratic by Western standards, but democratic states are by no means immune from terrorism. Indeed, in examining the most recent statistics, Rubin (1989) shows that the effects of terrorism, in the context of the total world population which die violently each year, are very limited. In 1986, 398 died and 574 were wounded and, of this small total, only some one-quarter suffered in the Middle East. Therefore, the impact is far more one of psychological than lethal warfare. As Rubin (1989) claims: 'Terrorism may well be the most overrated political phenomenon of our time'.

Thus, as perhaps the most closely monitored and minutely reported region of the world, the Middle East probably suffers unduly from publicity given to terrorist acts.

Being at the crossroads of the world, the Middle East has gained as a result of migrations over the centuries a more varied population than any region of similar size, globally. The complex mosaic of ethnic, religious, linguistic and cultural groups is bound to produce tensions. Whether these develop into terrorist movements is another question. Another issue is the vast discrepancy between the wealthy and the poor, whether at the level of the state or the individual. Obviously, the thinly populated oil-rich states can be contrasted with the densely populated developing countries of the region. However, such conditions occur in many other parts of the world and are in no sense particular to the Middle East. The reason that the Middle East seems so particularly prone to terrorism may be related to these factors but must also result from:

1   the establishment of Israel and its long-term support by the USA; and
2   the possession of such a high proportion of the global petroleum resources.

To gain publicity, usually the prime aim of the exercise (Clutterbuck 1994), to demonstrate their power, or to terrify certain sections of the population, terrorists have developed an armoury of tactics, including: political assassination of high profile figures; shooting; bombing; hostage taking and hijacking. Many of these tactics are designed for short-term gains, but hostage taking, for example the long drawn-out saga of the Beirut hostages, can be perpetrated with longer-term aims in mind, each success being marked by the liberation of a limited number of hostages. From the terrorist viewpoint, hijacking of airliners is particularly rewarding in that it generates vast amounts of media attention, often globally. A large airliner is likely to contain a wide variety of nationals, which ensures massive publicity. Examples include the Shi'a Fundamentalist hijackings of the TWA flight 147 from Athens to Beirut (1985) and of the Kuwaiti airliner (1988). The Lockerbie aircraft bombing (1988) still generates media attention and this is likely to be enhanced during the trials of suspects.

With the obvious exception of aerial hijacking, terrorism is normally focused on urban areas for a number of reasons. High concentrations of people and buildings ensure that for a modest terrorist input, the greatest possible damage is done and terror is spread more effectively and faster. This attracts greater media interest and frequently results in such increased security measures that the effects of any one activity are felt for a long time afterwards. Furthermore, within the urban fabric it is relatively easy for the perpetrators of any terrorism to escape. In the cities of developing countries there may well be ghetto 'no-go' areas in which terrorists are safe from security forces. In Western-style cities there is a multiplicity of high-grade targets and rapid movement is guaranteed.

It has been shown that there is a strong geographical basis for terrorism in that sites for activity are carefully selected so that maximum effect is gained. Furthermore, political, social and economic conditions may enhance the occurrence of terrorism. However, the effectiveness of the actions perpetrated depends very much upon perceptions. A relatively minor event can have a major political influence. Thus, in the interplay of geography and politics, terrorism is a classically geopolitical activity, a point enhanced by its international spread.

In all the countries of the Middle East, activities which can be labelled 'terrorism' have occurred. Only in two areas, the Occupied Territories and Lebanon, have terrorists (or freedom fighters) exercised real political power. The *intifada* which was initiated in the Gaza Strip in late 1987 was only possible as a result of the 'hearts and minds' approach practised by the PLO and the Palestine Communist Party (PCP). Although the PLO, based externally, sustained it, the *intifada* was based firmly in the Occupied Territories. Other groups such as Islamic Jihad co-operated with the local leadership but Hamas, an Islamist group, brought disharmony. Hamas achieved a relatively high profile, particularly amongst students, and, as long as the problems of Palestine remain unsettled, it is likely to recruit even more widely. Having been viewed by Israel as some sort of counter to the PLO, Hamas militancy is now a major concern.

During the 1980s, the conflict in Lebanon was increasingly influenced by Islamic Fundamentalist groups. In 1982, following the fundamentalist campaign in Syria, members of the Muslim Brotherhood moved to Tripoli and Sidon. Amal was founded in Baalbek and Hezbollah was established in the Bekaa Valley and Beirut. The Muslim Brotherhood is Sunni while the other two movements are both classified as Shi'a, Hezbollah being considered a particularly extreme group. Islamic Fundamentalism in Lebanon flowered after the impetus provided by the Iranian revolution. Iran provided direct support for Hezbollah which was opposed by Syria which supported Amal. One significant result was that from this time the West and Israel began to equate Fundamentalism with terrorism.

Conflict

The resultant of any of the geopolitical issues discussed in the preceding sections could be conflict. For elements of the macro-political agenda such as terrorism, arms smuggling, drug trafficking and the mass movements of refugees, there is already what might be described as a background low-intensity conflict. In this field, the difference between civilian operations involving such bodies as the police and customs and those of the military are becoming increasingly blurred.

Furthermore, as Salmon (1992) states: 'What is important is that we should realize that conflict lies at the heart of politics'. Disagreements or conflicts are resolved by the use of power and as a result, the values for society are allocated (Easton 1953). Internally, the use of such power is regulated by the government or regime. In contrast, internationally, since there is no world government or enforceable system of international law or underlying moral consensus, international relations becomes a delicate adjustment of power to power (Howard 1970).

However, as the world has moved towards a new global order and states have become increasingly interdependent, the use of power internationally has come to involve far more than mere military power. Security entails not merely the survival and territorial security of the state, but also the perpetuation of the values, patterns of social relations, lifestyles and varied other elements characteristic of that state (Schelling 1966). From studies of how states actually behave, many analysts discern that the aim tends to be security rather than hegemony and an increasing realization of the many competing demands placed upon their societies (Fox 1985).

This underlying complexity has been enhanced by the global geopolitical changes since 1989. One approach towards simplification is to analyse conflict in the Middle East in terms of levels or scales of conflict (Cottam 1989). Many of the geopolitical issues discussed which are potential triggers to conflict could be influential on more than one level, but in each case the most likely scale of disagreement can be identified. For example, problems over water supplies can be discerned at both the interstate and intrastate levels. However, it is disagreements over allocations between states, whether of surface water or of groundwater, which seem most likely to result in tensions.

Between 1989, the fall of the Berlin Wall and 1991, the disintegration of the Soviet Union, the confrontation between NATO and the Warsaw Pact ended. Since, in contrast to mainland Europe, the East–West boundary was unclear and indeed on several occasions changed in the Middle East, the effects of these events were somewhat muted. For example, the conflict between Israel and its Arab neighbours had throughout the Cold War period taken precedence over any other alignment. Moreover, the international system was defined by a bipolar structure and this overshadowed conflict at other levels. As a result, its removal has altered security assumptions. The old balance has been overturned, but a new one has yet to emerge. While the end of the East–West confrontation has offered new opportunities for international co-operation and has almost obliterated, at the present time, the possibility of global conflict, the probability of regional and small-scale crises has increased. Above all, the region in which there is a concentration of potentially volatile geopolitical factors is the Middle East.

This is borne out by the fact that since 1948 the Middle East has been the scene of continuous UN activity. Observer missions have been emplaced in: Palestine (UNTSO, 1948–present), Lebanon (UNOGIL, 1958), Yemen (UNYOM, 1963–64), Iran/Iraq (UNIMOG, 1988–90) and Iraq/Kuwait (UNIKOM, 1991–present). There is also a long history of peace-keeping and forces have been introduced as follows: Egypt/Israel (UNEFI, 1956–67), Cyprus (UNFICYP, 1964–present), Egypt/Israel (UNEFII, 1973–79), Syria/Israel (UNDOF, 1974–present) and Lebanon (UNIFIL, 1978–present).

At the present time, with one acknowledged superpower, the roles of the USA and the UN appear to have become almost interchangeable. Following the idea of the New World Order, the USA has identified its wider national security interests in the context of four areas being defined as critical to the growth of the world economy: North America, Western Europe, North-East

Asia and the Persian–Arabian Gulf (Anderson and Fenech 1994). The interest of the USA in the last area was, of course, amply demonstrated by the rapid response leading to Operation Desert Storm.

Nevertheless, while the Gulf states and oil are clearly high on the US security agenda, it seems unlikely that other areas of the Middle East would receive the same consideration. However, it is difficult to see, at the present time, any possibility of major power involvement in conflict within the Middle East. NATO appears to be taking on the mantle of the UN military force and there is no obvious opponent for the USA either in the Middle East or its environs. If the EU were to develop its own military arm or if the Russian Federation were to stabilize, the situation could change.

In the meantime, Fukuyama (1989) can merely postulate that forces resulting from nationalism or Fundamentalism are the only ones likely to oppose US hegemony. While the passing of the bipolar era has resulted the virtual disappearance of a threat at the global level, it has inevitably significantly reduced control at any level below that. Therefore, at the regional or inter-urban level and at the intra-urban or local level, conflict has become, if anything, more likely. The major security problems undoubtedly concern issues which could spill over national boundaries and become of regional importance. The danger of escalation would be enhanced by the large stockpiles of conventional weapons within the region. Despite the pious intents expressed by those Western powers who were involved in Operation Desert Storm, the arms build-up was continued at an even brisker pace, very soon after the conclusion of that conflict. Additionally, Israel is a nuclear power and the proliferation of such weapons is a possibility, particularly in the cases of Libya, Iraq and Iran. Furthermore, within the region is the greatest concentration of chemical weapons holders in the world: Syria, Iraq, Iran, Israel, Egypt and Libya. The existence of biological weapons can only be surmised.

Certain conflicts have, of course, already spilled over international boundaries. Actions against the Kurds have involved violations of the borders of Turkey, Iraq and Iran. Israel, Palestine, Lebanon and Syria are all enmeshed in a continuing confrontation. In the immediate area of the Middle East, there are obvious tensions in the Horn of Africa, while confrontation in the Caucasus and the Central Asian states, particularly Tajikistan, could spill over into the region. The most volatile state undoubtedly is Afghanistan. However, as competition between the states of the FSU to acquire economic outlets, including suitable pipeline and trans-shipment routes increases, this could produce further tensions. The competition for influence in the region between Turkey and Iran, long forecast, has so far failed to exacerbate a potentially fraught situation.

At the interstate level, much of the tension over the period since the Second World War can be attributed to the different effects of change within the region. Some states have remained traditional in their government and outlook, others have been highly nationalistic and even revolutionary. When leaders of the second group of countries have been religio-political the situation has been greatly complicated. Furthermore, those leaders who allied themselves with the West have increasingly been seen as less legitimate. Nasser and Musadiq remain heroes for their anti-Western stance. Sadat is characterized in many quarters as a traitor to the Arab cause. While ideological differences may give rise to tensions, the trigger for conflict is more likely to be related to terrorism, arms, drugs, boundary problems, refugee movements or water scarcity, all of which have already been discussed.

At the intrastate or local level, the key background factor producing tension has been the rate of development. Modern developments frequently sit uneasily alongside traditional practice. As educational levels have risen, so has the pressure for increased political participation. The rate and dimensions of change have compelled radical adjustments in terms of the composition of the governing élite, governmental and social institutions and prevailing norms (Cottam 1989). Such tensions have been enhanced, in some cases by religious differences and labour movements. Key factors are the proportion of the population which is Shi'a or Sunni and the percentage of the labour force which is expatriate.

As a result of all these geopolitical variables, a number of potential flashpoints can be identified. Politically, islands such as Abu Musa, the Tumbs, Warbah and the Hanish group have all been the centre of attention. Refugee camps and settlements in various parts of the Levant and Iran have been, in many cases, explosive. Water problems have arisen in the valleys of the Nile, the Tigris–Euphrates and the Jordan. International boundary issues have been highlighted in the Arabian Peninsula, and particularly between Israel and its neighbours.

Some of these disputes have now been settled. In late 1998, arbitration between Yemen and Eritrea over the southern Red Sea islands settled the question of sovereignty, awarding the Hanish Group to Yemen. Having given up its claim to the West Bank, Jordan, controversially as far as many in the Arab world are concerned, has settled its boundary with Israel. However, as settlement occurs in one place, so potential problems arise elsewhere. In the Nile catchment, the intention of Ethiopia to construct irrigation works on the upper Blue Nile has attracted belligerent comments from Egypt. The 'no fly zones', established under the aegis of the UN after the Gulf War (1991), have become the focus of attention and, allegedly in their defence the USA and UK have adopted a policy of regular bombing.

Some potential flashpoints appear to remain intractable. With only 58 km of coastline, Iraq relies for exports on two waterways, the Khor Zubair-Khor Abdullah and the Shatt al Arab, both shared with neighbours (Anderson 1993). International boundary concerns in the context of both of these waterways have resulted in recurrent conflicts. The Hawar Islands are vital as a result of the access they provide to seabed energy resources. Historically, owing to the largesse of the UK, Bahrain has a strong claim but geographically the islands can be considered as comprising part of the territorial integrity of Qatar.

The array of potential flashpoints throughout the Middle East needs to be appreciated in the context of the overall global location of the region. At the crossroads of the world, and having the major global concentration of proved oil reserves, the Middle East is a focus of routeways, particularly sea-lanes. Restrictions in the sea-lanes lead to a concentration of shipping and potential security problems. Within the Middle Eastern region, the Turkish Straits, the Strait of Hormuz and the Strait of Bab el Mandeb all constitute such choke points. Additionally, the Suez Canal is a man-made choke point. Further removed but related in that they exercise control over key parts of the global oil infrastructure are the Straits of Gibraltar and Malacca. With the exception of the Turkish Straits and the Suez Canal, jurisdiction over both of which is exercised by one country, all the choke points include the territorial waters of more than one state. As their locations are fixed by the configuration of the opposing coastlines and as shipping is likely to remain the major mode of transport for world trade, these straits are potentially permanent flashpoints.

Despite almost continuous conflict in and around Israel, and skirmishes and incursions in many other areas of the region, there has been only one period of fighting which was characterized as a war in the region during the decade of the 1990s. The Gulf War or, if the long running Iraq–Iran conflict of the 1980s is so designated, the second Gulf War, erupted following approximately six months of build-up in early 1991.

Following so closely upon the end of the Cold War when international hopes had risen that problems would be settled by negotiation rather than armed conflict, the Gulf War was a tragedy. The chronology of events is set out in detail in Anderson and Rashidian (1991) but the overall motives of the participants are likely to remain a cause for disagreement. Between 1982 and 1989 US–Iraq trade had grown from virtually nothing to $3.6 billion annually. Other European countries rushed to the Iraqi market and UK exports reached a peak of £450 million in 1989. Thus Iraq became one of the world's largest debtor nations with debts to the West of some $80 billion and total debts, including those to the Gulf states acquired during the Iran–Iraq war, amounting to $160 billion. Additionally, since 1985 the US Commerce Department had approved $750 million in exports of sensitive technology, much of it diverted to Iraq's nuclear chemical and missile programmes. German corporations supplied equipment for chemical and biological weapons and France supplied nuclear technology. Even when Iraq deployed nerve gas, the West remained unmoved. Two days before the invasion of Kuwait the USA formerly prohibited the sales of arms and technology. Thus it is reasonable to conclude that Iraq was encouraged in the development of its arsenal and in the strengthening of its position as the leading Arab if not regional state.

With vast debts and the requirement to rebuild its economy and infrastructure following the end of the war with Iran, Iraq was desperate for money from the sale of oil at a reasonable price, calculated to be about $18 a barrel. However, Kuwait and the UAE in particular had been exceeding OPEC quotas and the price had dropped to approximately $12. Thus strong feelings had already been engendered in Iraq about Kuwait and to these were added a variety of historical claims most of which could be argued in favour of either side.

Epilogue

The richness of Middle Eastern landscape and life is such that a volume like this can only hope to provide a few basic insights. The complexity can perhaps best be summarized by a series of paradoxes. Above all, within the apparent unity of the region there is great diversity. Culture, religion, language, climate, desert landscapes, the vital relationship with water, availability of oil money are all, to a greater or lesser degree, unifying factors. However, none is homogeneous or monolithic. All display diversity to the patient observer.

Within the complexity, there is often an awe-inspiring simplicity. In calligraphy, the text may appear complicated but the ornate flow of the letters encapsulates simplicity. The jumbled mass of a sand sea hides the simplicity of a single dune. The complexities of an increasingly urban lifestyle remain dependent upon one simple resource, water.

Geopolitical considerations of the Middle East tend to revolve around petroleum and the use of the oil weapon but, over time, the more significant resource is water. Pipelines transport oil across boundaries but the idea of the Peace Pipeline to pump water across the same boundaries is fraught with geopolitical and psychological implications. If the West Bank were chiefly significant for oil, it is possible to speculate that moves towards peace in the area would have been less controversial.

In the Middle East, the aspects of ancient culture are frequently seen juxtaposed with the most opulent modernity. At times, Bedouin encampments can be seen almost within the shadow of Kuwait City. Camels are transported to grazing in the back of a pickup vehicle. In the physical landscape, areas of extreme aridity surround some of the world's most sophisticated irrigation systems. Computer-controlled irrigation appears virtually alongside water-lifting devices which have been extant almost since the dawn of man. Development and underdevelopment, wealth and poverty, exist side by side.

The other abiding impression is of the dichotomy between centrality and transition. In so many ways the Middle East is at the centre of the world and yet it is also a zone of transition. Nowhere better exemplifies the transition from the North to the South. Kuwait and Bahrain, for example, are in so many ways of the developed world whereas Sudan typifies more than most countries underdevelopment.

Thus, although a common framework to life can be discerned, the main impression is of variety. There is greater variation in all the facets of human life in the Middle East than in any other area of comparable size.

Having been a centre of attention since civilization began, it seems most unlikely that the Middle East will ever be less than a focus in international relations (Lewis 1997). Location alone ensures the region will retain significance. This fact is merely underlined by the presence of globally dominant petroleum resources. Indeed, as the twenty-first century advances, dependence of the developed world in particular upon the Middle East for fuel minerals can only increase. Problems of water will loom larger and will result in more effective allocation. Thus, the region is likely to become more dependent upon imported food. In an era of globalization, resources such as petroleum which cannot, like finance, be moved freely round the world, provide a guarantee of security. However, at a time of resource geopolitics, they also elicit unwelcome attention.

It is not possible to come to any conclusion on a subject as diverse as the geography and geopolitics of the Middle East. All that can be said is that the region will retain its global importance and will continue to inspire life-long interest and devotion in discerning observers who attempt to understand and imbibe its culture.

References

Abella, M.I. (1995) Asian migrant and contract workers in the Middle East. In Cohen, R. (Ed.) *The Cambridge Survey of World Migration*. Cambridge: Cambridge University Press.

Abi-Aad, N. (1998) In Natural Gas in the Middle East: Status and Future Prospects. *OPEC Bulletin*, June.

Agnew, C. (1988) Soil hydrology in the Wahiba Sands. In Dutton, R.W. (Ed.) *The Scientific Results of The Royal Geographical Society's Oman Wahiba Sands Project 1985–1987*, Special Report No. 3. Muscat, Oman: Journal of Oman Studies.

Agnew, C. and Anderson, E. (1988) Dewfall and atmospheric conditions. In Dutton, R.W. (Ed.) *The Scientific Results of The Royal Geographical Society's Oman Wahiba Sands Project 1985–1987*. Special Report No. 3. Muscat, Oman: Journal of Oman Studies.

Agnew, C. and Anderson, E. (1992) *Water Resources in the Arid Realm*. London: Routledge.

Ajami, F. (1981) *The Arab Predicament: Arab Political Thought and Practice Since 1967*. New York: Cambridge University Press.

Alamgir, M. and Arora, P. (1991) *Providing Food Security for All*. London: Intermediate Technology Publications (International Fund for Agricultural Development (IFAD)).

Aldelman, H. (1995) The Palestinian Diaspora. In Cohen, R. (Ed.) *The Cambridge Survey of World Migration*. Cambridge: Cambridge University Press.

al-Otaiba, S. (1975) *OPEC and the Petroleum Industry*. New York: Wiley.

Amjad, R. (1989) *To the Gulf and Back*. New Delhi: ARTEP International Labour Organization.

Anderson, E.W. (1988) Preliminary dew measurements in the Eastern Prosopis Belt of the Wahiba Sands. In Dutton, R.W. (Ed.) *The Scientific Results of The Royal Geographical Society's Oman Wahiba Sands Project 1985–1987*, Special Report No. 3. Muscat, Oman: Journal of Oman Studies.

Anderson, E.W. (1992) Water conflict in the Middle East – a new initiative. *Jane's Intelligence Review* 4, 5, pp. 227–230.

Anderson, E.W. (1993) *An Atlas of World Political Flashpoints*. London: Pinter Reference.

Anderson, E.W. (1997) The *wied*: a representative Mediterranean landform. *GeoJournal* 41, 2, pp. 111–114.

Anderson, E.W. and Anderson L.D. (1998) *Strategic Minerals: Resource Geopolitics and Global Geo-Economics*. Chichester: Wiley.

Anderson, E.W. and Curry, W. (1987) *A View of Wadi Dayqah Gorge, Oman*. Muscat, Oman: Council for Conservation of Environment and Water Resources.

Anderson, E.W. and Fenech, D. (1994) New dimensions in Mediterranean security. In Gillespie, R. (Ed.) *Mediterranean Politics*, Vol. 1, pp. 9–21. London: Pinter.

Anderson, E.W., Gutmanis, I. and Anderson, L.D. (1999) *Economic Power in a Changing World*. London: Cassell.

Anderson, E.W. and Rashidian, K.H. (1991) *Iraq and the Continuing Middle East Crisis*. London: Pinter.

Anderson, R.S. Sorensen, M. and Willetts, B.B. (1991) A review of recent progress in our understanding of aeolian sediment transport. *Acta Mechanicia Supplement* 1, pp. 1–19.

Aronson, G. (1992) Hidden agenda: US-Israeli relations and the nuclear question. *Middle East Journal* 46, 4, pp. 617–630.

Beaumont, P., Blake, G.H. and Wagstaff, J.M. (1988) *The Middle East* (2nd edition). London: David Fulton Publishers.

Beydoun, Z.R. (1970) Southern Arabia and northern Somalia: comparative geology. *Philosophical Transactions Royal Society London* A267, pp. 267–292.

Beydoun, Z.R. (1982) The Gulf of Aden and NW Arabian Sea. In Nairn, A.E.M. and Stehli, F.G. (Eds) *The Ocean Basins and Margins 6: The Indian Ocean*. New York: Plenum.

Black, R. (1993) Geography and refugees: current issues. In Black, R. and Robinson, V. (Eds) *Geography and Refugees: Patterns and Processes of Change*. London: Belhaven Press.

Blake, G.H. (1992) International boundaries and territorial stability in the Middle East: an assessment. *GeoJournal* 28, 3. pp. 365–373.

Blake, G.H., Dewdney, J. and Mitchell, J. (1987) *The Cambridge Atlas of the Middle East and North Africa*. Cambridge: Cambridge University Press.

British Petroleum (1998) *BP Statistical Review of World Energy*.

Brough, S. (Ed.) (1989) *The Economist Atlas*. London: Hutchinson Business Books Ltd.

Brown, G.F. (1970) Eastern margin of the Red Sea and the coastal structures in Saudi Arabia. *Philosophical Transactions of the Royal Society A* 267, pp. 75–87.

Brown, L.R. (1984) Securing food supplies in state of the world food. In Starke, L. (Ed.) *State of the World 1984*. New York: W.W. Norton and Company.

Bunting, B.T. (1965) *The Geography of Soil*. London: Hutchison University Library.

Burdon, D.J. (1977) Flow of fossil groundwater. *Quarterly Journal Engineering Geology* 10, pp. 97–124.

Burdon, D.J. (1982) Hydrogeological conditions in the Middle East. *Quarterly Journal Engineering Geology* 15, pp. 71–82.

Bureau of Mines (1993) *Mineral Industries of the Middle East*. Washington DC: Department of the Interior, Bureau of Mines.

Bureau of Mines (1993) *Mineral Industries of Africa*. Washington DC: Department of the Interior, Bureau of Mines.

Butzer, K.W. (1975) Patterns of environmental change in the Near East during Late Pleistocene and Early Holocene times. In Wendorf, F. and Marks, A.E. (Eds) *Problems in Prehistory: North Africa and Levant*. Dallas: SMU Press.

Butzer, K.W. (1978) The Late Prehistoric environmental history of the Near East. In Brice, W.C. (Ed.) *The Environmental History of the Near and Middle East Since the Last Ice Age*. London: Academic Press.

Carpenter, C. (Ed.) (1991) *The Guinness World Data Book*. Enfield: Guinness Publishing.

Central Intelligence Agency (1998) *The World Fact Book 1997*. Washington DC:CIA.

Clutterbuck, R. (1994) *Terrorism in an Unstable World*. London: Routledge.

Cohen, R. (Ed.) (1995) *The Cambridge Survey of World Migration*. Cambridge: Cambridge University Press.

Coleman, R.G. (1993) *Geological Evolution of the Red Sea*. Oxford: Clarendon Press.

Cooke, R., Warren, A. and Goudie, A. (1993) *Desert Geomorphology*. London, UCL Press.

Copan, S.J. (1983) The Sahelian drought: social sciences and the political economy for underdevelopment. In Hewitt, K. (Ed.) *Interpretations of Calamity*. London: Allen and Unwin.

Cordesman, A.H. (1993) *After the Storm. The Changing Military Balance in the Middle East*. Boulder, Co: Westview Press.

Cottam, R.W. (1989) Levels of conflict in the Middle East. In Coffey, J.I. and Bonvicini, G. (Eds) *The Atlantic Alliance and the Middle East*. Pittsburgh: University of Pittsburgh Press.

Crossley, R., Watkins, C., Rave, M., Cripps, D., Carnell, A. and Williams, D. (1992) The sedimentary evolution of the Red Sea and Gulf of Aden. *Journal of Petroleum Geology* 15, 2, pp. 157–172.

Curzon, Lord (1907) *Frontiers: Romanes Lecture*. Oxford: Clarendon Press.

Dan, J., Yaalon, D.H., Koyumdjisky, H. and Raz, Z. (1976) *The Soils of Israel (with Map 1 : 500,000)*. The Volcani Centre, Bet Dugan Pamphlet No. 159.

Dempsey, M. (1983) *The Daily Telegraph Atlas of the Arab World*. London: Nomad Publishers.

Dostert, P.E. (1997) *Africa 1997*. Harper's Ferry, WV: Stryker-Post Publications.

Dregne, H.E. (1976) *Soils of Arid Regions*. Oxford: Elsevier.

Easton, D. (1953) *The Political System*. New York: Knopf.

Eisma, D. (1978) Stream deposition and erosion by the eastern shore of the Aegean. In Brice, W.C. (Ed.) *The Environmental History of the Near and Middle East Since the Last Ice Age*. London: Academic Press.

Esposito, J.L. (1992) *The Islamic Threat. Myth or Reality?* Oxford: Oxford University Press.

Field, J.O. (1993) From nutrition planning to nutrition management: the politics of action. In Pinstrup-Andersen, P. (Ed.) *The Political Economy of Food and Nutrition Policies*. Baltimore, MD: John Hopkins University Press.

Fisher, W.B. (1978) *The Middle East* (7th edition). London: Methuen.

FitzPatrick, E.A. (1986) *An Introduction to Soil Science* (2nd edition). Harlow: Longman Scientific and Technical.

Food and Agriculture Organization (1975) *FAO Production Yearbook 29*. Rome: FAO.

Fox, W.T.R. (1985) E. H. Carr and political realism: vision and revision. *Review of International Studies* 1, 1, pp. 1–16.

Frelick, B. (1992) Call them what they are: refugees. *World Refugee Survey*. Washington DC: US Committee for Refugees.

Fukuyama, F. (1989) The end of history? *National Interest* Summer, pp. 3–18.

Gal-Or, N. (1985) *International Cooperation to Suppress Terrorism*. New York: St Martin's Press.

Goldreich, Y. (1988) Temporal changes in the spatial distribution of rainfall in the central coastal plain of Israel. In Gregory, S. (Ed.) *Recent Climatic Change*. London: Belhaven Press.

Goudie, A. (Ed.) (1985) *Encyclopaedic Dictionary of Physical Geography*. Oxford: Blackwell.

Goudie, A. and Wilkinson, J. (1977) *The Warm Desert Environment*. London: Cambridge University Press.

Graf, W.L. (1988) *Fluvial Processes in Dryland Rivers*. London: Springer-Verlag.

Heathcote, R.L. (1983) *The Arid Lands: Their Use and Abuse*. London: Longman.

Hempton, M.R. (1987) Constraints on Arabian plate motion and extensional history of the Red Sea. *Tectonics* 6, p. 687.

Hersh, S.M. (1991) *The Samson Option*. New York: Random House.

Hewedy, A. (1989) *Militarization and Security in the Middle East: its Impact on Development and Democracy*. Tokyo: The UN University.

Hough, H. (1994) Israel's nuclear infra-structure. *Jane's Intelligence Review* 6, 11, pp. 508–511.

Howard, M. (1970) Military power and international order. In Garnett, J. (Ed.) *Theories of Peace and Security*. London: Macmillan.

Humphrey, M. (1993) Migrants, workers and refugees. The political economy of population movements in the Middle East. *Middle East Report* March–April.

Hunter, S. (1986) The Gulf economic crisis and its social and political consequences. *Middle East Journal* 40, 4, pp. 593–613.

Imeson, A.C. and Emmer, I.M. (1992) Implications of climatic change on land degradation in the Mediterranean. In Jeftić, L., Milliman, J.D. and Sestini, G. (Eds) *Climatic Change and the Mediterranean: Environmental and Societal Impacts of Climate Change and Sea-level Rise in the Mediterranean Region*. London: Routledge.

Inciardi, J.A. (1992) *The War on Drugs II. The Continuing Epic of Heroin, Cocaine, Crack, Crime, AIDS, and Public Policy*. London: Mayfield Publishing Company.

Issar, A.S. and Bruins, H.J. (1983) Special climatological conditions in the deserts of Sinai and the Negev during the Latest Pleistocene. *Palaeogeography, Palaeoclimatology, Palaeoecology* 43, pp. 63–72.

Jenny, H. (1941) *Factors of Soil Formation*. New York: McGraw-Hill.

Joffé, G. (1994) Territory, state and nation in the Middle East and North Africa. In Schofield, C.H. and Schofield, R.N. (Eds) *World Boundaries, Volume II, The Middle East and North Africa*. London: Routledge.

Jones, D.K.C., Cooke, R.U. and Warren, A. (1988) A terrain classification of the Wahiba Sands of Oman. In Dutton, R.W. (Ed.) *The Scientific Results of The Geographical Society's Oman Wahiba Sands Project 1985–1987*, Special Report No. 3. Muscat, Oman: Journal of Oman Studies.

Kay, M. (1986) *Surface Irrigation, Systems and Practice*. Cranfield: Cranfield Press.

Khoury, P.S. (1983) Islamic revivalism and the crisis of the secular state in the Arab world: an historical appraisal. In Ibrahim, I. (Ed.) *Arab Resources: The Transformation of a Society*. London: Croom Helm.

Kjellen, R. (1917) *Der Staat als Lebensform*. Berlin: K. Vowinckel.

Kolars, J. (1994) Problems of international river management: the case of the Euphrates. In Biswas, A.K. (Ed.) *International Waters of the Middle East from Euphrates–Tigris to Nile*. Oxford: Oxford University Press.

Köppen, W. (1931) *Die Klimate der Erde*. Berlin.

Lancaster, N. (1992) Arid geomorphology. *Progress in Physical Geography* 16, 4, pp. 489–495.

Le Houérou, H.N. (1992) Vegetation and land-use in the Mediterranean Basin by the year 2050: a prospective study. In Jeftič, L., Milliman, J.D. and Sestini, C (Eds) *Climatic Change and the Mediterranean: Environmental and Societal Impacts of Climate Change and Sea-level Rise in the Mediterranean Region.* London: Routlege.

Lewis, B. (1997) *Middle East.* London: Phoenix.

Lindh, G. (1992) Hydrological and water resources. Impact of climatic change. In Jeftič, L., Milliman, J.D. and Sestini, C. (Eds) *Climatic Change and the Mediterranean: Environmental and Societal Impacts of Climate Change and Sea-level Rise in the Mediterranean Region.* London: Routledge.

Lukman, R. (1998) Exploiting oil and gas opportunities in the OPEC member countries. *OPEC Bulletin* March.

Mackinder, H.J. (1904) The geographical pivot of history. *Geographical Journal* 23, pp. 412–437.

Macumber, P.G. and Head, M.J. (1991) Implications of the Wadi al-Hammeh sequences for the terminal drying of Lake Lisan, Jordan. *Palaeogeography, Paleoclimatology, Palaeoecology* 84, pp. 163–173.

Mansour, F. (1992) *The Arab World. Nation, State and Democracy.* Tokyo/London: UN University Press/Zed Books Limited.

Martinez, O.J. (1994) The dynamics of border interaction: new approaches to border analysis. In Schofield, C.H. (Ed.) *Global Boundaries, World Boundaries, Volume I.* London: Routledge.

Mather, J.R. (1974) *Climatology: Fundamentals and Applications.* New York: McGraw-Hill.

Maull, H. (1990) The arms trade with the Middle East and North Africa. In *The Middle East and North Africa* (37th edition). London: Europa Publications Limited.

Miller, A.G. and Cope, T.A. (1996) *Flora of the Arabian Peninsula and Socotra.* Edinburgh: Edinburgh University Press.

Moalla, S.N. and Pulford, I.D. (1995) Mobility of metals in Egyptian desert soils subject to inundation by Lake Nasser. *Soil Use and Management* 11, pp. 94–98.

Navias, M.S. (1993) Arms trade and arms control in the Middle East and North Africa since Operation Desert Storm. In *The Middle East and North Africa 1994* (40th edition). London: Europa Publications.

Newbigin, M.I. (1948) *Plant and Animal Geography.* New York: Dutton.

New York Protocol on Refugees 1967 (Amendment of UN Convention on Refugees 1951). *New Yorker, The.* Pfaff. January 28, 1991. Organization of African Unity Convention on Refugees 1967. Organization of American States: Cartagena Declaration 1985.

Parsa, A. (1978) *Flora of Iran.* Ministry of Science and Higher Education of Iran.

Penman, H.L. (1948) Natural evaporation from open water, bare soil and grass. *Proceedings of the Royal Society* Series A 193, pp. 120–145.

Peres, S. (1993) *The New Middle East.* New York: Henry Holt.

Peretz, D. (1978) *The Middle East Today* (3rd edition). New York: Holt, Rhinehart and Winston.

Pikkert, P. (1993) The longue durée: here today, here tomorrow. *Middle East International* 448, pp. 19–20.

Pye, K. and Tsoar, H. (1990) *Aeolian Sand and Sand Dunes.* London: Unwin Hyman.

Rashidian, K. (1991) *Fundamental Islam and Islamic States Stability.* University of Durham: unpublished paper.

Rees, J. (1990) *Natural Resources: Allocation, Economics and Policy* (2nd edition). London: Routledge.

Richards, A. and Waterbury, J. (1996) *A Political Economy of the Middle East*. Boulder, CO: Westview.

Roberts, N. (1982) Lake levels as an indicator of Near Eastern Palaeo-Climates: a preliminary appraisal. In Bintloff, D.L. and Van Zest, W. (Eds) *Paleoclimates, Paleoenvironments and Human Community in the Late Quaternary*. BAR: International Series 133.

Rubin, B. (1989) The political uses of terrorism in the Middle East. In Rubin, B. (Ed.) *The Politics Uses of Terrorism:Terror as a State and Revolutionary Strategy*. Washington DC: The John Hopkins University Press.

Russell, M.B. (1997) *The Middle East and South Asia 1997*. Harper's Ferry, WV: Stryker-Post Publications.

Ruthven, M. (1992) Islamic politics in the Middle East and North Africa. In *The Middle East and North Africa Regional Survey I*. London: Europa Publications Limited.

Sadowski, Y.M. (1993) *Scuds or Butter? The Political Economy of Arms Control in the Middle East*. Washington DC: The Brookings Institution.

Said, R. (1990) Geomorphology. In Said, R. (Ed.) *The Geology of Egypt*. Rotterdam: A. A. Balkema.

Salmon, T.C. (1992) The nature of international security. In Carey, R. and Salmon, T.C. (Eds) *International Security in the Modern World*. New York: St Martin's Press.

Sasson, A. (1990) *Feeding Tomorrow's World*. Paris: UNESCO/CTA.

Schelling, T.C. (1966) *Arms and Influence*. New Haven, CT: Yale University Press.

Schick, A.P. (1985) Water in arid lands. In Last, F.T., Hotz, M.C.B. and Bell, B.G. (Eds) *Land and its Uses – Actual and Potential*. London: Plenum Press.

Serjeant, R.P. (1994) The religions of the Middle East and North Africa. In *The Middle East and North Africa 1995*. London: Europa Publications Limited.

Shannon, P.M. and Naylor, D. (1989) *Petroleum Basin Studies*. London: Graham and Trotman.

Shuraydi, M.A. (1987) Pan-Arabism: a theory in practice. In Faris, H.A. (Ed.) *Arab Nationalism and the Future of the Arab World*. Massachusetts: Association of Arab-American University Graduates Monograph Series: No. 22.

Shuval, H.I. (1980) *Water Quality and Management Under Conditions of Scarcity*. New York: Academic Press.

Simmons, I.G. (1991) *Earth, Air and Water*. London: Edward Arnold.

Sluglett, P. and Farouk-Sluglett, M. (1996) *Guide to the Middle East*. London: Times Books.

Spector, L.S., McDonough, M.G. and Medeiros, E.S. (1995) *Tracking Nuclear Proliferation*. Carnegie Endowment for International Peace.

Sunday Times, The (1986) Reveal: The Secrets of Israel's Nuclear Arsenal, 5 October.

The International Institute for Strategic Studies (1997) *The Military Balance 1996/97*. London: Oxford University Press.

The International Institute for Strategic Studies (1998) *The Military Balance 1997/98*. London: Oxford University Press.

Thesiger, W. (1959) *Arabian Sands*. London: Longmans, Green.

Thornbury, W.D. (1954) *Principles of Geomorphology*. New York: Wiley.

Thornthwaite, C.W. (1948) An approach towards a rational classification of climate. *Geographical Review* 38, pp. 55–94.

Thornthwaite, C.W. (1954) The determination of potential evapotranspiration. *Publications in Climatology* 7, 1.

Tibi, B. (1981) *Arab Nationalism: A Critical Enquiry.* Edited and translated by Farouk-Sluglett, M. and Sluglett, P. New York: St Martin's Press.

Trewartha, G.T. (1964) *An Introduction to Climate.* New York: McGraw-Hill.

Tsoar, H. (1983) Dynamic processes acting on a longitudinal (*seif*) dune. *Sedimentology* 30, pp. 567–578.

UNESCO, Mallet, J. and Ghirardi, R. (1977) Map of the world distribution of arid regions. *MAB Technical Note 7.* Paris: UNESCO.

United Nations (1951) *Review of Economic Conditions in the Middle East.* New York: UN.

UN Convention on Refugees 1951. New York: UN.

Van Hear, N. (1995) Displaced people after the Gulf crisis. In Cohen, R. (Ed.) *The Cambridge Survey of World Migration.* Cambridge: Cambridge University Press.

Vita-Finzi, C. (1969) Late Quaternary continental deposits of central and western Turkey. *Man* 4, pp. 605–619.

Vita-Finzi, C. (1973) Supply of fluvial sediment to the Mediterranean during the last 20,000 years. In Stanley, D.J. (Ed.) *The Mediterranean Sea.* Cambridge: Cambridge University Press.

Ward, R.C. (1975) *Principles of Hydrology* (2nd edition). London: McGraw-Hill.

Warren, A. (1988) The dunes of the Wahiba Sands. In Dutton, R.W. (Ed.) *The Scientific Results of The Royal Geographical Society's Oman Wahiba Sands Project 1985–1987* Special Report No. 3. Muscat, Oman: Journal of Oman Studies.

Waterman, S. (1994) Boundaries and the changing world political order. In Schofield, C.H. and Schofield, R.N. (Eds) *World Boundaries, Volume II, The Middle East and North Africa.* London: Routledge.

Watson, A. (1990) The control of blowing sand and mobile desert dunes. In Goudie, A. (Ed.) *Techniques for Desert Reclamation.* Chichester: John Wiley & Sons Ltd.

Welty, G.A. (1987) Progressive versus reactive nationalism: the case of the Arab nation. In Faris, H.A. (Ed.) *Arab Nationalism and the Future of the Arab World.* Massachusetts: Association of Arab-American University Graduates Monograph Series: No. 22.

Wigley, T.M.I. (1992) Future climate of the Mediterranean Basin with particular emphasis on changes in precipitation. In Jeftič, L., Milliman, J.D. and Sestini, G. (Eds) *Climatic Change and the Mediterranean: Environmental and Societal Impacts of Climate Change and Sea-level Rise in the Mediterranean Region.* London: Routledge.

Wilkinson, J.C. (1977) *Water and Tribal Settlement in South-East Arabia: A Study of the Aflaj of Oman.* Oxford: Clarendon Press.

Wilkinson, P. (Ed.) (1981) *British Perspectives on Terrorism.* London: Allen & Unwin.

Yair, A. and Berkowicz, S.M. (1989) Climatic and nonclimatic controls of aridity: the case of the Northern Negev of Israel. *Catena Supplement 14: Arid and Semi-arid Environments.*

Yemen Observer, The (1999) Arabs face uncertain water future. 1, 9.

Zohary, M. (1962) *Plant Life of Palestine, Israel and Jordan.* New York: The Ronald Press Company.

Index

Note: *italicized* numbers indicate figures and tables

BUNKER HILL COMMUNITY COLLEGE

3 6189 00103174 8